DEC 20 1984

# ELECTROCHEMISTRY

*Calculations, Simulation, and Instrumentation*

COMPUTERS IN
CHEMISTRY AND INSTRUMENTATION

*edited by*

JAMES S. MATTSON      HARRY B. MARK, JR.      HUBERT C. MacDONALD, JR.

# ELECTROCHEMISTRY

*Calculations, Simulation, and Instrumentation*

*Edited by*

**JAMES S. MATTSON**
*Division of Chemical Oceanography*
*Rosensteil School of Marine*
*  and Atmospheric Sciences*
*University of Miami*
*Miami, Florida*

**HARRY B. MARK, JR.**
*Department of Chemistry*
*University of Cincinnati*
*Cincinnati, Ohio*

**HUBERT C. MacDONALD, JR.**
*Koppers Company, Inc.*
*Monroeville, Pennsylvania*

*MARCEL DEKKER, INC.,  New York    1972*

MARCEL DEKKER, INC.
*95 Madison Avenue, New York, New York 10016*


LIBRARY OF CONGRESS CATALOG CARD NUMBER: 76-189043
ISBN: 0–8247–1433–4


PRINTED IN THE UNITED STATES OF AMERICA

## INTRODUCTION TO THE SERIES

In the past decade, computer technology and design (both analog and digital) and the development of low cost linear and digital "integrated circuitry" have advanced at an almost unbelievable rate. Thus, computers and quantitative electronic circuitry are now readily available to chemists, physicists, and other scientific groups interested in instrument design. To quote a recent statement of a colleague, "the computer and integrated circuitry are revolutionizing measurement and instrumentation in science." In general, the chemist is just beginning to realize and understand the potential of computer applications to chemical research and quantitative measurement. The basic applications are in the areas of data acquisition and reduction, simulation, and instrumentation (on-line data processing and experimental control in and/or optimization in real time).

At present, a serious time lag exists between the development of electronic computer technology and the practice or application in the physical sciences. Thus, this series aims to bridge this communication gap by presenting comprehensive and instructive chapters on various aspects of the field written by outstanding researchers. By this means, the experience and expertise of these scientists is made available for study and discussion.

It is intended that these volumes will contain articles covering a wide variety of topics written for the nonspecialist but still retaining a scholarly level of treatment. As the series was conceived it was hoped that each volume (with the exception of Volume 1 which is an introductory discussion of basic principles and applications) would be devoted to one subject; for example, electrochemistry, spectroscopy, on-line analytical service systems. This format will be followed wherever possible. It soon became evident, however, that to delay publication of completed manuscripts while waiting to obtain a volume dealing with a single subject would be unfair to not only the authors but, more important, the intended audience. Thus, priority has been given to speed of publication lest the material become dated while awaiting publication. Therefore, some volumes will contain mixed topics.

The editors have also decided that submitted as well as the usual invited contributions will be published in the series. Thus, scientists who have recent developments and advances of potential interest should submit detailed outlines of their proposed contribution to one of the editors for consideration concerning suitability for publication. The articles should be imaginative, critical, and comprehensive survey topics in the field and/or other fields and which are written on a high level, that is, satisfying to specialists and nonspecialists alike. Parts of programs can be used in the text to illustrate special procedures and concepts, but, in general, we do not plan to reproduce complete programs themselves, as much of this material is either routine or represents a particular personality of either the author or his computer.

The Editors

# PREFACE

Although this volume of the series is devoted to the single discipline of Electrochemistry and, more specifically, discusses computer applications to various problems within the so-called field of electroanalytical chemistry, the material presented is actually general in nature and is useful to people in other areas of chemistry and also in many other experimental sciences. The principles and concepts of the computation methods, simulation approaches to mechanistic studies, and design of both simple and sophisticated instrumentation are perfectly general to similar problems in any of the physical sciences. Each of the chapters is written in such a way as to give clear and detailed descriptions of how the computer and/or computer-based electronic circuitry has been applied to each specific electrochemical problem. Thus, it would be an easy matter for other scientists to extrapolate this information to analogous problems in their areas. In other words, we intend for this volume to serve a dual purpose: to serve electrochemists in specific applications unique to their problems and to aid other scientists in the development and application of computer technology for their problems.

This volume of the series is subdivided into three parts which discuss the three basic applications of computers (digital, analog, and hybrid) to various problems in electrochemical studies. These three specific parts are the applications to: (i) complex and specialized calculations and data reduction, (ii) digital simulation (models) of complex electrochemical reaction mechanisms and kinetics, and (iii) new dimensions in electroanalytical instrumentation and measurement. It should be pointed out that this division of topics is not absolute, however, as several of the example problems in each specific category simultaneously employ the computer for operations in one other or both other applications. Thus, there is always present a certain amount of overlap in this field.

In the first part, which deals with calculations and data processing, Chapter 1 discussed the computer application to the problem of electro-capillary curve calculations. At first glance, one would expect that the treatment of data and calculation of results would require only a simple computation routine. However, this chapter points out clearly how the computer can "overdo" this job unless the operator is very knowledgeable about both the form of the system response and the operations being performed by the computer. This emphasis on the importance of the overall (or dual) competence of the operator in solving chemical and physical

v

experimental problems is very necessary at this early stage in the developmental applications of computers in science. Chapter 2 describes a variation to double layer capacitance calculations.

Chapters 3 to 6 describe the applications of data reduction techniques to various problems in electrochemical kinetic and mechanism studies. Chapter 6 is especially significant as it points out the reasons for carrying out data analysis in the Laplace plane rather than the real-time plane. This concept is relatively new and has the distinct advantage in electrochemical applications of not requiring the impulse generator (potentiostat, etc.) to operate in an ideal manner. Thus, useful kinetic data can be obtained at short times after initiation of the experiment where normal data analysis procedures break down. In addition, Laplace plane analysis allows more useful diagnostic criteria to be obtained than those available from classical techniques.

The second part of the volume deals with digital simulation of complex electrode reaction mechanisms. This is a relatively new tool in electrochemistry. Chapter 7 serves to introduce the technique and also brings the basic principles and applications up to date. Chapters 8 and 9 represent applications of these basic principles to two extremely complex systems: ring-disk electrode studies and electrochemiluminescence.

The third part discusses various aspects of the application of computers and computer-based electronic circuitry to instrumentation in electrochemistry. Chapter 10 is a detailed and comprehensive survey of the applications of <u>analog</u> operational amplifier circuitry to quantitative measurement. Although the operational amplifier-based instrumentation concept was initially proposed and demonstrated in the late 1950's and was a major factor in the "rebirth" of electroanalytical chemistry, this is the first complete review to be published. The rest of this part is devoted to the use of digital logic in electrochemical instrumentation. These applications are just recently being developed and surely will be responsible for a similar advance in new and quantitative applications of electrochemistry.

Chapter 11 serves as an introduction and "historic" review of the development of digital electrochemical instrumentation. Chapter 12 discusses classical as well as multiplexed ac polarography and the important new concept of "Fourier transform" electrochemistry which may open up a completely new area in the study of heterogeneous electron transfer kinetics. Chapter 13 covers the principles and applications of the built-in or on-line computer, which not only does data acquisition and reduction in real time but operates in a "feedback mode" to continuously optimize experimental parameters as the experiment is actually being run. It is obvious that these applications to instrumentation will greatly increase the efficiency and productivity of the scientists.

We wish to acknowledge the help and assistance of numerous colleagues

and graduate students for reading and commenting on most of the material
in this volume. We are very grateful for this help. Also philosophical and
technical discussions with Professor F. C. Anson and Professor Irving
Shain have been helpful in assembling this volume. Finally, we acknowledge
the special efforts of the individual authors, not only for their excellent
contributions, but also for their adherence to the schedule which allowed
prompt publication of a coherent volume.

April 1972            J. S. Mattson, Miami, Florida
                      H. C. MacDonald, Jr., Monroeville, Pennsylvania
                      H. B. Mark, Jr., Cincinnati, Ohio

# CONTRIBUTORS TO THIS VOLUME

DONALD G. DAVIS, Department of Chemistry, Louisiana State University in New Orleans, New Orleans, Louisiana

STEPHEN W. FELDBERG, Brookhaven National Laboratory, Upton, New York

HARVEY B. HERMAN, Department of Chemistry, University of North Carolina at Greensboro, Greensboro, North Carolina

HUBERT C. MacDONALD, JR., Koppers Company, Inc., Monroeville, Pennsylvania

J. T. MALOY, Department of Chemistry, West Virginia University, Morgantown, West Virginia

ROBERT F. MARTIN, Department of Chemistry, Vassar College, Poughkeepsie, New York

DAVID M. MOHILNER, Department of Chemistry, Colorado State University, Fort Collins, Colorado

PATRICIA R. MOHILNER, Department of Chemistry, Colorado State University, Fort Collins, Colorado

RICHARD S. NICHOLSON, National Science Foundation, Washington, D.C.

MICHAEL L. OLMSTEAD, Bell Telephone Laboratories, Murray Hill, New Jersey

ROBERT A. OSTERYOUNG, Department of Chemistry, Colorado State University, Fort Collins, Colorado

S. P. PERONE, Department of Chemistry, Purdue University, Lafayette, Indiana

ARTHUR A. PILLA, ESB Incorporated, Research Center, Yardley, Pennsylvania

KEITH B. PRATER, Department of Chemistry, University of Texas at El Paso, El Paso, Texas

RONALD R. SCHROEDER, Department of Chemistry, Wayne State University, Detroit, Michigan

DONALD E. SMITH, Department of Chemistry, Northwestern University, Evanston, Illinois

# CONTENTS

## PART I: CALCULATIONS

### A. INTRODUCTION AND THERMODYNAMIC CALCULATIONS

### B. ELECTROCHEMICAL KINETIC CALCULATIONS

PART II: SIMULATION METHODS

# ELECTROCHEMISTRY

*Calculations, Simulation, and Instrumentation*

Part I

# CALCULATIONS

A. Introduction and Thermodynamic Calculations

Chapter 1

# THERMODYNAMIC ANALYSIS OF ELECTROCAPILLARY DATA

Patricia R. Mohilner and David M. Mohilner

Department of Chemistry
Colorado State University
Fort Collins, Colorado 80521

## I.  INTRODUCTION

The perpetual dilemma of the experimenter — neither to interpret random errors as meaningful nor to obscure significant features of the data by excessive smoothing — is well illustrated in the application of the digital computer to the thermodynamic analysis of electrocapillary data. Before raw experimental measurements may be examined to determine if the data fit particular models, derived quantities must be calculated from the original

3

data. In this chapter, attention will be focussed on the problem of using a digital computer to convert raw data to quantities derived on the basis of fundamental thermodynamics; comparing derived results with theoretical models will not be done here.

Several different approaches to using a computer to calculate the quantities derived from experimental measurements will be presented and discussed. A preliminary report of this work has been published [1]. Before a computer method may be said to be fully validated, large quantities of data from a variety of sources must have been successfully processed. Because sufficient data from a multiplicity of laboratories have not been available, final conclusions will not be presented; instead, guidelines for the application of the digital computer to the thermodynamic analysis of electrocapillary data will be developed. Particular programs and subroutines which we have found useful are listed in Appendix I with brief comments designed to allow their incorporation into the data analysis schemes of other investigators.

## II.   THERMODYNAMIC BASIS

The thermodynamic theory of electrocapillarity is the subject of detailed discussion in several recent reviews [2-4]. For the purposes of this chapter it will suffice to present the equations which define the interrelationships of the thermodynamic variables. In order to direct the discussion to the basic concepts of computer analysis, only a simple electrochemical system will be considered and, consequently, simplified equations. More complex equations, pertinent to other types of electrochemical systems, may be found in the reviews cited and may be used as a basis for computer analysis by methods analogous to those developed later in this chapter.

The electrochemical system used throughout this chapter consists of an ideal polarized mercury electrode in contact with an aqueous solution containing a single salt and a single neutral organic sorbate. The temperature and external pressure are assumed to be held constant. Whenever more than one solution is considered, it is assumed that the activity of the electrolyte is held constant; the only composition variable is the activity of the neutral organic component. Under these assumptions, the electrocapillary equation describing the system is

$$d\gamma = -q^M dE - \Gamma_0 d\mu_0 \qquad (1)$$

where $\gamma$ is the interfacial tension, $q^M$ is the excess charge density on the mercury surface, E is the electrode potential, $\Gamma_0$ is the relative surface excess of the organic sorbate, and $\mu_0$ is its chemical potential in the solution.

Interfacial tension may be measured experimentally and $q^M$ and $\Gamma_0$

calculated from these measurements by

$$q^M = -\left(\frac{\partial \gamma}{\partial E}\right)_{\mu_0} \tag{2}$$

$$\Gamma_0 = -\left(\frac{\partial \gamma}{\partial \mu_0}\right)_E = \frac{-1}{RT}\left(\frac{\partial \gamma}{\partial \ln a_0}\right)_E \tag{3}$$

where $a_0$ is the activity of the organic sorbate in solution and R and T have their usual meaning. Differential capacitance, which is also experimentally accessible, is related to $q^M$ and $\gamma$ by

$$C = \left(\frac{\partial q^M}{\partial E}\right)_{\mu_0} = -\left(\frac{\partial^2 \gamma}{\partial E^2}\right) \tag{4}$$

where C is the differential capacitance.

Experimental measurements of interfacial tension must be processed by differentiation to obtain the derived quantities, $q^M$ and $\Gamma_0$, whereas measurements of differential capacitance must be integrated to obtain $q^M$.

## III. QUANTITIES DERIVED FROM INTERFACIAL TENSION

The analysis of data obtained by measuring interfacial tension is inherently more difficult than analysis of data obtained by measuring differential capacitance. The contrast arises because the mathematical processes involved, differentiation or integration, produce opposite effects in the presence of the random errors necessarily associated with experimentally determined numbers. Differentiation tends to magnify random errors; integration tends to smooth out these fluctuations. The simpler problem of integrating differential capacitance will be discussed in Section IV.

The primary purpose in developing a digital computer method of differentiation has been to replace graphical techniques which are tedious and not very accurate. The obvious advantage of the speed of computer processing in contrast to graphical analysis may be expected to become more important when the implementation of a proposed computer controlled-data acquisition system for measuring interfacial tension increases the speed with which data may be collected [5, 6].

In any method used to calculate derivatives from discrete numerical data, some type of curve-fitting procedure is required. In the case of interfacial tension data no algebraic equation relating the variables is known; theory provides only a differential equation (1). Therefore an approximation must be selected to represent the data in a manner such that derivatives may be calculated. Two principle approaches have been taken in choosing an approximation: (1) fitting an entire electrocapillary curve with a single function [7, 8] or (2) fitting an electrocapillary curve piecewise by moving a short segment of a polynomial through the curve. Both techniques will be discussed here, although the emphasis will be placed on the moving segment method.

## A.  Single Polynomial Techniques

When the relationship of $\gamma$ to E or to $\mu$ is approximated by fitting a single polynomial of degree n to an entire curve [7, 8], the problem of magnifying random errors does not usually arise because a least-squares fitting technique is used.  Data points in excess of (n + 1) contribute degrees of freedom for smoothing; the fitted values and derivatives at each point are influenced by all of the data.  This characteristic of single polynomial fitting is particularly attractive because it minimizes the influence of seriously erroneous values.

Because smoothing automatically accompanies differentiation in a single polynomial technique, the choice of n becomes critically important.  Using a polynomial of too low a degree may cause the fitted curve to smooth out significant features of the data, whereas too high a degree will cause the fitted curve to follow the fluctuations in the data which result from random error.  The degree of the polynomial may be preselected by the experimenter or selected by the computer according to a criterion included in the program. The latter approach has been recommended by Fawcett and Kent [7] as a result of a systematic study of the differentiation of  synthetic interfacial tension data which contained deliberately introduced random error of known standard deviation.

Fawcett and Kent used the decision-making capability of the computer to select the degree of the least squares polynomial by means of a statistical test.  The data were successively fit by polynomials of increasing degree until the test showed no significant decrease in the variance as a result of increased degree.  For the various sets of artificial data used, the degree of the  polynomial selected by the computer ranged from 3 to 12.  This study clearly shows that there is a hazard in using a preset arbitrary degree for the fitting polynomial.

Based on the results of their investigation, Fawcett and Kent recommended using a single least squares polynomial of computer-selected degree to represent interfacial tension data to be differentiated. They pointed out that this method requires the standard deviations of the measurements to be constant over the range of the data in order for the statistical test to be applicable.  These authors also noted that, when higher degree polynomials were required, often a term such as the 5th degree would not be significant whereas the 6th, 7th and 8th degree terms were.  To force the fitting process to continue beyond the lower degree their program  also required that  the estimated variance for the chosen fit be less than 1.25 times the variance for the error level in the dependent variable [7].

## B. Moving Polynomial Techniques

If a moving polynomial which fits the data exactly is used as the basis for differentiation, then a process of smoothing should precede the differentiation step. The simpler problem of differentiating already smooth data will be discussed first, followed by a description of smoothing by moving polynomials.

### 1. Differentiation

In contrast to single polynomial methods, the moving polynomial fitting does not assume that there exists one function which adequately describes the relationship between the variables over the entire range of the data. A much less restrictive assumption is made, namely, that a short interval of the data may be adequately represented by a polynomial, such as a cubic. The sample electrocapillary data in Fig. 1 illustrate an interval of four

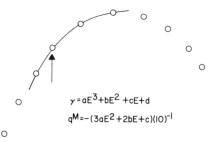

$$\gamma = aE^3 + bE^2 + cE + d$$
$$q^M = -(3aE^2 + 2bE + c)(10)^{-1}$$

Fig. 1. Illustration of four points of an electrocapillary curve fit by a cubic equation. Note that the slope of the curve is more faithful to the data at the interior points of the interval than at the ends.

points through which pass the curve representing the one, and only one, cubic equation of the form

$$\gamma = aE^3 + bE^2 + cE + d \tag{5}$$

which exactly fits these four points. The analytical formula for the derivative of a cubic equation is used to evaluate the charge density:

$$q^M = -(3aE^2 + 2bE + c)(10^{-1}) \tag{6}$$

where the minus sign arises from the thermodynamic relationship [Eq. (2)] and the factor $10^{-1}$ is inserted to express $\gamma$ in dyn/cm, E in V, and $q^M$ in $\mu C/cm^2$. Because the particular cubic equation used in this interval represents only these four data points well and, in fact, represents the slope of the curve well only at the interior points of the interval, the charge density is calculated from this equation only at an interior point (the second point is arbitrarily chosen). In order to evaluate the charge density at the next point in the data set, the computer is programmed to move the interval along the curve by dropping the first point from the interval, adding an additional point, calculating the coefficients of the new cubic equation which represents the new four-point interval, and then calculating the charge density at what has become the second point of the new interval. When this procedure has been repeated for all the data points, the interval has moved throughout the curve, and the charge density at each point has been calculated from the parameters of separate cubic equations (except for the end intervals). No assumptions have been made about an algebraic form for the entire electrocapillary curve.

Relative surface excesses may be evaluated in an entirely analogous manner by taking RT(ln a) as the independent variable instead of E. The subroutines for fitting a cubic equation to a four-point interval (CUBFIT), calculating charge density (QCAL), and calculating relative surface excess (XCAL) are listed in Appendix I.

The utility of the moving cubic fit method of differentiation was reported [1] based on a test case of 36 values of $q^M$ as a function of E for 0.1 M KCl [9]. In this test, the differential capacitance was calculated by differentiating the charge density. A median error of four parts per thousand was found for the values obtained when compared to the experimental differential capacitances from which $q^M$ had been derived [9]. A more complete set of data, 137 points, for the same system [10] was subjected to the same test. The agreement between calculated and measured capacitance was such that a conventional graph showing calculated points and the measured curve displays no visible discrepancies. Therefore, the errors are indicated by the frequency distribution chart in Fig. 2 where the relative errors are parts per 10,000! Clearly, the moving cubic fit method of differentiation is accurate when applied to suitable data.

An important conclusion is illustrated by these two tests which were conducted on data of identical quality. All other characteristics of the data being held constant, the accuracy of moving cubic fit differentiation is improved by increasing the number of points and decreasing their spacing.

## 2.   Smoothing

A moving polynomial may be used to smooth experimental data by a process similar to that described above for differentiation. The essential differences between moving polynomial smoothing and differentiation are: (1) more points are used per interval than are required for an exact fit of the data, thereby producing smoothing in a least squares sense, and (2) the smoothed value, rather than the derivative, is calculated at a central point in the interval.

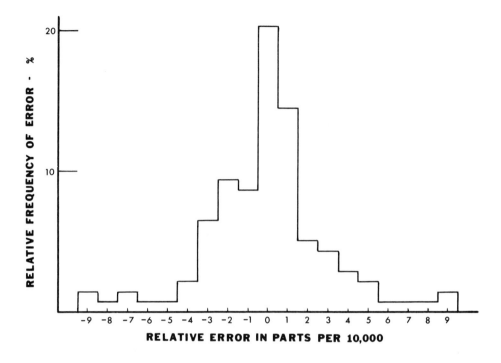

Fig. 2. Distribution of errors when 137 points of smooth data were differentiated by the moving cubic technique.

In applying the moving technique, our attention has centered on least squares polynomials of low degree. Several factors have led to an emphasis on low degree, and in particular to a substantial use of degree two. In a technique which requires the solution of a separate least squares problem for each point in the data, the increase in computation time for increasing degree of equation can rapidly become overwhelming, especially if the technique is applied repeatedly to refine the smoothing. A more fundamental reason for considering low degree equations is that, for the data of double layer interest, the curves are not expected to display detailed structure over short intervals of the independent variable. A higher degree polynomial, unless fit to a very large segment of the data, might well introduce artificial structure into the smooth data by overemphasizing erroneous points. The minimum number of points to be included in a segment for a single least squares calculation is established by the degree of the polynomial. If n is the degree of the polynomial, then a segment containing n + 1 points will be fit exactly with no smoothing. The number of points in excess of n + 1 provide the degrees of freedom for smoothing. Minimal smoothing is achieved with n + 2 points included in the interval. At the other extreme one segment may include all of the data points, in which case a moving technique is not being used. Because different sets of experimental data have different characteristics, no absolute guide can be given to selecting the number of points to include in a segment. The major factors to consider in choosing the number of points for a segment are: (1) the total number of points in the data set, (2) the intrinsic roughness, and (3) the amount of structure reasonably expected in the smooth curve.

These factors are interrelated but some generalizations may be made. The more points which are included in the total data set, especially if available at closely spaced intervals, the more effective the smoothing by a longer interval. If there are few data widely spaced, serious consideration should be given to using the single curve method rather than a moving technique. In considering the intrinsic roughness of the data and the amount of structure expected in the smooth curve, the scientific judgment of the electrochemist must be exercised. The rougher the data, the longer is the appropriate interval for smoothing. On the other hand, the more structure expected in the smooth curve, the shorter is the appropriate interval.

The capability of digital computers to make decisions may be utilized as a safeguard in selecting the number of points per segment in the case of interfacial tension curves. The differential capacitance, which is the negative of the second derivative of interfacial tension with respect to electrode potential [Eq. (4)], must be positive. (A negative capacitance is impossible because it violates the first law of thermodynamics.) Therefore, the smoothing routine can include a test of the sign of the second derivative for each interval with provision for adding additional points to the interval and resmoothing until a fitting conforming to the first law is obtained in that interval. In the case of smoothing with a moving quadratic function, the test is particularly simple and involves no additional calculations since the coefficient of the square term is one-half the second derivative. A subroutine (VLPOLY) which uses this principle is included in Appendix I.

Experience in applying moving smoothing techniques to a number of interfacial tension curves revealed that often one or two outlying points exerted a pronounced influence on the results. For this reason a data rejection routine was devised, based upon a rather long interval of a moving least squares quadratic fit. Standard deviations were calculated for each interval and these values averaged to give a "curve average standard deviation." Any point whose individual deviation from the fitted interval exceeded the "curve average standard deviation" multiplied by an input number was rejected. Typically, the input value was 2.5 or 3.0 (corresponding to approximate confidence levels of 95% and 99%, respectively). Using these values insured that only seriously outlying points would be rejected. The procedure is outlined in flow diagram form in Fig. 3 and a sample result of using this technique is given in Fig. 4. In this example charge density is plotted vs.

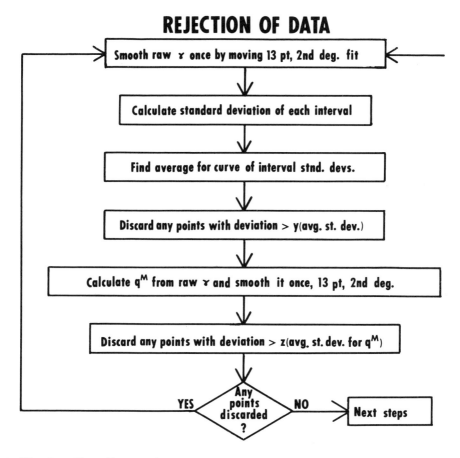

# REJECTION OF DATA

Smooth raw $\gamma$ once by moving 13 pt, 2nd deg. fit

Calculate standard deviation of each interval

Find average for curve of interval stnd. devs.

Discard any points with deviation > y(avg. st. dev.)

Calculate $q^M$ from raw $\gamma$ and smooth it once, 13 pt, 2nd deg.

Discard any points with deviation > z(avg. st. dev. for $q^M$)

Any points discarded ?

YES          NO

Next steps

Fig. 3.   Flow diagram for a routine to reject seriously outlying values in an interfacial tension curve. The routine is entered by the arrow at the upper right.

the electrode potential in order to make the errors visible. The interfacial tension data used were sufficiently good that a graphical display did not reveal suspect points. The curve is the charge density as calculated after smoothing the data. The individual points shown are charge densities calculated at the points which were subsequently rejected. All other charge densities lie on the curve. A subroutine (PITCH) which illustrates this method of data rejection is included in Appendix I.

Moving a four-point least squares quadratic interval through a curve (one degree of freedom smoothing in each interval) was usually found to give

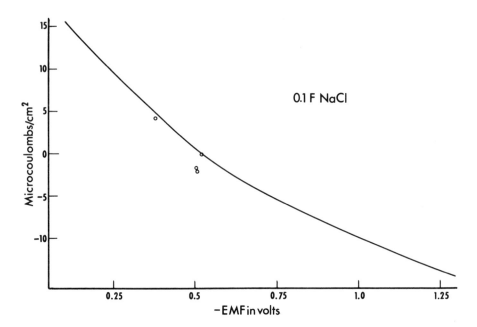

Fig. 4.  Charge density calculated from interfacial tension measurements
[13]. Points are for data rejected by computer routine diagrammed in Fig. 3.


insufficient smoothing as judged by the appearance of the charge density curve
calculated from the resultant "smooth" data. Repetitions of the smoothing
process improved the calculated charge density. The procedure for carrying
out these repetitions is described by the flow diagram in Fig. 5.

The decision box in Fig. 5, "Is criterion of smoothness satisfied?" is
the key to the entire process. Too few passes through the smoothing loop
will produce "smooth" data whose derivatives are so scattered as to contribute little understanding to the experimental question for which the data were
collected. Too many passes through the smoothing loop will systematically
shift the original data values and eventually erase the structure actually
present in the data. Precise meanings for "too few" and "too many" vary
from data set to data set, but generally less than 5 passes are too few and
more than 50 too many.

Several criteria for optimal smoothing, which will be described below,
have been used: (1) assignment of an arbitrary number of passes, (2) the
"$\epsilon$-test," (3) run statistics, and (4) increasing polynomial degree for each
successive pass.

The use of an arbitrary number of passes as a criterion for terminating

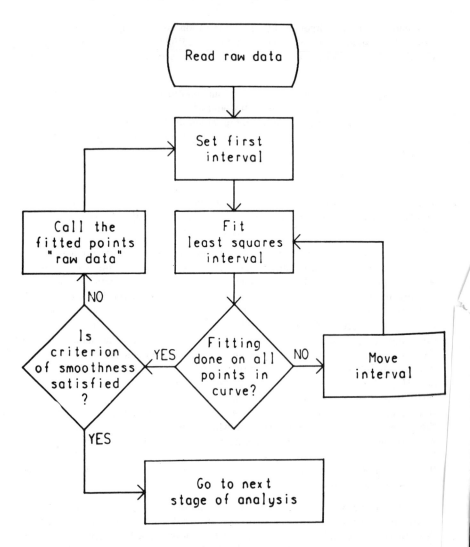

Fig. 5.   Flow diagram for the process of repetitively smoothing a curve by a moving interval of a low degree polynomial.

the smoothing process has little general utility, but can be used successful by the experimenter who has developed a "feeling" both for his data and for the smoothing process. An arbitrary number of iterations is useful in a program as an upper limit safeguard in case a set of data causes the program to go into an excessively long loop. The minor programming effort to incl

a counter of passes through the smoothing loop and testing that counter is
worthwhile, especially in a program which includes extensive data processing after the smoothing step. The upper limit value used should not exceed
50.

The $\epsilon$-test criterion was the first successful technique used to terminate
iterative smoothing in this work. It consists of an arbitrary input value, $\epsilon$,
which is used in a point-by-point comparison of the results of successive
passes of smoothing. Letting subscript i represent the data point and j represent the pass number, the test

$$\left| \gamma_{ij} - \gamma_{ij-1} \right| \leq \epsilon$$

is performed. If for any value of i the above relationship does not hold, an
additional pass of smoothing is initiated. When this relationship is satisfied
for all points, the smoothing process is terminated.

Useful results have been obtained from the $\epsilon$ test for smoothness. However, two serious disadvantages limit its general usefulness: first, the
smooth values produced are in no way tied to the original data, creating the
possibility of gross distortion; second, the selection of the input value for $\epsilon$
is totally arbitrary. The hazard of distorting the data is illustrated in Fig. 6
for a small section of a differential capacitance curve calculated by double
differentiation of experimental interfacial tension data. (The second derivative is used for illustrative purposes because the errors involved are too
small to be shown graphically for the original data or even for the charge
density.) The solid curve, displaying the general features of the characteristic "hump" was obtained by smoothing $\gamma$ using $\epsilon = 0.001$ dyn/cm, calculating charge density, smoothing $q^M$ using $\epsilon = 0.001$ $\mu C/cm^2$, and differentiating
these results. The dashed curve, in which the "hump" has become a plateau,
was obtained in the same way except that $\epsilon$ was one-half its former value.
Experience indicates that the best results are obtained using the $\epsilon$ test when
the value of $\epsilon$ is in the range of 1/10 to 1/20 of the uncertainty in the measured interfacial tension.

The criterion for terminating iterations of smoothing, which we have
designated "the run statistics criterion," was developed as a consequence of
the observations made during a detailed study of the $\epsilon$ test. By examining
smooth values obtained using progressively smaller values of $\epsilon$, it became
clear that initially the data were smoothed up or down in an apparently random fashion, but with additional iterations progressively longer sections of
a curve had all points "smoothed" upward while other sections were corrected
downward at all points. These results suggested that progressive distortion
of the data occurred as a function of increased iterations of smoothing and
led directly to the concept of run statistics.

"Run statistics" is a short term for the application of a statistical test
to an ordered sequence of objects of two types to determine if the particular

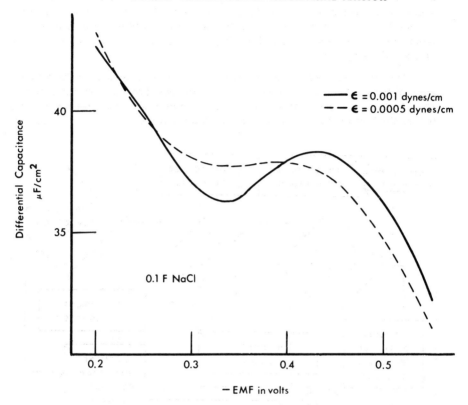

Fig. 6.    Magnified section of differential capacitance curves calculated from experimental interfacial tension measurements. See text for significance of the quantity $\epsilon$.

order at hand may be considered random at some specified probability level. In the application to data smoothing the two objects are the plus and minus signs associated with the deviations $\gamma_{smooth} - \gamma_{raw}$. The ordered sequence considered is taken from one end of the curve to the other. For example, a sequence such as + + - + - + - - + - appears random; + + + + + - - - - - is evidence of distortion. A run is defined as a group (one or more) of the same sign occurring together in the ordered sequence. Thus the first sequence above contains eight runs while the second has only two. Of course the intuitive feeling that the second sequence above is evidence of distortion cannot be translated directly into a computer program. However, a statistical test has been published [11] which does permit quantitative testing for

distortion defined in terms of a nonrandom arrangement of the ordered
sequence of the signs of the deviations. Appendix II on run statistics includes
a brief discussion of this statistical test as well as tables, calculated in this
laboratory, which are suitable for using this test in the context of smoothing
electrocapillary data.

The computer application of run statistics as a criterion for the termination of the iterative smoothing process is indicated by the flow diagram in
Fig. 7. It is assumed that one pass of smoothing has already been performed
before the test is started. Because the statistical test requires knowledge of

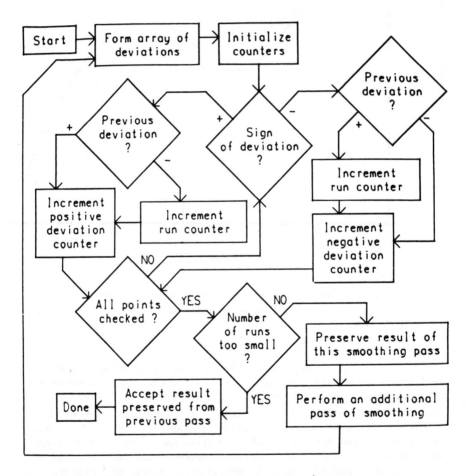

Fig. 7.    Flow diagram illustrating the application of the concept of run
statistics to terminating an iterative smoothing process. At the start it is
assumed that one pass of smoothing has been performed.

the total number of positive and negative deviations as well as the number of runs, three softwave counters are utilized in examining the deviations in order from one end of the curve. Since the identification of a run depends upon the sign of the deviation of the point at hand and the corresponding sign for the previous point, two flags are used to enable the computer to make the decisions labeled, "Previous deviation?" in Fig. 7. Details of the logic are illustrated by the listing of subroutine RUNCUR in Appendix I. The critical decision is, "Number of runs too small?" which asks if the number of runs counted in the deviations resulting from the current pass of smoothing is fewer than is consistent with a random arrangement.

As long as a random arrangement is indicated the smoothing process is continued; when a nonrandom arrangement of signs is detected the process is terminated and the smooth data are taken as those resulting from the previous pass (the last pass for which random order of deviations was indicated). Thus the procedure described here produces the maximum smoothing which does not introduce distortion as defined by the selection of the probability level (see Appendix II for discussion of probability level).

A fourth approach to iterative smoothing involves increasing the degree of the least squares polynomial on each successive pass. When combined with the use of a data rejection routine, this technique has produced satisfactory results [12].

## IV. INTEGRATION OF DIFFERENTIAL CAPACITANCE

A complete thermodynamic study of the electrical double layer at the electrode-solution interface includes measurements of the differential capacitance. The usual analysis of such measurements calls for the integration of these data. The process of integration presents no particular problem because differential capacitance data are generally of good quality, and integration — in contrast to differentiation — tends to smooth the data. The moving cubic technique applied to unsmoothed data has been used without difficulty to perform these integrations. A subroutine (GRATER) for integrating differential capacitance curves is included in Appendix I.

## V. THE ELECTROCAPILLARY SURFACE

The discussion to this point has considered interfacial tension and differential capacitance as functions of a single variable, primarily electrode potential. In a study of any particular system the real interest concerns functions of two variables, the electrode potential and the solution composition. In a typical investigation measurements are made on a series of solutions in which all composition variables are held constant except for the concentration of one component, e.g., the organic sorbate, whose concentration is different for each solution studied.

If the data were perfectly smooth, analysis could proceed simply, i.e., first consider $\gamma$ as a function of E alone and calculate the charge density for each solution. Then consider $\gamma$ as a function of $\mu$ alone and calculate the relative surface excess for each electrode potential. Since real data are not smooth, this procedure must be preceded by a smoothing step. A first approach at smoothing was to consider $\gamma$ as a function of a single variable, E, and smooth each curve. As a second step these smooth $\gamma$'s were treated as a function of a single variable, $\mu$, and again smoothed. While tolerable results could be obtained in this way, the inherent unity of the data as one function of two variables was ignored. This unity may be preserved by considering a functional form for $\gamma(E, \mu)$ which depicts the interfacial tension as a surface in a three-dimensional space extending over a plane whose points are the values of the independent variables E and $\mu$. One such functional form is

$$\gamma = aE^2 + b\mu^2 + cE\mu + dE + e\mu + f \tag{7}$$

where a, b, c, d, e, and f are constants.

The surface described by Eq. (7) is too simple to adequately describe any real family of electrocapillary curves, but it should be a reasonable approximation for any small patch of that surface. Thus, by using a moving technique in 3-space the complete set of electrocapillary curves for a system of interest may be analyzed as a unit. In practical application we have moved a patch over the surface by moving a rectangular area [the projection of the patch onto the (E, $\mu$) plane] through the (E, $\mu$) plane. In each position which the rectangular area occupies, the values of $\gamma$ corresponding to every (E, $\mu$) pair within that area are used to calculate a least squares fit to Eq. (7). The resultant coefficients, a...f, are used to calculate a smooth value for $\gamma$ at a central (E, $\mu$) point within the area. After a smooth value of $\gamma$ is calculated, the patch is moved and the process repeated until the entire surface has been smoothed. A program (SCONSF) and its associated subroutine (XYFIT2) for carrying out the surface smoothing procedure are included in Appendix I.

Smoothing a surface by the moving technique did not produce sufficient smoothing after one pass of the patch over the surface, just as in the case of one pass of an interval through a single curve. A repetitive technique was therefore adopted. At the present time, insufficient work has been done on the criterion for terminating the smoothing iterations. Useful results have been obtained from interfacial tension measurements on 5-chloro-1-pentanol in saturated NaF solution [13]. In this case the smoothing iterations were terminated arbitrarily after two passes of a 6 point × 6 point patch. This smoothing system is now developed only to the extent of treating data in which points have been taken at the same value of E for every solution. The values of q and $\Gamma$ are calculated from the same fitting which determines the smooth value of $\gamma$.

Application of the surface smoothing technique to differential capacitance data for the same chemical system [13] resulted in catastrophic failure
in the vicinity of the desorption peaks. Because differential capacitance data
can be integrated without smoothing, these data were analyzed by first
integrating each curve to charge density and then applying the surface
smoothing technique to the family of charge density curves. The necessity
for smoothing arose from the roughness of the data as a function of
concentration.

## VI.  DISCUSSION

The application of the digital computer to the thermodynamic analysis
of electrocapillary data can produce fruitful output or deception of the unwary.
The primary guideline in considering the results of any computer analysis
is that a computer cannot put out any better answers than are inherent in the
input data. Poor experimental data will not, by some digital magic, produce
good derived results. Given good experimental data, the computer can produce good derived results at significant savings in labor. Very early in our
work it became apparent that electrochemical judgment must be applied at all
stages of an experiment including questioning the significance of results
printed by a computer. If no programming errors have been made, there is
still no guarantee that results are meaningful. Differentiation of a quite
rough set of electrocapillary data could have led to a new theory of the electrical double layer to account for the resultant rising sawtooth charge density
curve; instead, these results suggested the need for smoothing experimental
data before differentiation.

The process used for smoothing raw data is critical to the entire
analysis of that data, because the analysis is actually performed on the
smooth data. The ideal smooth data faithfully represent all of the real features of the data and completely discard all of the random experimental error
associated with the measurements. The procedure to produce ideal smooth
data is not yet known. The techniques of single polynomial and moving fitting
described here are approaches toward that ideal. In any particular example,
the single polynomial method of fitting experimental data necessarily defines
$\gamma$ as a polynomial function of some particular degree in the independent
variable. This fact should be recalled when the derived results are considered, and the question should be asked, "Are the derived results a necessary
consequence of the original fitting of the data?" If so, of course, serious
questions arise as to the physical significance, if any, of those derived
results.

The moving technique does not force the data to fit a single polynomial
of some particular degree and, therefore, avoids the necessity of asking the
above question. Because the moving technique provides more variables in
defining the smoothing process, more care is required in its application.
Until large quantities of data collected in independent laboratories have been

analyzed, the guidelines given below must be considered as suggestions subject to modification as experience is amassed.

## A. Guidelines for Use of Moving Smoothing

1. Select a quadratic equation for the moving polynomial unless the data are expected to display a point of inflection. If the individual points are too widely separated to apply this guideline, the moving technique is probably not applicable, and more data are needed.

2. Use the shortest segment possible. For reasonably good interfacial tension data, use one degree of freedom per segment with the provision that an interval be extended if the sign of its second derivative implies a negative differential capacitance.

3. Terminate the iterative smoothing process by applying run statistics at the 0.20 probability level (good quality data) or the 0.10 level (medium quality data) or by the $\epsilon$ test using $\epsilon = 0.1x$ (uncertainty of the measurement) for medium quality data or $\epsilon = 0.05x$ (uncertainty of the measurement) for very high quality data.

4. Validate the output obtained. The first step in verifying that the smoothing process selected was appropriate is an inspection of the output for reasonableness in terms of known electrochemistry for the system investigated. Another simple verification method of the suitability of the choice of smoothing method is to repeat the process with a minor variation, e.g., reverse the order of the input data or use two degrees of freedom per segment, or terminate the iterations by $\epsilon$ test instead of run statistics. If the results obtained by the repetition do not agree, within experimental error, with those obtained the first time, the discrepancy must be resolved before the smooth data may be considered as a source of derived results.

## B. Advantages of Digital Computer Analysis

1. The speed of digital computers permits rapid analysis of data.

2. The data-handling capabilities of digital computers permit more data to be analyzed, thus leading to a more complete description of the electrochemical system.

3. Electrocapillary data can be analyzed as a function of two variables simultaneously.

## C. Disadvantage of Digital Computer Analysis

The psychological reaction to printed results can lead to a blind acceptance of a computer output as automatically valid. This disadvantage may be overcome only as the electrochemist understands what processing is actually being done to his experimental data when it is being treated by a particular computer program.

### ACKNOWLEDGMENT

This work was supported in part by the Air Force Office of Scientific Research, AFOSR(SRC)-OAR, USAF under Grants No. AF-AFOSR-68-1451 and No. AF-AFOSR-70-1887.

### APPENDIX I: FORTRAN LISTINGS

The routines listed here have been developed in the course of analyzing data in this laboratory. Although every routine presented here has been used successfully in the analysis of our data, the authors disclaim any responsibility for their performance in other situations. This disclaimer is given for several reasons, primarily as a warning that writing programs for a digital computer is an occupation beset with pitfalls, some of them slyly concealed. The reader may well be able to use these routines but should be aware that as yet unknown "bugs" may appear when applied to data which are in some significant (but not predictable) way different from those on which these routines have worked. Furthermore, in large computer systems every FORTRAN compiler has unique characteristics because local modifications are customarily introduced into the manufacturer's software. For example, at the Colorado State University Computer Center a new software system (CDC SCOPE 3.2) was introduced recently. We found that some programs which ran well under the old system (NCAR, which is the operation system for CDC 6600 and 6400 computers developed by the National Center for Atmospheric Research) did not even compile successfully under the new. Usually this difficulty was traced to a minor deviation from standard FORTRAN usage (e.g., accidentally introduced at the keypunch) which was tolerated by the former system. Most of the routines listed here have not yet been rerun under the new system. For the complete program listed here, special caution should be exercised in the matter of input/output statements which are usually slightly different from one computer system to another.

In using these routines, one should note that they have been developed in FORTRAN IV on a CDC 6400 computer which uses 14 or 15 decimal digits in single precision arithmetic. To implement these routines on computers using fewer digits in single precision arithmetic, many of the routines will have to be converted to double precision. Nonstandard subscripts (e.g., I+J-1) have been freely used because they are acceptable to CDC FORTRAN. To compile these routines under systems which do not accept this deviation from standard FORTRAN, it will be necessary to change the subscript notation. Although mixed mode arithmetic expressions have not been intentionally used, the reader should be alerted to the fact that CDC FORTRAN accepts these expressions; an accidental mixed mode expression may not have been detected.

Because these routines have been developed at various times for particular problems at hand, argument lists and sometimes COMMON statements
have been used as a means of transferring data between main programs and
subroutines. As a result, not all of these routines can be used in the same
program without introducing changes to make the method of transmission of
information mutually consistent.

CUBFIT (listing p. 25): This subroutine calculates the coefficients A,
B, C, D, of the cubic equation $S = AX^3 + BX^2 + CX + D$ which exactly fits the
four points, $S_i$, as a function of the corresponding four points, $X_i$. To use
this routine, the calling program must contain the COMMON block /CUBE/.
Before calling CUBFIT the calling program must place the coordinates ($X_i$,
$S_i$) for the four points to be fit into the designated words of /CUBE/. After
a call to CUBFIT is executed, the coefficients are available to the calling
program in words A, B, C, D of /CUBE/.

QCAL (listing p. 26): This subroutine calculates the charge density
from one electrocapillary curve. The calling program must have the first
150 words of blank COMMON arranged to correspond to the blank COMMON
list of QCAL, i.e., the array of electrode potentials first, the smooth
interfacial tension next, and the array into which the calculated charge densities will be placed last. The calling program must also include the
COMMON block /CALC/. Its first word must be the number of points in the
electrocapillary curve. The remaining eight words of the block are not used
in QCAL. Upon execution of a call to QCAL, the calculated charge densities
are available to the calling program in the third array of blank COMMON.

XCAL (listing p. 26): This subroutine calculates relative surface excess
for a curve of smooth interfacial tension as a function of RTlna for the
organic sorbate. The variable ACTLOG must be preset to RTlna by the
calling program (where a is the activity and R and T have their usual significance). Necessary communication with the calling program is provided
through COMMON statements. Blank COMMON contains, in order, the
arrays ACTLOG, smooth interfacial tension, and relative surface excess,
the first two to be supplied by the calling program, the last available to the
calling program after execution of a CALL XCAL statement. The calling
program must supply the number of points in the curve as the first word of
COMMON block /CALC/, the remaining eight words being irrelevant to
XCAL. Finally, the calling program must supply the temperature in degree
Celsius as the 36th word of COMMON block /HEAD/, the other words in that
block being irrelevant to XCAL.

PITCH (listing p. 27): This subroutine rejects outlying data points. It
is included here only to illustrate the technique used and is incompatible as
it stands with the previous routines. The COMMON block labeled /CALC/ is
not the same as the one in QCAL or XCAL. PITCH calls a subroutine

QUADSM which is not included here. Its functions are to calculate the coefficients A, B, C of the least squares quadratic function which fits one interval containing INTSZE number of points; to calculate the integer MIDPT which is the subscript of the middle point in the interval; and to calculate YORIG, a number used in normalizing the data. The calling program must supply the following information in the argument list in the CALL statement: INTSZE = number of points in the long interval to be used; S = multiplier of the "curve average standard deviation"; and M = a code to indicate if data rejection is to be applied to interfacial tension (M = 1) or to charge density (M = 2). The calling program must preset the following values in COMMON block /CALC/: NPTS = number of points in the curve; EMF = the array of electrode potential; SURF = the array of interfacial tension; and CHARGE = the array of charge density (need not be supplied if rejection of interfacial tension points is requested). After execution of a call to PITCH, the following information is available to the calling program via COMMON block /CALC/: NPTS = the number of points remaining in the curve after points have been rejected, and the arrays EMF and SURF or CHARGE now containing only the remaining points.

VLPOLY (listing p. 28): This subroutine performs one pass of moving least squares smoothing on a curve, fitting the moving interval by a polynomial of any desired degree. It may also be used for fitting a single least squares polynomial to an entire curve by letting the number of points in the interval (INTSZE) equal the number of points in the curve (NPTS). The arguments to be supplied by the calling routine are: NDEG = degree of polynomial, INTSZE = number of points to be included in the moving interval (not less than NDEG + 2), XVAL = array of independent variable values (e. g., electrode potentials), YVAL = array of dependent variable values (e. g., interfacial tension), IVL = a code controlling testing of the second derivative (= 1 means, test sign of second derivative; = 0 means, do not), and NPTS = number of points in the curve. Upon return to the calling program, the array YVAL contains the fitted values of the independent variable.

MATINV (listing p. 31): This subroutine inverts a matrix. It was taken from the literature [14] and modified to operate in our computation environment. The modifications consisted of correcting misprints and inserting additional CONTINUE statements to permit compilation under CDC SCOPE Fortran Extended. The argument A is the matrix, and N is its dimension. The inverted matrix is available in A upon return from this subroutine.

RUNCUR (listing p. 32): This routine smooths a single curve by an iterative process of moving interval fitting using the run statistics criterion for terminating the iterations. Before calling RUNCUR the calling program must preset the following variables in COMMON block /KURSM/: NPTS = number of points in curve; ISYS = a code to treat a special case which arose in this laboratory (in normal usage ISYS = 0); CONF = the cumulative normal (Gaussian) distribution value for the probability level being used; and

NCUT = the maximum number of smoothing iterations to be allowed. In calling RUNCUR the following arguments must be supplied in the CALL statement: XVAL = array of electrode potentials; RAWY = array of interfacial tension data; SMY = an array into which RUNCUR will place the smooth values; LLIM = the array dimensioned 50 × 25 which contains the run statistics data (e.g., Table 1 or Table 2 in Appendix II); NDEG = degree of the moving polynomial to be used; and IVL = a code to designate if the second derivation test is to be made (= 0 means, no test; = 1 means, test and extend interval if needed). After executing a call to RUNCUR the smooth values are available to the calling program in the array designated SMY in the argument list. In addition the number of iterations of the smoothing process is returned as NPASS in COMMON block /KURSM/. The integer KEY in that block is set to zero if the smoothing process was normal. If IVL is set to zero this routine should be applicable to data other than interfacial tension.

GRATER (listing p. 34): This subroutine integrates a curve of the dependent variable YG, a function of XG, to produce the function G at the points XG. Integration is done from the point CX where the value of the integral is CG. The N is the number of points in the curve. The calling routine must supply N, XG, YG, CX, and CG. Upon return to the calling routine, G is available to that calling routine.

SCONSF (listing p. 35): This program performs the surface smoothing described in this chapter. The subroutine XYFIT2, described below, is required by this program. Execution of this program produces an output listing for each solution with columns containing electrode potential, experimental interfacial tension, smooth interfacial tension, charge density, and relative surface excess. The variable names for the input data are: ALPH = an array of alphanumeric data which is printed out in the heading; NKURVE = number of interfacial tension curves making up the set of data; NEMF = number of points in each curve (same for all curves); NPATCH = size of moving patch; NLOOP = arbitrary number of times the patch is to be moved over the entire surface; CONC = activity (or concentration) of organic sorbate for which the interfacial tension curve is being read by the current pass of the loop which terminates at statement 20; EMF = electrode potential; and SURF = interfacial tension. This routine presently contains a known but uncured "bug." In the case where equally spaced values of EMF are used and the particular value zero is one potential, then using NPATCH = an odd number produces peculiar results in the vicinity of EMF = 0. Or, if equally spaced values of EMF are used, if there is no EMF = 0, if there are positive and negative EMF's of equal magnitude, and if NPATCH is even, peculiar results occur in the region of EMF = 0. The reason for this peculiarity is not now known, but, in those cases where it might be a problem, the difficulty may be circumvented by using an even value of NPATCH if EMF = 0 is a data point or an odd value if it is not.

XYFIT2 (listing p. 38): This subroutine performs a second degree least squares fit of a function of two variables for a set of, at most, 121 points. The number of points (NPTS) and their coordinates (X, Y, Z) are transmitted to XYFIT2 through COMMON block /TRIDIM/, and the coefficients of the least squares surface are made available to the calling program through the same block.

```
      SUBROUTINE CUBFIT                                         CUBFT 01
      DIMENSION Y(4), DEN(4)                                    CUBFT 02
      COMMON/CUBE/X(4),S(4),A,B,C,D                             CUBFT 03
      Y(1) = S(1)                                               CUBFT 04
      Y(2) = -S(2)                                              CUBFT 05
      Y(3) = S(3)                                               CUBFT 06
      Y(4) = -S(4)                                              CUBFT 07
      BA = X(2)-X(1)                                            CUBFT 08
      CA = X(3)-X(1)                                            CUBFT 09
      DA = X(4)-X(1)                                            CUBFT 10
      CB = X(3)-X(2)                                            CUBFT 11
      DB = X(4)-X(2)                                            CUBFT 12
      DC = X(4)-X(3)                                            CUBFT 13
      DEN(1) = BA*CA*DA                                         CUBFT 14
      DEN(2) = BA*CB*DB                                         CUBFT 15
      DEN(3) = DC*CA*CB                                         CUBFT 16
      DEN(4) = DC*DA*DB                                         CUBFT 17
      GB = X(2)                                                 CUBFT 18
      GC = X(3)                                                 CUBFT 19
      GD = X(4)                                                 CUBFT 20
      J = 1                                                     CUBFT 21
      A = 0.                                                    CUBFT 22
      B = 0.                                                    CUBFT 23
      C = 0.                                                    CUBFT 24
      D = 0.                                                    CUBFT 25
 8001 WA = GB + GC + GD                                         CUBFT 26
      WE = GB*GC + GB*GD + GC*GD                                CUBFT 27
      WI = -Y(J)/DEN(J)                                         CUBFT 28
      A = A + WI                                                CUBFT 29
      WI = Y(J)(WA/DEN(J)                                       CUBFT 30
      B = B + WI                                                CUBFT 31
      WI = -Z(J)*WE/DEN(J)                                      CUBFT 32
      C = C + WI                                                CUBFT 33
      WI = Y(J)*GB*GC*GD/DEN(J)                                 CUBFT 34
      D = D + WI                                                CUBFT 35
      GO TO (8002,8003,8004,8005),J                             CUBFT 36
 8002 GB = X(1)                                                 CUBFT 37
      J = 2                                                     CUBFT 38
      GO TO 8001                                                CUBFT 39
 8003 GC = X(2)                                                 CUBFT 40
      J = 3                                                     CUBFT 41
      GO TO 9001                                                CUBFT 42
 8004 GD = X(3)                                                 CUBFT 43
      J = 4                                                     CUBFT 44
      GO TO 8001                                                CUBFT 45
 8005 RETURN                                                    CUBFT 46
      END                                                       CUBFT 47
```

```
      SUBROUTINE QCAL                                         QCAL   01
C   VERSION OF 10/20/67
      COMMON EMF(50),SURF(50),Q(50)
      COMMON/CUBE/X(4),Y(4),A,B,C,D                           QCAL   03
      COMMON/CALC/NEMF,EPS,OPT(7)
      DO 5312 M=1,4                                           QCAL   05
      X(M)=EMF(M)                                             QCAL   06
 5312 Y(M)=SURF(M)                                            QCAL   07
      CALL CUBFIT                                             QCAL   08
      Q(1)=(-3.*A*EMF(1)*EMF(1)-2.*B*EMF(1)-C)/10.            QCAL   09
      Q(2)=(-3.*A*EMF(2)*EMF(2)-2.*B*EMF(2)-C)/10.            QCAL   10
      L=NEMF-3                                                QCAL   11
      DO 5314 I=2,L                                           QCAL   12
      J=I+1                                                   QCAL   13
      DO 5313 M=1,4                                           QCAL   14
      MM=M+I-1                                                QCAL   15
      X(M)=EMF(MM)                                            QCAL   16
 5313 Y(M)=SURF(MM)                                           QCAL   17
      CALL CUBFIT                                             QCAL   18
 5314 Q(J)=(-3.*A*EMF(J)*EMF(J)-2.*B*EMF(J)-C)/10.            QCAL   19
      J=NEMF-1                                                QCAL   20
      Q(J)=(-3.*A*EMF(J)*EMF(J)-2.*B*EMF(J)-C)/10.            QCAL   21
      J=NEMF                                                  QCAL   22
      Q(J)=(-3.*A*EMF(J)*EMF(J)-2.*B*EMF(J)-C)/10.            QCAL   23
      RETURN                                                  QCAL   24
      END                                                     QCAL   25


      SUBROUTINE XCAL                                         XCAL   01
C   VERSION OF 10/20/67                                       XCAL   02
      COMMON ACTLOG(50),SURF(50),GAM(50)                      XCAL   03
      COMMON/CALC/NCONC,EPS,OPT(7)                            XCAL   04
      COMMON/HEAD/STUFF(38)                                   XCAL   05
      COMMON/CUBE/X(4),Y(4),A,B,C,D                           XCAL   06
      EXCESS(ARG)=-(3.*A*ARG*ARG+2.*B*ARG+C)/(8.315E+07*ABTEMP)   XCAL   07
      ABTEMP=STUFF(36)+273.16                                 XCAL   08
      DO 5512 M=1,4                                           XCAL   09
      X(M)=ACTLOG(M)                                          XCAL   10
 5512 Y(M)=SURF(M)                                            XCAL   11
      CALL CUBFIT                                             XCAL   12
      DO 5513 J=1,2                                           XCAL   13
 5513 GAM(J)=EXCESS(ACTLOG(J))                                XCAL   14
      L=NCONC-3                                               XCAL   15
      DO 5515 I=2,L                                           XCAL   16
      J=I+1                                                   XCAL   17
      DO 5514 M=1,4                                           XCAL   18
      MM= M+I-1                                               XCAL   19
      X(M)=ACTLOG(MM)                                         XCAL   20
 5514 Y(M)=SURF(MM)                                           XCAL   21
      CALL CUBFIT                                             XCAL   22
 5515 GAM(J)=EXCESS(ACTLOG(J))                                XCAL   23
      GAM(NCONC-1)=EXCESS(ACTLOG(NCONC-1))                    XCAL   24
      GAM(NCONC)=EXCESS(ACTLOG(NCONC))                        XCAL   25
      RETURN                                                  XCAL   26
      END                                                     XCAL   27
```

```
      SUBROUTINE PITCH(INTSZE,S,M)
C VERSION OF 11/4/68
C M=1 MEANS DO PITCHING ON ARRAY 2 IN /CALC/
C M=2 MEANS DO PITCHING ON ARRAY 3 IN /CALC/
      DIMENSION MARK(100),TRIAL(30),DVINT(30),DEV(100),HOLD(100)
      COMMON DUMMY(2000)
      COMMON/CALC/NPTS,EMF(100),SURF(100),CHARGE(100),INT,MIDPT,A,B,C,
     1 YORIG,X(30),Y(30),KOUNT,JUNK(33)
      FIT(ARG)=A*ARG*ARG+B*ARG+C+YORIG
      DO 1 I = 1, NPTS
1     MARK(I)=0
      N=NPTS-INTSZE+1
      GRNDSM=0.
      IF(M.EQ.2) GO TO 101
      IEXIT=1
3     DO 20 INT=1,N
      CALL QUADSM(INTSZE)
      SUMSQ=0.
      DO 5 I = 1, INTSZE
      TRIAL(I)=FIT(X(I))
      DVINT(I)=TRIAL(I)-SURF(INT+I-1)
5     SUMSQ=SUMSQ+DVINT(I)*DVINT(I)
      FLN=INTSZE-1
      STDEV=SQRT(SUMSQ/FLN)
      GRNDSM=GRNDSM+STDEV
      IF(INT.NE.1) GO TO 7
      IX=MIDPT-1
      DO 6 I = 1, IX
6     DEV(I)=DVINT(I)
7     DEV(INT+MIDPT-1)=DVINT(MIDPT)
20    CONTINUE
      IX=MIDPT+1
      DO 21 I = IX,INTSZE
21    DEV(NPTS-INTSZE+I)=DVINT(IX)
      FLN=N
      STDEV=GRNDSM/FLN
      TEST=S*STDEV
      DO 22 I = 1, NPTS
      IF(ABS(DEV(I)).LT.TEST) GO TO 22
      MARK(I)=1
22    CONTINUE
      NKEEP=NPTS
      KOUNT=0
      GO TO (23,103),IEXIT
23    DO 40 I = 1, NKEEP
      IF(MARK(I).EQ.0) GO TO 40
      J=I-KOUNT
      WRITE(6,51)EMF(J),DEV(J),S,STDEV
51    FORMAT(1X,*EMF = *,F7.4,* DROPPED FOR RESID = *,E11.4/1X,*GREATER
     1THAN *,F3.1,* TIMES CURVE AVG. ST. DEV. OF *,E11.4)
      NPTS=NPTS-1
      KOUNT=KOUNT+1
      IF(J.ER.NPTS+1) GO TO 40
      DO 30 K=J,NPTS
      SURF(K)=SURF(K+1)
30    EMF(K)=EMF(K+1)
40    CONTINUE
      RETURN
101   DO 102 I = 1, NPTS
      HOLD(I)=SURF(I)
102   SURF(I)=CHARGE(I)
      IEXIT=2
      GO TO 3
103   DO 104 I = 1, NPTS
104   SURF(I)=HOLD(I)
      GO TO 23
      END
```

```
      SUBROUTINE VLPOLY(NDEG, INTSZE, XVAL, YVAL, IVL, NPTS)
C  VERSION OF 12/22/69
C  MOVING LEAST SQUARES INTERVAL FIT OF A COMPLETE CURVE BY A
C  POLYNOMIAL OF DEGREE NDEG USING AN INTERVAL SIZE OF INTSZE POINTS
C  PER INTERVAL.  ONE CALL OF ROUTINE FITS WHOLE CURVE WITH FITTED
C  VALUES RETURNED IN YVAL WHERE RAW DATA WAS.  IF INTSZE=NPTS, A SINGLE
C  FIT IS DONE ON THE WHOLE CURVE.  IVL = 0 MEANS USE FIXED LENGTH,
C  = 1 MEANS STRETCH INTERVAL TO INSURE 2ND DERIVATIVE IS NEGATIVE.
C  ON RETURN ROUTINE SETS IVL=0 IF NO INTERVAL WAS STRETCHED AND =1 IF
C  ONE OR MORE INTERVALS WERE EXTENDED
C  USES SUBROUTINE MATINV
C
      DIMENSION XVAL(1),YVAL(1),XMAT(50,50)
      COMMON/MATRIX/A(50),SUMXY(50),SUMX(100)
C
C  DIMENSIONS LIMIT EXECUTION TO 49TH DEGREE - TO CHANGE LIMIT XMAT
C  MUST BE DIMENSIONED (NDEG+1) X (NDEG+1), A AND SUMXY (NDEG+1),
C  SUMX(2*NDEG) AND MATINV SUBROUTINE DIMENSION MUST MATCH THAT OF XMAT
C  FIRST TEST FOR SUFFICIENT POINTS
C
      IF(INTSZE.GE.NDEG+2) GO TO 11
      WRITE(6,5) INTSZE, NDEG
   5  FORMAT(1X,*LEAST SQUARES FIT OF*,I3,1X,*POINT INTERVAL BY POLYNOMI
     1AL OF DEGREE*,I3,1X,*IMPOSSIBLE*)
      STOP
  11  LIM=2*NDEG
      LIMY=NDEG+1
C
C  NEXT SET UP LOOP TO MOVE INTERVAL THRU CURVE - KOUNT WILL RECORD
C  EXTENSIONS OF INTERVALS IF DESIRED AND NECESSARY.
C  FIRST TEST IF INTERVAL EXTENSION MAKES SENSE, IF NOT SET IVL=0
C
      KOUNT=0
      NEND=NPTS-INTSZE+1
      IF(INTSZE+4.LE.NPTS) GO TO 13
      IVL=0
  13  DO 100 INT=1, NEND
C
C  INT IS THE INDEX OF THE FIRST POINT OF THE INTERVAL CURRENTLY BEING
C  FITTED.
C
      IF(IVL.EQ.0) GO TO 15
C
C  THE NEXT SET OF DEFINITIONS ARE NEEDED ONLY IF INTERVAL EXTENSION
C  IS A POSSIBILITY
C
      INDEXL=INT
      INDEXR=INT+INTSZE-1
      JINT=INTSZE
  15  EN=INTSZE
C
C  CALCULATIONS WILL BE DONE WITH DATA NORMALIZED ABOUT MEANS OF X AND
C  Y.  FIRST FIND MEANS
C
```

```
      SX=0.
      SY=0.
      DO 20 I=1,INTSZE
      SX=SX+XVAL(INT+I-1)
 20   SY=SY+YVAL(INT+I-1)
      XBAR=SX/EN
      YBAR=SZ/EN
C
C   NEXT FORM SUMS FOR LEAST SQUARES SOLUTION
C
      DO 25 I=1,LIM
 25   SUMX(I)=0.
      DO 30 I=1,LIMY
 30   SUMXY(I)=0.
      DO 45 I=1, INTSZE
      XX=XVAL(INT+I-1)-XBAR
      YY=YVAL(INT+I-1)-YBAR
      DO 35 J=1,LIM
 35   SUMX(J)=SUMX(J)+XX**J
      SUMXY(1)=SUMXY(1)+YY
      DO 40 J=2,LIMY
 40   SUMXY(J)=SUMXY(J)+XX**(J-1)*YY
 45   CONTINUE
C
C   SUMS FORMED - SET UP MATRIX XMAT
C
 48   XMAT(1,1)=EN
      DO 50 I=2,LIMY
 50   XMAT(I,1)=SUMX(I-1)
      DO 55 J=2, LIMY
      DO 55 I=1,LIMY
 55   XMAT(I,J)=SUMX(I+J-2)
C
C   NEXT INVERT THE MATRIX WITH INVERSE PLACED IN XMAT IN LIEU OF
C   ORIGINAL MATRIX
C
      CALL MATINV(XMAT,LIMY)
C
C   NEXT CALCULATE ARRAY OF COEFFICIENTS
C
      DO 60 I=1,LIMY
      A(I)=0.
      DO 60 J=1, LIMY
 60   A(I)=A(I)+XMAT(I,J)*SUMXY(J)
C
C   TEST 2ND DERIVATIVE IF DESIRED
C
      IF(IVL.EQ.0) GO TO 82
C
C   TEST IS DONE AT EVERY POINT IN INTERVAL.  FAILURE AT ANY POINT
C   TERMINATES ROUTINE AND CAUSES INTERVAL EXTENSION
C
      DO 80 I=1,INTSZE
      SUM=0.
      NN=NDEG-1
      DO 62 J=1, NN
      FLJ=J
```

```
   62    SUM= SUM + FLJ*(FLJ+1.)*A(J+2)*(XVAL(INT+I-1)-XBAR)**(J-1)
         IF(SUM.LT.0.) GO TO 80
C
C  ROUTINE FOR INTERVAL EXTENSION ADAPTED FROM VLCUBE OF 9/20/68
C
         KOUNT=KOUNT+1
         IF(INDEXL.EQ.1) GO TO 72
         IOUT=1
   64    JINT=JINT+1
         INDEXL=INDEXL-1
         XX=XVAL(INDEXL)-XBAR
         YY=YVAL(INDEXL)-YBAR
   66    DO 68 J=1, LIM
   68    SUMX(J)=SUMX(J)+XX**J
         SUMXY(1)=SUMXY(1)+YY
         DO 70 J=2,LIMY
   70    SUMXY(J)=SUMXY(J)+XX**(J-1)*YY
         EN=JINT
         GO TO(72,48,48), IOUT
   72    IOUT=2
         IF(INDEXR.EQ.NPTS) GO TO 74
         JINT=JINT+1
         INDEXR=INDEXR+1
         XX=XVAL(INDEXR)-XBAR
         YY=YVAL(INDEXR)-YBAR
         GO TO 66
   74    IF(INDEXL.EQ.1) GO TO 76
         IOUT=3
         GO TO 64
   76    WRITE(6,78)
   78    FORMAT(1X,*VLPOLY COVERED WHOLE CURVE*)
         STOP
   80    CONTINUE
C
C  AT THIS POINT INTERVAL IS FITTED - NOW CALCULATE SMOOTH POINTS.
C
   82    IF(INT.NE.1) GO TO 90
         MID=INTSZE/2
         DO 85 I=1, MID
         XA=XVAL(I)-XBAR
         YVAL(I)=A(1)+YBAR
         DO 85 J=2,LIMY
   85    YVAL(I)=YVAL(I)+A(J)*XA**(J-1)
   90    I=INT+MID
         XA=XVAL(I)-XBAR
         YVAL(I)=A(1)+YBAR
         DO 92 J=2,LIMY
   92    YVAL(I)=YVAL(I)+A(J)*XA**(J-1)
  100    CONTINUE
C
C  NOW INTERVAL HAS MOVED THRU CURVE,  IT REMAINS TO CALCULATE RIGHT
C  END POINTS
C
         MID=2+ INTSZE/2
         DO 105 I= MID,INTSZE
         J=NPTS-INTSZE+I
         XA=XVAL(J)-XBAR

         YVAL(J)=A(1)+YBAR
         DO 105 K=2, LIMY
  105    YVAL(J)=YVAL(J)+A(K)*XA**(K-1)
C
C  THE LAST THING TO DO IS SET IVL TO SHOW IF ANY INTERVALS WERE
C  EXTENDED
C
         SUMXY(1)=XBAR
         SUMXY(2)=YBAR
         IF(IVL.EQ.0) RETURN
         IF(KOUNT.EQ.0) GO TO 110
         IVL=1
         RETURN
  110    IVL=0
         RETURN
         END
```

```
      SUBROUTINE MATINV(A,N)
C  VERSION OF 12/22/69
C  TAKEN FROM DICKSON, THE COMPUTER AND CHEMISTRY, FREEMAN, SAN
C  FRANCISCO, 1968, P. 142.
C
      DIMENSION A(50,50), IPV(50,3)
C  INITIALIZATION
      DO 1 J=1,N
    1 IPV(J,3)=0
      DO 3 I=1,N
C  SEARCH FOR PIVOT ELEMENT
      AMAX=0.
      DO 15 J=1,N
      IF(IPV(J,3)-1)7,15,7
    7 DO 6 K=1,N
      IF(IPV(K,3)-1)9,6,9
    9 IF(AMAX-ABS(A(J,K)))11,6,6
   11 IROW=J
      ICOLUM=K
      AMAX=ABS(A(J,K))
    6 CONTINUE
   15 CONTINUE
      IPV(ICOLUM,3)=IPV(ICOLUM,3)+1
      IPV(I,1)=IROW
      IPV(I,2)=ICOLUM
C  INTERCHARNGE ROWS TO PUT PIVOT ELEMENT ON DIAGONAL
      IF(IROW-ICOLUM ) 16,17,16
   16 DO 20 L=1,N
      SWAP=A(IROW,L)
      A(IROW,L)=A(ICOLUM,L)
   20 A(ICOLUM,L)=SWAP
C  DIVIDE PIVOT ROW BY PIVOT ELEMENT
   17 PIVOT=A(ICOLUM,ICOLUM)
      A(ICOLUM,ICOLUM)=1.0
      DO 23 L=1,N
   23 A(ICOLUM,L)=A(ICOLUM,L)/PIVOT
C  REDUCE THE NONPIVOT ROWS
      DO 3 L1=1,N
      IF(L1-ICOLUM)26,3,26
   26 T=A(L1,ICOLUM)
      A(L1,ICOLUM)=0.0
      DO 29 L=1,N
   29 A(L1,L)=A(L1,L)-A(ICOLUM,L)*T
    3 CONTINUE
C  INTERCHANGE THE COLUMNS
      DO 31 I=1,N
      L=N-I+1
      IF(IPV(L,1)-IPV(L,2)) 34,31,34
   34 JROW=IPV(L,1)
      JCOLUM=IPV(L,2)
      DO 32 K=1,N
      SWAP=A(K,JROW)
      A(K,JROW)=A(K,JCOLUM)
      A(K,JCOLUM)= SWAP

   32 CONTINUE
   31 CONTINUE
      RETURN
      END
```

```
          SUBROUTINE RUNCUR(XVAL,RAWY,SMY,LLIM,NDEG,IVL)
C     VERSION OF 12/22/69
C     TO SMOOTH A SINGLE CURVE BY SUCCESSIVE PASSES OF (NDEG+2) POINT
C     MOVING INTERVAL OF A POLYNOMIAL OF DEGREE NDEG.
C     STOP ITERATIONS ON THE BASIS OF NUMBER OF RUNS IN SIGN OF DEVIATIONS
C
          DIMENSION XVAL(1), RAWY(1), SMY(1), LLIM(50,25),SAVE(100),COMPAR(1
         100)
          COMMON/KURSM/ NPTS, NPASS, KEY, ISYS, CONF, NCUT
          IVSAVE=IVL
          NPASS=0
          KEY=0
          DO 1 I=1, NPTS
    1     SMY(I)=RAWY(I)
          INTSZE=NDEG+2
    3     CALL VLPOLY(NDEG,INTSZE,XVAL,SMY,IVL,NPTS)
          IF(IVSAVE.EQ.0) GO TO 5
          IF(IVL.EQ.0) GO TO 5
          NPASS=NPASS+1
          IVL=1
          IF(NPASS.LT.NCUT) GO TO 3
          KEY=2
          RETURN
    5     IF(ISYS.EQ.0) GO TO 9
          DO 7 I=1,NPTS
          SAVE(I)=SMY(I)
    7     COMPAR(I)=SMY(I)
          GO TO 13
    9     DO 11 I=1, NPTS
          SAVE(I)=SMY(I)
   11     COMPAR(I)=RAWY(I)
   13     NPASS=NPASS+1
          IF(NPASS.LT.NCUT) GO TO 15
          KEY=3
          RETURN
   15     IF(IVSAVE.EQ.1) IVL=1
          CALL VLPOLY(NDEG,INTSZE,XVAL,SMY,IVL,NPTS)
          IF(IVSAVE.EQ.0) GO TO 21
          IF(IVL.EQ.0) GO TO 21
          NPASS=NPASS+1
          DO 17 I=1, NPTS
   17     SAVE(I)=SMY(I)
          IF(NPASS.LT.NCUT) GO TO 15
          KEY=4
          RETURN
   21     NPLUS=0
          NMINUS=0
          NRUN=0
          KPLUS=0
          KMIN=0
          DO 31 I=1, NPTS
          DIFF=SMY(I)-COMPAR(I)
          IF(DIFF) 23,31,27
   23     IF(KMIN.EQ.1) GO TO 25
```

```
      KPLUS=0
      KMIN=1
      NRUN=NRUN+1
25    NMINUS=NMINUS+1
      GO TO 31
27    IF(KPLUS.EQ.1) GO TO 29
      KMIN=0
      KPLUS=1
      NRUN=NRUN+1
29    NPLUS=NPLUS+1
31    CONTINUE
      IF(NPLUS.GT.25.AND.NMINUS.GT.25) GO TO 45
      IF(NPLUS.GT.50.OR.NMINUS.GT.50) GO TO 43
      IF(NPLUS.LT.NMINUS) GO TO 41
32    KUTOFF=LLIM(NPLUS,NMINUS)
      IF(KUTOFF.NE.0) GO TO 33
      KEY=1
      RETURN
33    IF(NRUN.LE.KUTOFF) GO TO 37
      DO 35 I=1,NPTS
35    SAVE(I)=SMY(I)
      GO TO 13
37    DO 39 I=1, NPTS
39    SMY(I)=SAVE(I)
      KEY=0
      RETURN
41    KEEP=NPLUS
      NPLUS=NMINUS
      NMINUS=KEEP
      GO TO 32
43    KEY=5
      RETURN
45    ENP=NPLUS
      ENM=NMINUS
      TWONN=2.*ENP*ENM
      ENEN=ENP+ENM
      KUTOFF=TWONN/ENEN+1.-CONF*SQRT(TWONN*(TWONN-ENEN)/(ENEN*ENEN*
     1 (ENEV-1.)))
      GO TO 33
      END
```

```
      SUBROUTINE GRATER(N,XG,YG,G,CX,CG)
C  VERSION OF 7/16/70
C  TO CALCULATE N POINTS OF G, THE INTEGRAL OF YG(XG) WHERE CG IS THE
C  VALUE OF THE INTEGRATION CONSTANT AT XG=CX.  XG MUST BE STRICTLY
C  INCREASING.  NOTE THAT IF YG = CHARGE, THE RESULTANT G MUST BE
C  MULTIPLIED BY -10 TO BE INTERFACIAL TENSION.
C  USES SUBROUTINE CUBFIT.
      COMMON /CUBE/ X(4),Y(4),A,B,C,D
      DIMENSION XG(1), YG(1), G(1)
      GRAL(ARG)=ARG*(D+(ARG*(C/2.+ARG*(B/3.+ARG*A/4.))))
      DO 1 I = 1,N
      IF(CX.LE.XG(I)) GO TO 4
1     CONTINUE
2     WRITE(6,3) CX
3     FORMAT(1X,*INTEGRATION CONSTANT NOT IN INTERIOR OF CURVE*,E12.5)
      STOP
4     IF(I.LE.3.OR.I.GE.N-2) GO TO 2
      DO 5 J=1,4
      X(J)=XG(I+J-3)
5     Y(J)=YG(I+J-3)
      CALL CUBFIT
      G(I)=GRAL(XG(I))-GRAL(CX)+CG
      G(I-1)=GRAL(XG(I-1))-GRAL(CX)+CG
      L=N-3
      M=I-1
      DO 7 K=M,L
      DO 6 J=1,4
      X(J)=XG(K+J-1)
6     Y(J)=YG(K+J-1)
      CALL CUBFIT
7     G(K+2)=G(K+1)+GRAL(XG(K+2))-GRAL(XG(K+1))
      G(N)=G(N-1)+GRAL(XG(N))-GRAL(XG(N-1))
8     DO 9 J=1,4
      X(J)=XG(M+J-3)
9     Y(J)=YG(M+J-3)
      CALL CUBFIT
      G(M-1)=G(M)+GRAL(XG(M-1))-GRAL(XG(M))
      IF(M.EQ.3) GO TO 10
      M=M-1
      GO TO 8
10    G(1)=G(2)+GRAL(XG(1))-GRAL(XG(2))
      RETURN
      END
```

```
      PROGRAM SCONSF
C  VERSION OF 7/30/69
C  TO TAKE A SET OF ELECTROCAPILLARY CURVES AND SMOOTH THEM BY FITTING
C  A MOVING 2ND DEGREE PATCH TO THE ELECTROCAPILLARY SURFACE
C  DATA MUST BE ON A REGULAR GRID WITH NO MISSING POINTS
C  GRID INTERVALS NEED NOT BE CONSTANT
      COMMON ALPH(16),EMF(50),YMU(50),SURF(50,50),SMOOTH(50,50),CHARGE(5
     10,50),XCESS(50,50),CONC(50),RAWSUR(50,50)
      COMMON /TRIDIM/ NPTS,X(121),Y(121),Z(121),A,B,C,D,E,F
      FIT(ARG1,ARG2)=A*ARG1*ARG1+B*ARG2*ARG2+C*ARG1*ARG2+D*ARG1+E*ARG2+F
      QFIT(ARG1,ARG2)=-0.1*(2.*A*ARG1+C*ARG2+D)
      XFIT(ARG1,ARG2)=-(2.*B*ARG2+C*ARG1+E)
      R=8.31436E+07
      T=298.18
      READ(5,1) (ALPH(I),I=1,16)
  1   FORMAT(8A10)
C  ALPHAMERIC HEADING INFO TAKES TWO CARDS
      WRITE(6,2)
  2   FORMAT(1H1,*SELF-CONSISTANT SURFACE ANALYSIS, PROGRAM SCONSF 7/29/
     169*/)
      WRITE(6,3)(ALPH(I),I=1,16)
  3   FORMAT(1X,8A10)
      READ(5,6)NKURVE,NEMF,NPATCH,NLOOP
  6   FORMAT(4I5)
      NP=(NPATCH-NPATCH/2)-1
      DO 20 N=1, NKURVE
      READ(5,10)CONC(N)
  10  FORMAT(E12.5)
C  YMU IS SOLUTE ACTIVITY FOR A SINGLE CURVE AND BECOMES CHEMICAL
C  POTENTIAL
      YMU(N)=R*T*ALOG(CONC(N))
      READ(5,12)(EMF(I),SURF(I,N),I=1,NEMF)
  12  FORMAT(2F10.5)
  20  CONTINUE
C  THIS COMPLETES INPUT DATA
      DO 25 I=1,NEMF
      DO 25 J=1,NKURVE
  25  RAWSUR(I,J)=SURF(I,J)
      DO 72 LOOP=1,NLOOP
C  MOVING PATCH WILL BE NPATCH X NPATCH (LIMIT NPATCH = 11)
      NPTS=NPATCH*NPATCH
      DO 30 I=1, NPATCH
      DO 30 J=1, NPATCH
      L=J+NPATCH*(I-1)
      X(L)=EMF(I)
      Y(L)=YMU(J)
  30  Z(L)=SURF(I,J)
      CALL XYFIT2
      NPTC=NPATCH/2+1
      DO 31 I = 1, NPTC
      DO 31 J=1,NPTC
      SMOOTH(I,J)=FIT(EMF(I),YMU(J))
      CHARGE(I,J)=QFIT(EMF(I),YMU(J))
  31  XCESS(I,J)=XFIT(EMF(I),YMU(J))
```

```
C   TOP LEFT CORNER DONE - NEXT TOP EDGE
        NPCM=NPATCH-1
        NPCH=NPATCH+1
        DO 40 N=NPCH,NEMF
        IS=N-NPCM
        DO 35 I = IS,N
        DO 35 J=1,NPATCH
        L=J+NPATCH*(I-IS)
        X(L)=EMF(I)
        Y(L)=YMU(J)
 35     Z(L)=SURF(I,J)
        CALL XYFIT2
        DO 40 J=1,NPTC
        SMOOTH(N-NP,J)=FIT(EMF(N-NP),YMU(J))
        CHARGE(N-NP,J)=QFIT(EMF(N-NP),YMU(J))
 40     XCESS(N-NP,J)=XFIT(EMF(N-NP),YMU(J))
        NN=NEMF-NP +1
        DO 41 I=NN,NEMF
        DO 41 J=1,NPTC
        SMOOTH(I,J)=FIT(EMF(I),YMU(J))
        CHARGE(I,J)=QFIT(EMF(I),YMU(J))
 41     XCESS(I,J)=XFIT(EMF(I),YMU(J))
C   THIS COMPLETES TOP EDGE
        NK=NKURVE-1
        DO 60 M= NPCH,NK
        JS=M-NPCM
        DO 45 I=1,NPATCH
        DO 45 J=JS,M
        L=J-JS+1+NPATCH*(I-1)
        X(L)=EMF(I)
        Y(L)=YMU(J)
 45     Z(L)=SURF(I,J)
        CALL XZFIT2
        DO 46 I=1,NPTC
        SMOOTH(I,M-NP)=FIT(EMF(I),YMU(M-NP))
        CHARGE(I,M-NP)=QFIT(EMF(I),YMU(M-NP))
 46     XCESS(I,M-NP)=XFIT(EMF(I),YMU(M-NP))
        DO 55 N=NPCH,NEMF
        IS=N-NPCM
        DO 50 I=IS,N
        DO 50 J=JS,M
        L=J-JS+1+NPATCH*(I-IS)
        X(L)=EMF(I)
        Y(L)=YMU(J)
 50     Z(L)=SURF(I,J)
        CALL XYFIT2
        SMOOTH(N-NP,M-NP)=FIT(EMF(N-NP),YMU(M-NP))
        CHARGE(N-NP,M-NP)=QFIT(EMF(N-NP),YMU(M-NP))
 55     XCESS(N-NP,M-NP)=XFIT(EMF(N-NP),YMU(M-NP))
        DO 60 I= NN,NEMF
        SMOOTH(I,M-NP)=FIT(EMF(I),YMU(M-NP))
        CHARGE(I,M-NP)=QFIT(EMF(I),YMU(M-NP))
 60     XCESS(I,M-NP)=XFIT(EMF(I),YMU(M-NP))
C   ONLY BOTTOM EDGE LEFT TO DO
        JS=NKURVE-NPCM
        DO 65 I=1,NPATCH
        DO 65 J=JS,NKURVE
```

```
         L=J-JS+1+NPATCH*(I-1)
         X(L)=EMF(I)
         Y(L)=YMU(J)
  65     Z(L)=SURF(I,J)
         CALL XZFIT2
         DO 66 I=1,NPTC
         JI=NKURVE-NP
         DO 66 J=JI,NKURVE
         SMOOTH(I,J)=FIT(EMF(I),YMU(J))
         CHARGE(I,J)=QFIT(EMF(I),YMU(J))
  66     XCESS(I,J)=XFIT(EMF(I),YMU(J))
         DO 70 N=NPCH,NEMF
         IS=N-NPCM
         DO 68 I=IS,N
         DO 68 J=JS,NKURVE
         L=J-JS+1+NPATCH*(I-IS)
         X(L)=EMF(I)
         Y(L)=YMU(J)
  68     Z(L)=SURF(I,J)
         CALL XYFIT2
         DO 70 J=JI,NKURVE
         SMOOTH(N-NP,J)=FIT(EMF(N-NP),YMU(J))
         CHARGE(N-NP,J)=QFIT(EMF(N-NP),YMU(J))
  70     XCESS(N-NP,J)=XFIT(EMF(N-NP),YMU(J))
         DO 71 I=NN,NEMF
         DO 71 J=JI,NKURVE
         SMOOTH(I,J)=FIT(EMF(I),YMU(J))
         CHARGE(I,J)=QFIT(EMF(I),YMU(J))
  71     XCESS(I,J)=XFIT(EMF(I),YMU(J))
         DO 72 I=1,NEMF
         DO 72 J=1,NKURVE
  72     SURF(I,J)=SMOOTH(I,J)
C  CALCULATION COMPLETE
         DO 100 N=1,NKURVE
         WRITE(6,73) CONC(N)
  73     FORMAT(1H1,*FOR CONCENTRATION = *,E12.5,* MOLES/LITER*)
         WRITE(6,75) YMU(N)
  75     FORMAT(1X,*FOR CHEMICAL POTENTIAL = *,E12.5)
         WRITE(6,76)NPATCH,NPATCH,NLOOP
  76     FORMAT(1X,*USING *,I2,* X *,I2,* PATCH MOVED OVER SURFACE *,I2,* T
        1IMES*)
         WRITE(6,78)
  78     FORMAT(3X,*EMF*,7X,*RAW*,5X,*SMOOTH*,4X,*CHARGE*,6X,*EXCESS*)
         WRITE(6,80)(EMF(I),RAWSUR(I,N),SMOOTH(I,N),CHARGE(I,N),XCESS(I,N),
        1I=1,NEMF)
 100     CONTINUE
  80     FORMAT(1X,F7.4,2(2X,F8.3),2X,F8.4,2X,E12.5)
         STOP
         END
```

```
      SUBROUTINE XYFIT2
C  VERSION OF 12/18/68
C  TO FIT A PATCH OF THE GENERAL 2ND DEGREE LEAST SQUARES SURFACE
C  OF Z OVER X AND Y
      COMMON /TRIDIM/ NPTS,X(121),Y(121),Z(121),A,B,C,D,E,F
C  THE VALUES OF X AND Y MAY BE REDUNDANT, I.E. FOR 3 POINTS EACH ON
C  3 CURVES AT THE SAME EMFS THE ARRAYS X AND Y WOULD BE  EMF(1),EMF(2)
C  EMF(3),EMF(1),EMF(2),EMF(3),EMF(1),EMF(2),EMF(3),MU(1),MU(1),MU(1),
C  MU(2),MU(2),MU(2),MU(3),MU(3),MU(3)
C
C  THE VALUES OF THE COEFFICIENTS ARE RETURNED AS A,B,C,D,E,F ACCORDING
C  TO Z=AX**2+BY**2+CXY+DX+EY+F
C
C  INITIALIZE SUMS
      SX=0.
      SY=0.
      SZ=0.
      SZX=0.
      SZY=0.
      SXY=0.
      SZXY=0.
      SX2=0.
      SY2=0.
      SZX2=0.
      SZY2=0.
      SX2Y2=0.
      SX2Y=0.
      SXY2=0.
      SX3=0.
      SY3=0.
      SX3Y=0.
      SXY3=0.
      SX4=0.
      SY4=0.
C  THEN FORM SUMS
      DO 10 I = 1, NPTS
      SX=SX+X(I)
      SY=SY+Y(I)
      SZ=SZ+Z(I)
      SZX=SZX+Z(I)*X(I)
      SZY=SZY+Z(I)*Y(I)
      SXY=SXY+X(I)*Y(I)
      SZXY=SZXY+Z(I)*X(I)*Y(I)
      XSQ=X(I)*X(I)
      SX2=SX2+XSQ
      YSQ=Y(I)*Y(I)
      SY2=SY2+YSQ
      SZX2=SZX2+Z(I)*XSQ
      SZY2=SZY2+Z(I)*YSQ
      SX2Y2=SX2Y2+XSQ*YSQ
      SX2Y=SX2Y+XSQ*Y(I)
      SXY2=SXY2+X(I)*YSQ
      SX3=SX3+X(I)*XSQ
      SY3=SY3+Y(I)*YSQ
```

```
          SX3Y=SX3Y+X(I)*XSQ*Y(I)
          SXY3=SXY3+X(I)*YSQ*Y(I)
          SX4=SX4+XSQ*XSQ
   10     SY4=SY4+YSQ*YSQ
          EN=NPTS
C   START CALCULATION OF FIRST COEFFICIENT
          SOVT=SX2Y/SX2
          ALPHA=SY3-SY2*SOVT
          BETA=SXY2-SXY*SOVT
          GAMMA=SXY-SX*SOVT
          DELTA=SY2-SY*SOVT
          EPS=SY-EN*SOVT
          GOVT=SZX2/SX2
          ETA=SZY2-SY2*GOVT
          EMOVT=SX4/SX2
          PSI=SX2Y2-SY2*EMOVT
          POVT=SX2Y2/SX2
          RHO=SY4-SY2*POVT
          ROVT=SX3/SX2
          CAPLAM=SXY2-SY2*ROVT
          QOVT=SX3Y/SX2
          YA=SXY3-SZ2*QOVT
          CLOVAL=CAPLAM/ALPHA
          SMLAM=SX2Y-SXY*ROVT-BETA*CLOVAL
          TVERT=SX2-SX*ROVT-GAMMA*CLOVAL
          BACKE=SXY-SY*ROVT-DELTA*CLOVAL
          BACKF=SX-EN*ROVT-EPS*CLOVAL
          ETOVAL=ETA/ALPHA
          RHOVAL=RHO/ALPHA
          YAOVAL=YA/ALPHA
          BEOVAL=BETA/ALPHA
          DOWNE=SZX2-SXY*GOVT-BETA*ETOVAL
          CHAIR=SXY3-SXY*POVT-BETA*RHOVAL
          SIGMA=SX2Z2-SXY*GOVT-BETA*YAOVAL
          SMMU=SX3Y-SXY*EMOVT-PSI*BEOVAL
          CAPSIG=SZX-SX*GOVT-GAMMA*ETOVAL-TVERT*DOWNE/SMLAM
          OMEGA=SXY2-SX*POVT-GAMMA*RHOVAL-TVERT*CHAIR/SMLAM
          CAPDEL=SX2Y-SX*QOVT-GAMMA*YAOVAL-TVERT*SIGMA/SMLAM
          BACKC=SXY2-SY*QOVT-DELTA*YAOVAL-BACKE*SIGMA/SMLAM
          BACKEP=SXY-EN*QOVT-EPS*YAOVAL-BACKF*SIGMA/SMLAM
          UVERT=SX3-SX*EMOVT-GAMMA*PSI/ALPHA-TVERT*SMMU/SMLAM
          AN1=SZZ-SY *GOVT-DELTA*ETOVAL-BACKE*DOWNE/SMLAM-BACKC*CAPSIG/CAPDE
         1L
          AN2=SY2-EN*POVT-EPS*RHOVAL-BACKF*CHAIR/SMLAM-BACKEP*OMEGA/CAPDEL
          AN3=SZ-EN*GOVT-EPS*ETOVAL-BACKF*DOWNE/SMLAM-BACKEP*CAPSIG/CAPDEL
          AN4=SY3-SY*POVT-DELTA*RHOVAL-BACKE*CHAIR/SMLAM-BACKC*OMEGA/CAPDEL
          AD1=SX2Y-SY*EMOVT-DELTA*PSI/ALPHA-BACKE*SMMU/SMLAM-BACKC*UVERT/CAP
         1DEL
          AD3=SX2-EN*EMOVT-EPS*PSI/ALPHA-BACKF*SMMU/SMLAM-BACKEP*UVERT/CAPDE
         1L
C   GET FIRST COEFFICIENT
          A=(AN1(AN2-AN3*AN4)/(AD1*AN2-AD3*AN4)
C   ON TO SECOND
          B=(AD1*AN3-AD3*AN1)/(AD1*AN2-AD3*AN4)
C   ON TO THIRD
          SMA=SX2Y-SY*EMOVT-DELTA*PSI/ALPHA-BACKE*SMMU/SMLAM
          SMB=SX2-EN*EMOVT-EPS*PSI/ALPHA-BACKF*SMMU/SMLAM
```

```
      CN1=SY3-SY*POVT-DELTA*RHOVAL-BACKE*CHAIR/SMLAM-SMA*OMEGA/UVERT
      CN2=SZ-EN*GOVT-EPS*ETOVAL-BACKF*DOWNE/SMLAM-SMB*CAPSIG/UVERT
      CN3=SY2-EN*POVT-EPS*RHOVAL-BACKF*CHAIR/SMLAM-SMB*OMEGA/UVERT
      CN4=SZY-SY*GOVT-DELTA*ETOVAL-BACKE*DOWNE/SMLAM-SMA*CAPSIG/UVERT
      CD2=SXY-EN*QOVT-EPS*YAOVAL-BACKF*SIGMA/SMLAM-SMB*CAPDEL/UVERT
      CD4=SXZ2-SY*QOVT-DELTA*YAOVAL-BACKE*SIGMA/SMLAM-SMA*CAPDEL/UVERT
      C=(CN1*CN2-CN3*CN4)/(CN1*CD2-CN3*CD4)
C  ON TO FOURTH
      POVM=SX2Y2/SX4
      SMD=SY4-SX2Y2*POVM
      SME=SXY3-SX3Y*POVM
      SMF=SXZ2-SX3*POVM
      SMG=SY3-SX2Y*POVM
      SMJ=SY2-SX2*POVM
      SMK=SZY2-SZX2*POVM
      QOVM=SX3Y/SX4
      EOVD=SME/SMD
      SMQ=SX2Y2-SX3Y*QOVM-SME*EOVD
      SMR=SX2Y-SX3*QOVM-SMF*EOVD
      SMS=SXY2-SX2Y*QOVM-SMG*EOVD
      SMT=SXY-SX2*QOVM-SMJ*EOVD
      SMX=SZXY-SZX2*QOVM-SMK*EOVD
      ROVM=SX3/SX4
      FOVD=SMF/SMD
      ROVQ=SMR/SMQ
      SMY=SZX-SZX2*ROVM-SMK*FOVD-SMX*ROVQ
      SMZ=SXY-SX2Y*ROVM-SMG*FOVD-SMS*ROVQ
      AVERT=SX-SX2*ROVM-SMJ*FOVD-SMT*ROVQ
      AKET=SY-SX2Y*SX2/SX4-SMG*SMJ/SMD-SMS*SMT/SMQ
      ABRA=EN-SX2*SX2/SX4-SMJ*SMJ/SMD-SMT*SMT/SMQ
      BRACE=SX2-SX3*ROVM-SMF*FOVD-SMR*ROVQ
      DN1=SZZ-SX2Y*SZX2/SX4-SMG*SMK/SMD-SMS*SMX/SMQ-AKET*SMY/AVERT
      DN2=SY-SX2*SX2Y/SX4-SMJ*SMG/SMD-SMS*SMT/SMQ-ABRA*SMZ/AVERT
      DN3=SZ-SX2*SZX2/SX4-SMJ*SMK/SMD-SMT*SMX/SMQ-ABRA*SMY/AVERT
      DN4=SY2-SX2Y*SX2Y/SX4-SMG*SMG/SMD-SMS*SMS/SMQ-AKET*SMZ/AVERT
      DD1=SMZ-AKET*BRACE/AVERT
      DD3=AVERT-ABRA*BRACE/AVERT
      D=(DN1(DN2-DN3*DN4)/(DD1*DN2-DD3*DN4)
C  ON TO FIFTH
      E=(DD1*DN3-DD3*DN1)/(DD1*DN2-DD3*DN4)
C  ON TO LAST
      FN1=SY2-SX2Y*SX2Y/SX4-SMG*SMG/SMD-SMS*SMS/SMQ-SMZ*SMZ/BRACE
      FN2=SZ-SX2*SZX2/SX4-SMJ*SMK/SMD-SMT*SMX/SMQ-AVERT*SMY/BRACE
      FN3=SY-SX2*SX2Y/SX4-SMJ*SMG/SMD-SMS*SMT/SMQ-AVERT*SMZ/BRACE
      FN4=SZY-SX2Y*SZX2/SX4-SMG*SMK/SMD-SMS*SMX/SMQ-SMZ*SMY/BRACE
      FD2=EN-SX2*SX2/SX4-SMJ*SMJ/SMD-SMT*SMT/SMQ-AVERT*AVERT/BRACE
      F=(FN1*FN2-FN3*FN4)/(FN1*FD2-FN3*FN3)
      RETURN
      END
```

## APPENDIX II: RUN STATISTICS

The concept of the statistical test for a nonrandom arrangement of positive and negative deviations may be grasped by considering a fixed number of positive and negative deviations and letting u be the number of runs in the signs of the deviations when they are ordered from one end of a curve to the other. A critical number of runs, $u'$, exists for any given probability level, $\delta$, such that

$$Pr\{u \leq u'\} = \delta$$

which is read, "The probability that u is less than or equal to $u'$ is $\delta$." The value of $\delta$ is a function of $u'$ and the fixed number of positive and negative deviations. Combinatorial formulas for calculating $\delta$ are given by Swed and Eisenhart [11] along with tables of the largest integer, $u'$, consistent with values of $\delta$ = 0.005, 0.01, 0.025, and 0.05; additional tables for $\delta$ = 0.10 and 0.20 have been calculated from the formulas of Swed and Eisenhart and are included here as Tables 1 and 2.

To illustrate the use of these tables, the sequences of signs given previously, (a) ++-+-+--+- and (b) +++++-----, may be tested. One counts the number of plus signs (five in each case), the number of minus signs (five in each case), and the number of runs (eight in case a, two in case b) and selects a probability level (0.20 for example). Referring to a table of critical values of $u'$ (Table 2 for example), enter the column M for the number of plus signs and the row N for the number of minus signs. For $\delta$ = 0.20 and 5 plus and 5 minus signs, the critical value of the number of runs is 4. Thus we conclude that case a is consistent with random deviations but that in case b the 2 runs are too few, compared with the critical value of 4, to represent random deviations.

Had a probability level of 0.10 been selected (Table 1) in the above example, the critical value of $u'$ would have been 3 and the conclusions would have been unchanged. The significance of the probability level in the smoothing problem is that a lower probability level permits more iterations of smoothing before termination. Experience suggests that $\delta$ = 0.20 is appropriate when the data set contains 30-40 points and smaller values of $\delta$ are suitable for larger data sets.

In using Tables 1 and 2 it should be noted that the values 0 entered in the table have no statistical significance. They have been entered only to provide a complete 2-dimensional array for the computer and are interpreted by the program as indicating a nonrandom case (see listing of subroutine RUNCUR). Since the table is symmetric, a full array is not used; the cases omitted are treated by interchanging counts of positive and negative deviations. For those cases where the number of plus and minus signs both exceed 25, a normal approximation for the critical value may be used [11]. The formula for its calculation is included in the listing of RUNCUR in Appendix I.

## TABLE 1
### Critical Values of Number of Runs, u; for Probability Level $\delta = 0.10$

| N\M | 1 | 2 | 3 | 4 | 5 | 6 | 7 | 8 | 9 | 10 | 11 | 12 | 13 | 14 | 15 | 16 | 17 | 18 | 19 | 20 | 21 | 22 | 23 | 24 | 25 |
|---|---|---|---|---|---|---|---|---|---|---|---|---|---|---|---|---|---|---|---|---|---|---|---|---|---|
| 1 | 0 | 0 | 0 | 0 | 0 | 0 | 0 | 0 | 0 | 0 | 0 | 0 | 0 | 0 | 0 | 0 | 0 | 0 | 0 | 0 | 0 | 0 | 0 | 0 | 0 |
| 2 | 0 | 0 | 0 | 0 | 2 | 2 | 2 | 2 | 2 | 2 | 2 | 2 | 2 | 2 | 2 | 2 | 2 | 2 | 3 | 3 | 3 | 3 | 3 | 3 | 3 |
| 3 | 0 | 0 | 2 | 2 | 2 | 2 | 3 | 3 | 3 | 3 | 3 | 3 | 3 | 3 | 4 | 4 | 4 | 4 | 4 | 4 | 4 | 4 | 4 | 4 | 4 |
| 4 | 0 | 0 | 2 | 2 | 3 | 3 | 3 | 3 | 4 | 4 | 4 | 4 | 4 | 4 | 4 | 5 | 5 | 5 | 5 | 5 | 5 | 5 | 5 | 5 | 5 |
| 5 | 0 | 2 | 2 | 3 | 3 | 3 | 4 | 4 | 4 | 5 | 5 | 5 | 5 | 5 | 5 | 6 | 6 | 6 | 6 | 6 | 6 | 6 | 6 | 6 | 6 |
| 6 | 0 | 2 | 2 | 3 | 3 | 4 | 4 | 5 | 5 | 5 | 5 | 6 | 6 | 6 | 6 | 6 | 7 | 7 | 7 | 7 | 7 | 7 | 7 | 8 | 8 |
| 7 | 0 | 2 | 3 | 3 | 4 | 4 | 5 | 5 | 5 | 6 | 6 | 6 | 7 | 7 | 7 | 7 | 7 | 8 | 8 | 8 | 8 | 8 | 8 | 8 | 8 |
| 8 | 0 | 2 | 3 | 3 | 4 | 5 | 5 | 5 | 6 | 6 | 7 | 7 | 7 | 7 | 7 | 7 | 8 | 8 | 8 | 9 | 9 | 9 | 9 | 9 | 9 |
| 9 | 0 | 2 | 3 | 4 | 4 | 5 | 5 | 6 | 6 | 7 | 7 | 7 | 8 | 8 | 8 | 9 | 9 | 9 | 9 | 10 | 10 | 10 | 10 | 10 | 10 |
| 10 | 0 | 2 | 3 | 4 | 5 | 5 | 6 | 6 | 7 | 7 | 8 | 8 | 8 | 9 | 9 | 9 | 10 | 10 | 10 | 10 | 10 | 11 | 11 | 11 | 11 |
| 11 | 0 | 2 | 3 | 4 | 5 | 5 | 6 | 7 | 7 | 8 | 8 | 9 | 9 | 9 | 10 | 10 | 10 | 10 | 11 | 11 | 11 | 11 | 12 | 12 | 12 |
| 12 | 0 | 2 | 3 | 4 | 5 | 6 | 6 | 7 | 7 | 8 | 9 | 9 | 9 | 10 | 10 | 10 | 11 | 11 | 11 | 12 | 12 | 12 | 12 | 13 | 13 |
| 13 | 0 | 2 | 3 | 4 | 5 | 6 | 7 | 7 | 8 | 8 | 9 | 9 | 10 | 10 | 11 | 11 | 11 | 12 | 12 | 12 | 13 | 13 | 13 | 13 | 14 |
| 14 | 0 | 2 | 3 | 4 | 5 | 6 | 7 | 7 | 8 | 9 | 9 | 10 | 10 | 11 | 11 | 11 | 12 | 12 | 13 | 13 | 13 | 14 | 14 | 14 | 14 |
| 15 | 0 | 2 | 4 | 4 | 5 | 6 | 7 | 7 | 8 | 9 | 10 | 10 | 11 | 11 | 12 | 12 | 12 | 13 | 13 | 13 | 14 | 14 | 14 | 15 | 15 |
| 16 | 0 | 2 | 4 | 5 | 6 | 6 | 7 | 7 | 9 | 9 | 10 | 10 | 11 | 11 | 12 | 12 | 13 | 13 | 14 | 14 | 14 | 15 | 15 | 15 | 16 |
| 17 | 0 | 2 | 4 | 5 | 6 | 6 | 7 | 8 | 9 | 10 | 10 | 11 | 11 | 12 | 12 | 13 | 13 | 14 | 14 | 15 | 15 | 15 | 16 | 16 | 16 |
| 18 | 0 | 2 | 4 | 5 | 6 | 7 | 8 | 8 | 9 | 10 | 10 | 11 | 12 | 12 | 13 | 13 | 14 | 14 | 15 | 15 | 15 | 16 | 16 | 17 | 17 |
| 19 | 0 | 3 | 4 | 5 | 6 | 7 | 8 | 8 | 9 | 10 | 11 | 11 | 12 | 13 | 13 | 14 | 14 | 15 | 15 | 16 | 16 | 16 | 17 | 17 | 17 |
| 20 | 0 | 3 | 4 | 5 | 6 | 7 | 8 | 9 | 10 | 10 | 11 | 12 | 12 | 13 | 13 | 14 | 15 | 15 | 16 | 16 | 16 | 17 | 17 | 18 | 18 |
| 21 | 0 | 3 | 4 | 5 | 6 | 7 | 8 | 9 | 10 | 10 | 11 | 12 | 13 | 13 | 14 | 14 | 15 | 15 | 16 | 16 | 17 | 17 | 18 | 18 | 19 |
| 22 | 0 | 3 | 4 | 5 | 6 | 7 | 8 | 9 | 10 | 11 | 11 | 12 | 13 | 14 | 14 | 15 | 15 | 16 | 16 | 17 | 17 | 18 | 18 | 19 | 19 |
| 23 | 0 | 3 | 4 | 5 | 6 | 7 | 8 | 9 | 10 | 11 | 12 | 12 | 13 | 14 | 14 | 15 | 16 | 16 | 17 | 17 | 18 | 18 | 19 | 19 | 20 |
| 24 | 0 | 3 | 4 | 5 | 6 | 7 | 8 | 9 | 10 | 11 | 12 | 13 | 13 | 14 | 15 | 15 | 16 | 17 | 17 | 18 | 18 | 19 | 19 | 20 | 20 |
| 25 | 0 | 3 | 4 | 5 | 6 | 8 | 8 | 9 | 10 | 11 | 12 | 13 | 14 | 14 | 15 | 16 | 16 | 17 | 17 | 18 | 19 | 19 | 20 | 20 | 21 |
| 26 | 0 | 3 | 4 | 5 | 6 | 8 | 8 | 10 | 10 | 11 | 12 | 13 | 14 | 15 | 15 | 16 | 17 | 17 | 18 | 18 | 19 | 19 | 20 | 21 | 21 |
| 27 | 0 | 3 | 4 | 5 | 6 | 8 | 9 | 10 | 11 | 12 | 12 | 13 | 14 | 15 | 16 | 16 | 17 | 18 | 18 | 19 | 19 | 20 | 20 | 21 | 21 |
| 28 | 0 | 3 | 4 | 6 | 6 | 8 | 9 | 10 | 11 | 12 | 13 | 13 | 14 | 15 | 16 | 16 | 17 | 18 | 18 | 19 | 20 | 20 | 21 | 21 | 22 |
| 29 | 0 | 3 | 4 | 6 | 7 | 8 | 9 | 10 | 11 | 12 | 13 | 14 | 14 | 15 | 16 | 17 | 17 | 18 | 19 | 19 | 20 | 21 | 21 | 22 | 22 |
| 30 | 0 | 3 | 4 | 6 | 7 | 8 | 9 | 10 | 11 | 12 | 13 | 14 | 15 | 15 | 16 | 17 | 18 | 18 | 19 | 20 | 20 | 21 | 21 | 22 | 23 |
| 31 | 0 | 3 | 4 | 6 | 7 | 8 | 9 | 10 | 11 | 12 | 13 | 14 | 15 | 16 | 16 | 17 | 18 | 19 | 19 | 20 | 21 | 21 | 22 | 22 | 23 |
| 32 | 0 | 3 | 4 | 6 | 7 | 8 | 9 | 10 | 11 | 12 | 13 | 14 | 15 | 16 | 17 | 17 | 18 | 19 | 20 | 20 | 21 | 22 | 22 | 23 | 23 |
| 33 | 0 | 3 | 4 | 6 | 7 | 8 | 9 | 10 | 11 | 12 | 13 | 14 | 15 | 16 | 17 | 18 | 18 | 19 | 20 | 21 | 21 | 22 | 23 | 23 | 24 |
| 34 | 0 | 3 | 4 | 6 | 7 | 8 | 9 | 10 | 12 | 12 | 14 | 14 | 15 | 16 | 17 | 18 | 19 | 19 | 20 | 21 | 22 | 22 | 23 | 23 | 24 |
| 35 | 0 | 3 | 4 | 6 | 7 | 8 | 9 | 10 | 12 | 12 | 14 | 14 | 16 | 16 | 17 | 18 | 19 | 20 | 20 | 21 | 22 | 22 | 23 | 24 | 24 |
| 36 | 0 | 3 | 4 | 6 | 7 | 8 | 10 | 10 | 12 | 13 | 14 | 15 | 16 | 16 | 17 | 18 | 19 | 20 | 21 | 21 | 22 | 23 | 23 | 24 | 25 |
| 37 | 0 | 3 | 4 | 6 | 7 | 8 | 10 | 10 | 12 | 13 | 14 | 15 | 16 | 17 | 18 | 18 | 19 | 20 | 21 | 22 | 22 | 23 | 24 | 24 | 25 |
| 38 | 0 | 3 | 4 | 6 | 7 | 8 | 10 | 11 | 12 | 13 | 14 | 15 | 16 | 17 | 18 | 19 | 20 | 20 | 21 | 22 | 23 | 23 | 24 | 25 | 25 |
| 39 | 0 | 3 | 4 | 6 | 7 | 8 | 10 | 11 | 12 | 13 | 14 | 15 | 16 | 17 | 18 | 19 | 20 | 20 | 21 | 22 | 23 | 24 | 24 | 25 | 26 |
| 40 | 0 | 3 | 4 | 6 | 7 | 8 | 10 | 11 | 12 | 13 | 14 | 15 | 16 | 17 | 18 | 19 | 20 | 21 | 22 | 22 | 23 | 24 | 25 | 25 | 26 |
| 41 | 0 | 3 | 4 | 6 | 7 | 8 | 10 | 11 | 12 | 13 | 14 | 15 | 16 | 17 | 18 | 19 | 20 | 21 | 22 | 22 | 23 | 24 | 25 | 26 | 26 |
| 42 | 0 | 3 | 4 | 6 | 7 | 8 | 10 | 11 | 12 | 13 | 14 | 15 | 16 | 17 | 18 | 19 | 20 | 21 | 22 | 23 | 24 | 24 | 25 | 26 | 26 |
| 43 | 0 | 3 | 4 | 6 | 7 | 8 | 10 | 11 | 12 | 13 | 14 | 16 | 16 | 18 | 18 | 20 | 20 | 21 | 22 | 23 | 24 | 24 | 25 | 26 | 27 |
| 44 | 0 | 3 | 4 | 6 | 7 | 8 | 10 | 11 | 12 | 14 | 14 | 16 | 17 | 18 | 19 | 20 | 20 | 21 | 22 | 23 | 24 | 25 | 26 | 26 | 27 |
| 45 | 0 | 3 | 4 | 6 | 8 | 8 | 10 | 11 | 12 | 14 | 15 | 16 | 17 | 18 | 19 | 20 | 21 | 22 | 22 | 23 | 24 | 25 | 26 | 26 | 27 |
| 46 | 0 | 3 | 4 | 6 | 8 | 9 | 10 | 11 | 12 | 14 | 15 | 16 | 17 | 18 | 19 | 20 | 21 | 22 | 23 | 24 | 24 | 25 | 26 | 27 | 28 |
| 47 | 0 | 3 | 4 | 6 | 8 | 9 | 10 | 11 | 12 | 14 | 15 | 16 | 17 | 18 | 19 | 20 | 21 | 22 | 23 | 24 | 24 | 25 | 26 | 27 | 28 |
| 48 | 0 | 3 | 4 | 6 | 8 | 9 | 10 | 11 | 12 | 14 | 15 | 16 | 17 | 18 | 19 | 20 | 21 | 22 | 23 | 24 | 25 | 26 | 26 | 27 | 28 |
| 49 | 0 | 3 | 4 | 6 | 8 | 9 | 10 | 12 | 12 | 14 | 15 | 16 | 17 | 18 | 19 | 20 | 21 | 22 | 23 | 24 | 25 | 26 | 27 | 27 | 28 |
| 50 | 0 | 3 | 4 | 6 | 8 | 9 | 10 | 12 | 13 | 14 | 15 | 16 | 17 | 18 | 19 | 20 | 21 | 22 | 23 | 24 | 25 | 26 | 27 | 28 | 28 |

TABLE 2
Critical Values of Number of Runs, u; for Probability Level $\delta = 0.20$

| N\M | 1 | 2 | 3 | 4 | 5 | 6 | 7 | 8 | 9 | 10 | 11 | 12 | 13 | 14 | 15 | 16 | 17 | 18 | 19 | 20 | 21 | 22 | 23 | 24 | 25 |
|---|---|---|---|---|---|---|---|---|---|---|---|---|---|---|---|---|---|---|---|---|---|---|---|---|---|
| 1 | 0 | 0 | 0 | 0 | 0 | 0 | 0 | 0 | 0 | 0 | 0 | 0 | 0 | 0 | 0 | 0 | 0 | 0 | 0 | 0 | 0 | 0 | 0 | 0 | 0 |
| 2 | 0 | 0 | 2 | 2 | 2 | 2 | 2 | 2 | 3 | 3 | 3 | 3 | 3 | 3 | 3 | 3 | 3 | 3 | 3 | 3 | 3 | 3 | 3 | 3 | 3 |
| 3 | 0 | 2 | 2 | 3 | 3 | 3 | 3 | 3 | 4 | 4 | 4 | 4 | 4 | 4 | 4 | 4 | 4 | 4 | 4 | 4 | 4 | 4 | 4 | 4 | 4 |
| 4 | 0 | 2 | 3 | 3 | 3 | 4 | 4 | 4 | 4 | 4 | 5 | 5 | 5 | 5 | 5 | 5 | 5 | 6 | 6 | 6 | 6 | 6 | 6 | 6 | 6 |
| 5 | 0 | 2 | 3 | 3 | 4 | 4 | 5 | 5 | 5 | 5 | 5 | 6 | 6 | 6 | 6 | 6 | 6 | 6 | 7 | 7 | 7 | 7 | 7 | 7 | 7 |
| 6 | 0 | 2 | 3 | 4 | 4 | 5 | 5 | 5 | 6 | 6 | 6 | 6 | 7 | 7 | 7 | 7 | 7 | 8 | 8 | 8 | 8 | 8 | 8 | 8 | 8 |
| 7 | 0 | 2 | 3 | 4 | 5 | 5 | 5 | 6 | 6 | 7 | 7 | 7 | 7 | 8 | 8 | 8 | 8 | 8 | 9 | 9 | 9 | 9 | 9 | 9 | 9 |
| 8 | 0 | 2 | 3 | 4 | 5 | 5 | 6 | 6 | 7 | 7 | 8 | 8 | 8 | 8 | 9 | 9 | 9 | 9 | 10 | 10 | 10 | 10 | 10 | 10 | 10 |
| 9 | 0 | 3 | 4 | 4 | 5 | 6 | 6 | 7 | 7 | 8 | 8 | 8 | 9 | 9 | 9 | 10 | 10 | 10 | 10 | 10 | 11 | 11 | 11 | 11 | 11 |
| 10 | 0 | 3 | 4 | 4 | 5 | 6 | 7 | 7 | 8 | 8 | 9 | 9 | 9 | 10 | 10 | 10 | 11 | 11 | 11 | 11 | 12 | 12 | 12 | 12 | 12 |
| 11 | 0 | 3 | 4 | 5 | 5 | 6 | 7 | 8 | 8 | 9 | 9 | 9 | 10 | 10 | 11 | 11 | 11 | 11 | 12 | 12 | 12 | 12 | 13 | 13 | 13 |
| 12 | 0 | 3 | 4 | 5 | 6 | 6 | 7 | 8 | 8 | 9 | 9 | 10 | 10 | 11 | 11 | 12 | 12 | 12 | 13 | 13 | 13 | 13 | 14 | 14 | 14 |
| 13 | 0 | 3 | 4 | 5 | 6 | 7 | 7 | 8 | 9 | 9 | 10 | 10 | 11 | 11 | 12 | 12 | 13 | 13 | 13 | 13 | 14 | 14 | 14 | 15 | 15 |
| 14 | 0 | 3 | 4 | 5 | 6 | 7 | 8 | 8 | 9 | 10 | 10 | 11 | 11 | 12 | 12 | 13 | 13 | 13 | 14 | 14 | 14 | 15 | 15 | 15 | 16 |
| 15 | 0 | 3 | 4 | 5 | 6 | 7 | 8 | 9 | 9 | 10 | 11 | 11 | 12 | 12 | 13 | 13 | 14 | 14 | 14 | 15 | 15 | 15 | 16 | 16 | 16 |
| 16 | 0 | 3 | 4 | 5 | 6 | 7 | 8 | 9 | 10 | 10 | 11 | 12 | 12 | 13 | 13 | 14 | 14 | 15 | 15 | 15 | 16 | 16 | 16 | 17 | 17 |
| 17 | 0 | 3 | 4 | 5 | 6 | 7 | 8 | 9 | 10 | 11 | 11 | 12 | 13 | 13 | 14 | 14 | 15 | 15 | 15 | 16 | 16 | 17 | 17 | 17 | 18 |
| 18 | 0 | 3 | 4 | 6 | 6 | 8 | 8 | 9 | 10 | 11 | 12 | 12 | 13 | 13 | 14 | 15 | 15 | 15 | 16 | 16 | 17 | 17 | 18 | 18 | 18 |
| 19 | 0 | 3 | 4 | 6 | 7 | 8 | 9 | 10 | 10 | 11 | 12 | 13 | 13 | 14 | 14 | 15 | 15 | 16 | 16 | 17 | 17 | 18 | 18 | 19 | 19 |
| 20 | 0 | 3 | 4 | 6 | 7 | 8 | 9 | 10 | 10 | 11 | 12 | 13 | 13 | 14 | 15 | 15 | 16 | 16 | 17 | 17 | 18 | 18 | 19 | 19 | 19 |
| 21 | 0 | 3 | 4 | 6 | 7 | 8 | 9 | 10 | 11 | 12 | 12 | 13 | 14 | 14 | 15 | 16 | 16 | 17 | 17 | 18 | 18 | 19 | 19 | 20 | 20 |
| 22 | 0 | 3 | 4 | 6 | 7 | 8 | 9 | 10 | 11 | 12 | 13 | 13 | 14 | 15 | 15 | 16 | 17 | 17 | 18 | 18 | 19 | 19 | 20 | 20 | 21 |
| 23 | 0 | 3 | 4 | 6 | 7 | 8 | 9 | 10 | 11 | 12 | 13 | 14 | 14 | 15 | 16 | 16 | 17 | 18 | 18 | 19 | 19 | 20 | 20 | 21 | 21 |
| 24 | 0 | 3 | 4 | 6 | 7 | 8 | 9 | 10 | 11 | 12 | 13 | 14 | 15 | 15 | 16 | 17 | 17 | 18 | 19 | 19 | 20 | 20 | 21 | 21 | 22 |
| 25 | 0 | 3 | 4 | 6 | 7 | 8 | 9 | 10 | 11 | 12 | 13 | 14 | 15 | 16 | 16 | 17 | 18 | 18 | 19 | 19 | 20 | 21 | 21 | 22 | 22 |
| 26 | 0 | 3 | 4 | 6 | 7 | 8 | 10 | 10 | 12 | 12 | 13 | 14 | 15 | 16 | 17 | 17 | 18 | 19 | 19 | 20 | 20 | 21 | 21 | 22 | 23 |
| 27 | 0 | 4 | 5 | 6 | 7 | 8 | 10 | 11 | 12 | 13 | 14 | 14 | 15 | 16 | 17 | 18 | 18 | 19 | 20 | 20 | 21 | 21 | 22 | 22 | 23 |
| 28 | 0 | 4 | 5 | 6 | 7 | 8 | 10 | 11 | 12 | 13 | 14 | 15 | 16 | 16 | 17 | 18 | 19 | 19 | 20 | 21 | 21 | 22 | 22 | 23 | 23 |
| 29 | 0 | 4 | 5 | 6 | 8 | 8 | 10 | 11 | 12 | 13 | 14 | 15 | 16 | 16 | 17 | 18 | 19 | 20 | 20 | 21 | 21 | 22 | 23 | 23 | 24 |
| 30 | 0 | 4 | 5 | 6 | 8 | 9 | 10 | 11 | 12 | 13 | 14 | 15 | 16 | 17 | 18 | 18 | 19 | 20 | 20 | 21 | 22 | 22 | 23 | 24 | 24 |
| 31 | 0 | 4 | 5 | 6 | 8 | 9 | 10 | 11 | 12 | 13 | 14 | 15 | 16 | 17 | 18 | 19 | 19 | 20 | 21 | 21 | 22 | 23 | 23 | 24 | 25 |
| 32 | 0 | 4 | 5 | 6 | 8 | 9 | 10 | 11 | 12 | 13 | 14 | 15 | 16 | 17 | 18 | 19 | 20 | 20 | 21 | 22 | 22 | 23 | 24 | 24 | 25 |
| 33 | 0 | 4 | 5 | 6 | 8 | 9 | 10 | 11 | 12 | 14 | 14 | 16 | 16 | 17 | 18 | 19 | 20 | 21 | 21 | 22 | 23 | 23 | 24 | 25 | 25 |
| 34 | 0 | 4 | 5 | 6 | 8 | 9 | 10 | 11 | 12 | 14 | 15 | 16 | 16 | 18 | 18 | 19 | 20 | 21 | 22 | 22 | 23 | 24 | 24 | 25 | 26 |
| 35 | 0 | 4 | 5 | 6 | 8 | 9 | 10 | 12 | 12 | 14 | 15 | 16 | 17 | 18 | 18 | 19 | 20 | 21 | 22 | 23 | 23 | 24 | 25 | 25 | 26 |
| 36 | 0 | 4 | 5 | 6 | 8 | 9 | 10 | 12 | 13 | 14 | 15 | 16 | 17 | 18 | 19 | 20 | 20 | 21 | 22 | 23 | 24 | 24 | 25 | 26 | 26 |
| 37 | 0 | 4 | 5 | 6 | 8 | 9 | 10 | 12 | 13 | 14 | 15 | 16 | 17 | 18 | 19 | 20 | 21 | 22 | 22 | 23 | 24 | 25 | 25 | 26 | 27 |
| 38 | 0 | 4 | 5 | 6 | 8 | 9 | 10 | 12 | 13 | 14 | 15 | 16 | 17 | 18 | 19 | 20 | 21 | 22 | 23 | 23 | 24 | 25 | 26 | 26 | 27 |
| 39 | 0 | 4 | 5 | 6 | 8 | 9 | 10 | 12 | 13 | 14 | 15 | 16 | 17 | 18 | 19 | 20 | 21 | 22 | 23 | 24 | 24 | 25 | 26 | 27 | 27 |
| 40 | 0 | 4 | 5 | 6 | 8 | 9 | 10 | 12 | 13 | 14 | 15 | 16 | 17 | 18 | 19 | 20 | 21 | 22 | 23 | 24 | 25 | 25 | 26 | 27 | 28 |
| 41 | 0 | 4 | 5 | 6 | 8 | 9 | 10 | 12 | 13 | 14 | 15 | 16 | 18 | 18 | 20 | 20 | 21 | 22 | 23 | 24 | 25 | 26 | 26 | 27 | 28 |
| 42 | 0 | 4 | 5 | 6 | 8 | 10 | 10 | 12 | 13 | 14 | 16 | 16 | 18 | 19 | 20 | 21 | 22 | 22 | 23 | 24 | 25 | 26 | 27 | 27 | 28 |
| 43 | 0 | 4 | 5 | 6 | 8 | 10 | 11 | 12 | 13 | 14 | 16 | 17 | 18 | 19 | 20 | 21 | 22 | 23 | 24 | 24 | 25 | 26 | 27 | 28 | 28 |
| 44 | 0 | 4 | 5 | 6 | 8 | 10 | 11 | 12 | 13 | 14 | 16 | 17 | 18 | 19 | 20 | 21 | 22 | 23 | 24 | 25 | 26 | 26 | 27 | 28 | 29 |
| 45 | 0 | 4 | 5 | 6 | 8 | 10 | 11 | 12 | 14 | 14 | 16 | 17 | 18 | 19 | 20 | 21 | 22 | 23 | 24 | 25 | 26 | 27 | 27 | 28 | 29 |
| 46 | 0 | 4 | 5 | 6 | 8 | 10 | 11 | 12 | 14 | 14 | 16 | 17 | 18 | 19 | 20 | 21 | 22 | 23 | 24 | 25 | 26 | 27 | 28 | 28 | 29 |
| 47 | 0 | 4 | 5 | 6 | 8 | 10 | 11 | 12 | 14 | 15 | 16 | 17 | 18 | 19 | 20 | 21 | 22 | 23 | 24 | 25 | 26 | 27 | 28 | 29 | 29 |
| 48 | 0 | 4 | 5 | 6 | 8 | 10 | 11 | 12 | 14 | 15 | 16 | 17 | 18 | 20 | 20 | 22 | 22 | 24 | 24 | 25 | 26 | 27 | 28 | 29 | 30 |
| 49 | 0 | 4 | 5 | 6 | 8 | 10 | 11 | 12 | 14 | 15 | 16 | 17 | 18 | 20 | 20 | 22 | 23 | 24 | 25 | 26 | 26 | 27 | 28 | 29 | 30 |
| 50 | 0 | 4 | 5 | 6 | 8 | 10 | 11 | 12 | 14 | 15 | 16 | 17 | 18 | 20 | 21 | 22 | 23 | 24 | 25 | 26 | 27 | 28 | 28 | 29 | 30 |

## REFERENCES

[1].    D. M. Mohilner and P. R. Mohilner, J. Electrochem. Soc., 115:261 (1968).

[2].    D. M. Mohilner, "The Electrical Double Layer. Part I. Elements of Double Layer Theory," in Electroanalytical Chemistry, Vol. 1 (A. J. Bard, ed.), Marcel Dekker, Inc., New York, 1966, pp. 241-409.

[3].    P. Delahay, Double Layer and Electrode Kinetics, Interscience, New York, 1965.

[4].    M. A. V. Devanathan and B. V. K. S. R. A. Tilak, Chem. Rev., 65:635 (1965).

[5].    J. Lawrence and D. M. Mohilner, J. Electrochem. Soc., 118:259 (1971).

[6].    J. Lawrence and D. M. Mohilner, J. Electrochem. Soc., 118: 1596 (1971).

[7].    W. R. Fawcett and J. E. Kent, Can. J. Chem., 48:47 (1970).

[8].    R. G. Barradas and F. M. Kimmerle, Can. J. Chem., 45:109 (1967).

[9].    D. C. Grahame, J. Am. Chem. Soc., 71:2975 (1949).

[10].   D. C. Grahame, M. A. Poth, and J. I. Cummings, ONR Tech. Rept. No. 7, Dec. 1951, Project NR 051-150.

[11].   F. S. Swed and C. Eisenhart, Ann. Math. Stat., 14:66 (1943).

[12].   P. R. Mohilner and D. M. Mohilner, American Chemical Society, Southwest Regional Meeting, Dec. 1968, Abstr. No. 5.

[13].   K. Doblhofer and D. M. Mohilner, J. Phys. Chem., 75: 1698 (1971).

[14].   T. R. Dickson, The Computer and Chemistry, W. H. Freeman, San Francisco, 1968, p. 142.

Chapter 2

# A PROGRAM TO CALCULATE THE RELATIVE SURFACE EXCESS OF A SUBSTANCE ADSORBED AT THE SURFACE OF A MERCURY ELECTRODE FROM DIFFERENTIAL CAPACITANCE DATA

Hubert C. MacDonald, Jr.
Koppers Company, Inc.,
Monroeville, Pennsylvania, 15146

## I. INTRODUCTION

The measurement of the differential capacitance of a solution is a convenient, sensitive, and relatively rapid method for the determination of specific adsorption. It is also possible to arrive at the relative surface excess of the adsorbate at the electrode surface from differential capacitance data. These calculations, although not difficult, are tedious and time consuming. It is for this reason that this program was developed.

The apparatus and experimental procedures have been described elsewhere [1, 2]. Mohilner and Mohilner [3] give a brief literature survey of the various computer methods which have been developed for the treatment of electrocapillary data. Since these topics appear elsewhere, they will not be discussed further here.

There are two steps in the calculation of the surface concentration [4]. First, the electrode charge must be calculated according to the following equation:

$$q = \int_{E_1}^{E_2} CdE + const$$

where q is the charge at the surface of the electrode, C is the measured differential capacitance, E is the electrode potential, and const is the integration constant. The integration constant is determined by noting that q is zero at electrocapillary maximum (ECM), the point of zero charge. The determination of the ECM, however, requires further experimental work and, for the purposes here, it is convenient to use the quantity $q^*$ defined by

$$q^* = q - const.$$

This is permissible because the derivative of the charge will be used in future calculations and the derivative of the left-hand side of the above equation is identical with the derivative of the right-hand side. The second step is the calculation of the surface excess, $\Gamma$. This is computed according to

$$\Gamma = \frac{1}{2.3RT} \int_{E_1}^{E_2} \left(\frac{\partial q^*}{\partial \log c}\right)_E dE$$

where R and T have their usual thermodynamic meaning, E is the potential, and c is the concentration of the surface-active agent. This equation is obtained from the Gibbs adsorption equation. The customary approximations have been made to perform the calculation. The quantity $(\partial q^*/\partial \log c)$ is the slope of the charge curves as a function of the concentration for a given constant potential.

Obviously, then, in order to calculate the surface excess of an adsorbed species, it is necessary to perform an integration, a differentiation, and a second integration. Before these operations are described, a discussion of the data, software, and hardware is needed to the extent that it will aid one in using this program.

## II. EXPERIMENTAL

The data that are recorded during the experiment are the capacities. The differential capacitance is obtained by dividing the capacity by the electrode area. A program can be written to perform this calculation and this is, of course, an elementary matter. It is convenient, however, to put the output of this program onto a device (card punch, disk storage, etc.) which can be readily used as an input source to the program for the calculation of the surface excess. The integration subroutine requires that the intervals between the points of a differential capacitance curve be equal. It was found that if a 25-mV interval was used, satisfactory results were obtained. The smoothing function, as it is implemented here, also requires the data points to be equally spaced. There must be differential capacitance curves for at least four different concentrations of adsorbate including a curve of the supporting electrolyte and no adsorbate. It is necessary that the measurements be carried out at potentials sufficiently cathodic to insure complete desorption of material in the solutions with the highest concentration of adsorbate. If this condition is not satisfied, then integration must be made about the ECM. The program, then, must be modified to perform the integration in this manner and, in addition, further experimental work is necessary to determine the potential of the ECM.

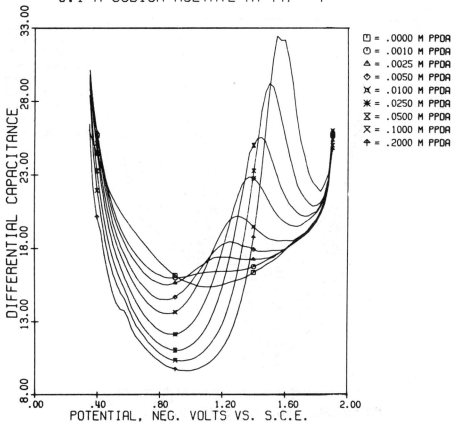

Fig. 1. Results of differential capacitance measurements.

### III. PROGRAM

This program is complete in itself but, depending on one's particular situation, there may be routines available which aid in the interpretation of results. Plots generated on the printer, for example, were useful in checking the differential capacitance curves for points in error. Such routines are generally available to all users of a computer at any given installation or can be easily added as a subroutine. Also, the software for programming a Calcomp plotter was available for generating plots of the data (see Figs. 1-3).

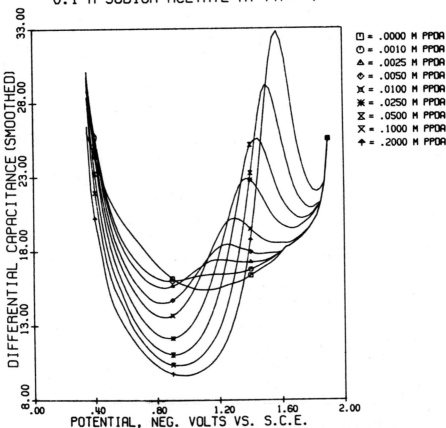

Fig. 2.  Differential capacitance data which have been smoothed by the smoothing function.

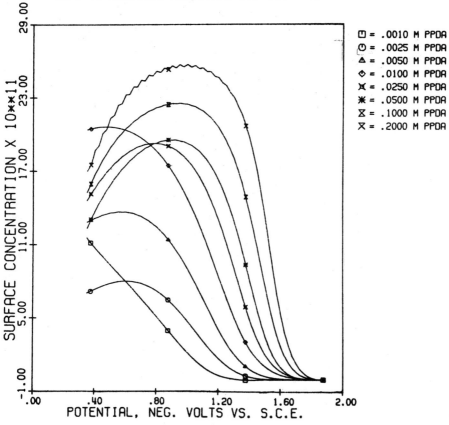

Fig. 3. Surface concentrations as a function of potential calculated from the differential capacitance curves in Fig. 2.

This program is written in standard Fortran IV Level G and it ran on an IBM System 360/67. It was found that the single precision of this system was sufficient to give good results. If this program proves to exceed the core storage of any particular system, it is quite convenient to reduce it to three separate programs performing the task of each of the subroutines.

A listing of the programs appears in the Appendix as the programs were actually compiled and run. The statement numbers on the far left are numbers supplied by the compiler and refer only to Fortran statements and not to comments. These are the numbers which will be referred to below. A familiarity with programming will be assumed so that this discussion will be concerned only with an explanation of what the variables represent and suggestions on how to prepare the input deck.

Statements 18 and 19 allow for the identification of a calculation with two title cards of up to 72 characters each. Statement 20 provides the program with:

NCONC, the number of differential capacitance curves to be processed;
NPOT, the number of potentials at which measurements were made;
SMTH1 and SMTH3, constants which control the calling of the smoothing function (this particular compiler was forgiving and allowed SMTH1 and SMTH3 to be read in under an integer format);
IPLI, IPL2, and IPL3, constants which control the generation of a printer plot.

The values of these last five constants must be nonzero in order to omit calling the function over which they have control. For this reason it is suggested that the following five cards be added to the program after Statement 21 and removed only when these features wish to be used:

SMTH1 = 1.0
SMTH3 = 1.0
IPL1   = 1
IPL2   = 1
IPL3   = 1

The next data to be read in are the molar concentrations of the surface agent; the order must be from the lowest concentration, which is 0.0 $\underline{M}$, to the highest. Next, the potentials are required, starting with the most cathodic and omitting the sign. For the calculations, POT(I) will be changed to -POT(I) by the program.

The value 1/RT is now calculated and the position of the decimal point for this constant is adjusted by the factor $10^5$ so that the final results will have the proper magnitude.

Only the differential capacitance data remain to be entered. The differential capacitance is placed in a vector that is NPOT long. It is obvious that all the capacitance points must be of the same spacing and that there must be the same number of points for each set of capacitance. The data are placed into a scratch vector so that calculations may be accomplished. The operations of smoothing, if desired, and integrating are performed and the results are returned to the array SAVE (I, J) for later use.

The smoothing function, SMOOTH, is of standard type [5] and is based on a centered least-squares approximation of a section of the curve. The section of the curve taken in this instance is a five-point segment. The integration formula, SIMP, uses Simpson's Rule to calculate the area under the first two and last two points. This subroutine requires that the data points be equally spaced.

If statements 45, 63, and 85 are left in the program, it is suggested that the following subroutine be added to avoid an error return because of an undefined function.

```
      SUBROUTINE PLTR (NC, A, C, IL)
C     NOT FINISHED
      RETURN
      END
```

The value $\partial q*/\partial \log c$ is now calculated by the subroutine THREER. THREER calculates the derivative by fitting a least-squares polynomial using three points taken from the array SAVE (I, J) at a constant value of the electrode potential. The derivative is thus calculated and multiplied by $1/RT \times 10^5$, and the calculated value is stored in the array SAVE (I, J).

The values obtained from THREER are now integrated with respect to the potential. At this point it is not advisable to call on the smoothing function. This option is included, however, to provide for further refinement in the data. The values returned by SIMP are the surface concentrations of the adsorbed species. Statement 77 or 80 is designed so that this output will eventually appear on a line printer. Statement 89, however, is intended to send the results to temporary storage or a card punch so that they may be used as data for other programs. For example, one may wish to make computer plots of the surface concentration. This is the most convenient method of transferring the data.

Figures 1 and 2 are differential capacitance curves and Fig. 3 is the surface concentration calculated by the program. The curves in Fig. 2 have been smoothed by the smoothing function. For the region in which there is no adsorption, the points have been averaged so that the value of $\partial q*/\partial \log c$ is zero. The significance of these curves has been discussed elsewhere [2].

It is suggested that the user check the results that this program provides by a manual calculation. Such a caution is well heeded in the use of any program to avoid subtle pitfalls which are impossible to detect by any other manner. In addition, this will provide the user with first-hand knowledge of the capabilities of this program for this calculation.

## ACKNOWLEDGMENT

The author acknowledges Mr. Robert LaBudde for his assistance in developing the computer programs, the University of Michigan for providing computer time, and the National Science Foundation for providing financial assistance.

## APPENDIX

```
0001      DIMENSION POT(99), Y(99), YY(99), CONC(13), XCONC(12)
0002      DIMENSION SAVE(99,13), A(99), TITLE(36), C(12), P(99)
      C
      C
0003 1000 FORMAT (18A4)
0004 1005 FORMAT (1H1, (/,30X, 18A4) )
0005 1010 FORMAT (20I4)
0006 1015 FORMAT (1H-,20X, 'NCONC = ', I2, 5X, 'NPOT = ', I2, 5X, 'SMTH1 = '
     1, I2, 5X, 'SMTH2 DOES NOT EXIST YET', 5X, 'SMTH3 = 'I2, /, 1H0, 20
     2X, 'IPL1 = ', I2, 5X, 'IPL2 = ', I2, 5X, 'IPL3 = ', I2)
0007 1020 FORMAT (10F8.0)
0008 1025 FORMAT (1H-, 20X, 'CONC(1) = ', F4.2, 5X, 'CONC(',I2,') = ', F5.3
     1/, 21X,'POT(1) = ', F5.3, 5X,'POT(', I2, ') = ', F5.3, /,1H0,
     2 20X, '1/RT = ', F10.6)
0009 1030 FORMAT (1H1, 30X, 'FOR CONCENTRATION = ', F6.4, '    MOLES/LITER',/
     1,1H0, 18X, 1H+, 4X, 'INPUT', 10X, 'SMOOTHED', 7X, 'INTEGRAL - FULL
     2', /, 7X, 'POTENTIAL', 3X, 1H+, 5X, 'DATA', 11X, 'VALUES', 8X, 'SI
     3MPSONS''S RULE', /, 5X, 70(1H+), 2(/, 19X, 1H+ ) )
0010 1035 FORMAT (8X, F7.3, 4X, 1H+, F9.4, 5X, F13.4, /, (8X, F7.3, 4X, 1H+,
     1 F9.4, 5X, F13.4, 8X, F10.4) )
0011 1040 FORMAT (8X, F7.3, 4X, 1H+, F9.4, /, (8X, F7.3, 4X, 1H+, F9.4, 26X,
     1 F10.4) )
0012 1045 FORMAT (1H+,24X,'FOR POTENTIAL = ',F7.3,' VOLTS VS. S.C.E.',/,1H0,
     14X, 'LOG BASE E', 4X, 1H+,4X, 'INPUT', 20X,'DIFFERENTIAL - THREE
     2', /, 4X, 'OF THE CONC.', 3X, 1H+, 5X, 'DATA', 27X,'POINT RULE',
     3 /, 5X, 70(1H+), 2(/, 19X, 1H+) )
0013 1050 FORMAT (7X, F8.4, 4X, 1H+, F9.5, 25X, F15.7)
0014 1060 FORMAT (1H6)
0015 1070 FORMAT (10F8.4)
0016 1080 FORMAT (1H2, 20X, '      IN THE CALCULATION OF THE SURFACE CONCENTR
     1ATION THE NUMBERS LABELLED ''INPUT DATA'' ARE', /, 1H0, 20X,
     2'THE DERIVATIVES OF THE CHARGE WITH RESPECT TO THE LOG OF THE CONC
     3ENTRATION, I.E. D Q*/D LN C,'/, 1H0, 20X, 'AND THE INTEGRAL IS TH
     4E SURFACE CONTRATION AND THE UNITS ARE MOLES/(CM*CM) X 10**11')
0017 1090 FORMAT (1H1)
```

```
0018   1100   READ (5, 1000, END=1900) (TITLE(I), I = 1, 36)
0019          WRITE (6, 1005) (TITLE(I), I = 1, 36)
0020          READ (5, 1010) NCONC, NPOT, SMTH1, SMTH3, IPL1, IPL2, IPL3
0021          WRITE (6, 1015)NCONC, NPOT, SMTH1, SMTH3, IPL1, IPL2, IPL3
0022          READ (5, 1020) (CONC(I), I = 1, NCONC)
0023          READ (5, 1020) (POT(I), I = 1, NPOT)
       C
       C      CALCULATE 1/RT
       C
0024          RTEMP = 1.0 E+05 / (8.31432 * 298.16)
       C
       C      WRITE SOME OF THE INPUT DATA
       C
0025          WRITE (6, 1025) CONC(1), NCONC, CONC(NCONC), POT(1), NPOT,
              1 POT(NPOT), RTEMP
0026          WRITE (6, 1080)
       C
       C      CHECK TO MAKE SURE VECTORS WILL NOT BE OVER RUN
       C
0027          IF (NCONC .GT. 13 .OR. NPOT .GT. 99) GO TO 1100
       C
       C      SET THE POTENTIALS EQUAL TO THEIR NEGATIVE FOR CALCULATIONS
       C
0028          DO 1110 I = 1, NPOT
0029          P(I) = POT(I)
0030   1110   POT(I) = -POT(I)
       C
       C      CALCULATE CHARGE CURVES USING SIMPSON'S RULE
       C
0031   1200   DO 1260 I = 1, NCONC
```

```
0032        READ (5, 1020) (Y(J), J = 1, NPOT)
0033        WRITE (6, 1030) CONC(I)
0034        IF (SMTH1 .NE. 0) GO TO 1210
0035        CALL SMOOTH (Y, YY, NPOT)
0036        CALL SIMP (YY, NPOT, POT, A)
0037        WRITE (6, 1035) POT(1), Y(1), YY(1), (POT(J), Y(J), YY(J),
           1 A(J), J = 2, NPOT)
0038        GO TO 1220
0039  1210  CALL SIMP (Y, NPOT, POT, A)
0040        WRITE (6, 1040) POT(1), Y(1), (POT(J), Y(J), A(J), J = 2, NPOT)
0041  1220  DO 1230 J = 2, NPOT
0042  1230  SAVE(J, I) = A (J)
      C
      C     PLOT POINTS HERE IF IPL1 IS ZERO
      C
0043        IF (IPL1 .NE. 0) GO TO 1260
0044        IL = 1
0045        CALL PLTR (NPOT, A, P, IL)
0046  1260  CONTINUE
0047        WRITE (6, 1090)
0048        NC = NCONC - 1
0049        DO 1300 I = 1, NC
0050  1300  C(I) = CONC(I + 1)
0051        XCONC(I) = ALOG (CONC(I + 1) )
0052        DO 1360 I = 2, NPOT
0053        WRITE (6, 1045) POT(I)
0054        DO 1310 J = 1, NC
0055  1310  Y(J) = SAVE (I, J+1)
0056        CALL THREER (XCONC, Y, NC, A)
0057        WRITE (6, 1050) (XCONC(J), Y(J), A(J), J = 1, NC)
0058        IF (IPL2 .NE. 0) WRITE (6, 1060)
0059        DO 1320 J = 1, NC
0060  1320  SAVE(I, J) = A(J) * RTEMP
      C
      C     PRODUCE A PRINTER PLOT IF IPL2 IS ZERO
      C
0061        IF (IPL2 .NE. 0) GO TO 1360
0062        IL = 2
0063        CALL PLTR (NC, A, C, IL)
0064  1360  CONTINUE
```

```
0065          WRITE (6, 1090)
       C
0066          NP = NPOT - 1
0067          DO 1410 I = 1, NP
0068          P(I) = P(I + 1)
0069   1410   POT(I) = POT(I + 1)
0070          DO 1460 I = 1, NC
0071          WRITE (6, 1030) CONC(I+1)
0072          DO 1420 J = 2, NPOT
0073   1420   Y(J-1) = SAVE(J, I)
0074          IF (SMTH3 .NE. 0) GO TO 1440
0075          CALL SMOOTH (Y, YY, NP)
0076          CALL SIMP (YY, NP, POT, A)
0077          WRITE (6, 1035) POT(1), Y(1), YY(1), (POT(J), Y(J), YY(J),
0078        1 A(J), J = 2, NP)
0079          GO TO 1450
0080   1440   CALL SIMP (Y, NP, POT, A)
0081          WRITE (6, 1040) POT(1), Y(1), (POT(J), Y(J), A(J), J = 2, NP)
0082   1450   DO 1455 J = 1, NP
0083   1455   SAVE(J, I) = A(J)
       C
       C      PRODUCE A PRINTER PLOT IF IPL3 ZERO
       C
       C
0083          IF (IPL3 .NE. 0) GO TO 1460
0084          IL = 3
0085          CALL PLTR (NP, A, P, IL)
0086   1460   CONTINUE
0087          WRITE (6, 1090)
       C
       C
0088          DO 1500 I = 1, NC
0089   1500   WRITE (7, 1070) (SAVE(J, I), J =1, NP)
0090          GO TO 1100
       C
0091   1900   STOP
0092          END
```

```
0001        SUBROUTINE SIMP(Y,N,X,A)
0002        DIMENSICN Y(100), X(100), A(100)
0003        A(1)=0.
0004        IF (N-3) 200,50,50
0005   50   DX1 = X(2) - X(1)
0006        DX2 = X(3) - X(2)
0007        DY1 = Y(2) - Y(1)
0008        DY2 = Y(3)-Y(2)
0009        SDX1 = DX1*DX1
0010        SDX2 = DX2*DX2
0011        CDX1 = DX1*SDX1
0012        CDX2 = DX2 *SDX2
0013        SUM = DX1+DX2
0014        DENOM = CDX1*DX2*SUM
0015        A(2) = Y(2)*DX1 -(0.5*SDX1*(SDX1*DY2+SDX2*DY1)- 0.333333*CDX1* (
0016       1DX1*DY2-DX2*DY1))/DENOM
0017        DO 100 I=3,N
0018        A(I) = A(I-2) + Y(I-1)*SUM+(0.5*(SDX2-SDX1)*(SDX1*DX2+SDX2*DY1)+
0019       1(CDX1+CDX2)*0.333333*(DX1*DY2-DX2*DY1))/DENOM
0020        IF (I.EQ.N) GO TO 100
0021        DX1 = DX2
0022        SDX1 = SDX2
0023        CDX1 = CDX2
0024        DX2 = X(I+1) -X(I)
0025        SDX2 = DX2*DX2
0026        CDX2 = DX2*SDX2
0027        DY1 = DY2
0028        DY2 = Y(I+1) -Y(I)
0029        SUM = DX1+DX2
0030        DENOM = CDX1*DX2*SUM
0031  100   CONTINUE
0032  200   RETURN
0033        END
```

```
0001        SUBROUTINE THREER (X,Y,N,A)
0002        DIMENSION X(100), Y(100), A(100)
0003        IF (N-3) 200,50,50
0004  50    DX1 = X(2) - X(1)
0005        DX2 = X(3) - X(2)
0006        DY1 = Y(2)-Y(1)
0007        DY2 = Y(3)-Y(2)
0008        SDX1 = DX1*DX1
0009        SDX2 = DX2*DX2
0010        DENOM = SDX1*DX2+SDX2*DX1
0011        A(1) = ((SDX1*DY2+SDX2*DY1)-2.*DX1*(DX1*DY2-DX2*DY1))/DENOM
0012        M = N-1
0013        DO 100 I=2,M
0014        A(I) = (SDX1*DY2+SDX2*DY1)/DENOM
0015        IF (I.EQ.M) GO TO 100
0016        DX1 = DX2
0017        SDX1 = SDX2
0018        DY1 = DY2
0019        DX2 = X(I+2) - X(I+1)
0020        SDX2 = DX2*DX2
0021        DY2 = Y(I+2)-Y(I+1)
0022        DENOM = SDX1*DX2+SDX2*DX1
0023  100   CONTINUE
0024        A(N) = ((SDX1*DY2+SDX2*DY1+2.*DX2*(DX1*DY2-DX2*DY1))/DENOM
0025  200   RETURN
0026        END
```

```
0001          SUBROUTINE SMOOTH(Y,YY,N)
0002          DIMENSION Y(100),YY(10C)
0003          YY(1)= Y(1)
0004          IF (N.LT.2) RETURN
0005          YY(2)= Y(2)
0006          YY(N-1)= Y(N-1)
0007          YY(N) = Y(N)
0008          IF (N.LT.5) RETURN
0009          M= N-2
0010          DO 100 I=3,M
0011  100     YY(I) = 0.02857143*(-3.*Y(I-2)+12.*Y(I-1)+17.*Y(I)+12.*Y(I+1)-3.*Y
             1(I+2))
0012          RETURN
0013          END
```

## REFERENCES

[1]. L. R. McCoy and H. B. Mark, Jr., J. Phys. Chem., 73:953 (1969).

[2]. (a) H. C. MacDonald, Jr. and H. B. Mark, Jr., J. Phys. Chem., submitted for publication. (b) H. C. MacDonald, Jr., Doctoral Dissertation, University of Michigan, 1969.

[3]. D. M. Mohilner and P. R. Mohilner, J. Electrochem. Soc., 115:261 (1968).

[4]. D. M. Mohilner, Electroanalytical Chemistry, Vol. 1 (A. J. Baird, ed.), Marcel Dekker, New York, 1966, p. 241.

[5]. F. B. Hildebrand, Introduction to Numerical Analysis, McGraw-Hill, New York, 1956.

B. Electrochemical Kinetic Calculations

Chapter 3

# APPLICATION OF COMPUTERS TO
# SOLUTION OF ORGANIC ELECTRODE REACTION MECHANISMS

Harvey B. Herman

Department of Chemistry
University of North Carolina at Greensboro
Greensboro, North Carolina, 27412

## I. INTRODUCTION

Adding or removing electrons electrochemically from stable organic molecules can produce instability. Chemists are using this phenomenon as a convenient method for understanding decay mechanisms of unstable molecules [1]. Excitation of an electrode system causes a response: a "normal" response when no chemical reactions are involved (or in some cases extremely fast reactions), and a "perturbed" response when chemical reactions are important in the time span of measurement. The relationship between the two different responses can in many cases be used to calculate the rate constant(s) involved in the chemical reaction. For example, the current response of an electrode system can be measured after a step potential change. Under certain conditions, the normal response would predict a constant product of current and the square root of time. Whenever the electron transfer product is unstable in solution and homogeneously reacts to regenerate the starting material, the current-root-time product is no longer constant. In this case the ratio of the two current-root-time products can be used to calculate the rate constant with the aid of a table or graph. Frequently the numbers used in the table or graph were generated by a computer program working with analytical solutions of Fick's Law. The mathematical methods used in electrochemistry, e.g., Laplace transforms, quite often arrive at equations which are not useful unless worked on further with numerical methods. Computers can be extremely helpful in this area.

When more conventional mathematical procedures fail, computers can be used to solve Fick's Law directly by finite difference methods. In recent years interest in these methods has increased as more complicated problems have been approached. Happily, computers at the same time have become faster, making these methods really practical.

The measurement of rate constants is only the first step in any study. Other numbers and correlations must be squeezed out of the raw data. The "best" numbers are found by regression analysis, linear and nonlinear. Anyone who has done this with a desk calculator, for example, can understand how tedious it can be and can appreciate the utilization of computers for these calculations.

The present chapter deals with the two basic applications of computers to the understanding of organic electrode mechanisms: (1) rate constant measurement and (2) regression analysis. I hope that the discussion of these applications will be helpful to others interested in this field, particularly neophytes.

## II. RATE CONSTANT DETERMINATION

### A. Analytical Solution Necessary

Among the more common transient electrochemical methods for the determination of homogeneous chemical reaction rates concomitant with electron transfer are chronoamperometry and chronopotentiometry [2]. Most of the simple linear diffusion equations have been solved and are available in the literature. The following implicit equation in $k\tau$ appears frequently [3] in discussions of the EC mechanism (electron transfer-chemical reaction)

$$O + ne = R \xrightarrow{k} Y \tag{1}$$

and current-reversal chronopotentiometry,

$$\mathrm{erf}[k^{1/2}(\tau + t)^{1/2}] - 2\,\mathrm{erf}[(k\tau)^{1/2}] = 0. \tag{2}$$

After measuring transition times the chemist is faced with a choice. He can use a small graph in the original article or prepare his own working curve to determine the rate constant. Many workers have adopted the latter course to improve the precision of the final result. Working curves, plots, and tables are a convenient bridge between experimental numbers and the rate constant. Digital computer solutions of equations like (2) are an obvious aid in the preparation of extensive tables and even graphs. Two methods which can be used to solve these implicit analytical solutions are the "brute-force" [4] and Newton-Raphson methods [5]. Occasionally tables can be prepared by algebraic manipulation of an equation, but more often, methods similar to the ones cited are essential.

## 1. Brute-Force Method

In practice this method is very simple to use and will work with many of the equations of interest to electrochemists. Usually these equations have only one real positive root and it is possible to choose an initial guess less than this value. Under these conditions convergence (defined below) will normally occur, albeit slowly. The method has been severely criticized because of this slow convergence feature [6]. However, I have only abandoned it when computer time became excessive.

The brute-force method is based on the well-known observation that an equation of the form $y = f(x)$ changes sign when a root is passed. Equation (2) can be rearranged to this form and made dimensionless. Thus,

$$y = \text{erf}[(k\tau)^{1/2}(1 + t/\tau)^{1/2}] - 2\,\text{erf}[(k\tau)^{1/2}(t/\tau)^{1/2}] \tag{3}$$

This equation is a function of the two dimensionless variables $k\tau$ and $t/\tau$. If we choose a value for one, $k\tau$, we can solve for the other, $t/\tau$. In this case zero can be an initial guess for $t/\tau$ after specifying $k\tau$. In other cases a very small nonzero value must be used as zero can be a trivial solution or a singularity. The initial sign of y is saved and a second guess, say 1, is attempted. Change of sign indicates that the root has been bracketed. A third guess, 0.1, can then be tried. Each time the root is passed the increments are decreased by a factor of 10. If the root has not been passed the increment is simply added to the last guess. In this manner we can bracket the root $t/\tau$ as closely as we choose and the method is said to have converged.

The above procedure is illustrated in a block diagram, Fig. 1. A FORTRAN program, which compiles a working table of $t/\tau$ for values of $k\tau$ from 0.01 to 3.00, is shown in Fig. 2. Statement number 10 can be any other equation, electrochemical or otherwise, as long as the above restrictions are kept in mind. Equations with more than one positive root can be accommodated with adjustments in the initial guess and incrementing variable. Modifications in the program are necessary for negative roots. Also, the number of iterations should be limited in case of nonconvergence for any reason. In the program illustrated TAU is $k\tau$ and X is initially kt. The ratio $t/\tau$ is calculated by dividing X by TAU before printing.

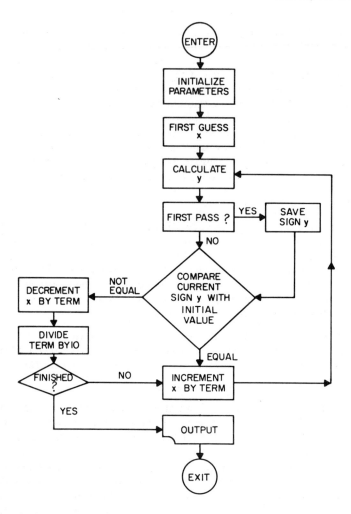

Fig. 1.  Block diagram of brute-force procedure to calculate a positive root of an arbitrary function.

## 2.  Newton-Raphson Method

The rate of convergence can be improved markedly if the Newton-Raphson method is used.  The objective is the same as was discussed previously, namely, to prepare a table of $t/\tau$ vs. $k\tau$.  A possible disadvantage of the method is that a knowledge of the function's derivative is required for solution.  The derivative can, of course, be evaluated numerically but in the example discussed below it is explicitly determined.

```
C      THIS PROGRAM CALCULATES TRANSITION RATIOS FOR CURRENT REVERSAL
C      CHRONOPOTENTIOMETRY WITH A FOLLOWING CHEMICAL REACTION.
C      VALUES OF K*TAU (TAU) ARE GENERATED BY THE PROGRAM AND THE RATIO OF
C      TAU2 TO TAU1 (X) IS SOLVED BY AN ITERATIVE METHOD
       WRITE (6,999)
999    FORMAT (1H1)
       DO 500 JJ=1,300
       TAU=(JJ)/100.
       NOSIG=3
       X=0.0
       M=1
       LA=0
       R=2.0
10     Y=ERF(SQRT(TAU+X))-R*ERF(SQRT(X))
96     IF (M-1)   300,100,102
100    Z=Y
       M=M+1
102    IF (Z)   98,200,99
98     IF (Y)   71,200,73
99     IF (Y)   73,200,71
71     X=X+10.0**(-LA)
       GO TO 10
73     X=X-10.0**(-LA)
       LA=LA+1
199    IF ((X/10.0**(-LA))-10.0**(NOSIG))   71,200,200
200    X=X/TAU
       WRITE (6,900) R,TAU,X
500    CONTINUE
900    FORMAT (3F10.5)
       GO TO 920
300    WRITE (6,905)
905    FORMAT (2X,5HERROR)
920    STOP
       END
```

Fig. 2. FORTRAN program listing to calculate chronopotentiometric transition time ratios when electrode process is perturbed by following chemical reaction. Procedure illustrated in Fig. 1 was used.

The Newton-Raphson method is described in several numerical-analysis and programming texts [7, 8]. The equation which we wish to solve is again written in the form, $y = f(x)$. The derivative of the function, e.g., Eq. (3), is evaluated. Thus,

$$d(y) = d(erf[(k\tau)^{1/2}(1 + t/\tau)^{1/2}]) - 2d(erf[(k\tau)^{1/2}(t/\tau)^{1/2}]) \qquad (4)$$

using

$$d(erf[(ax)^{1/2}]) = \frac{a}{\pi^{1/2}[ax]^{1/2}} \exp(-ax). \qquad (5)$$

An initial guess is made for x(kt) and the value of y and dy calculated. The current value of x is corrected with the equation,

$$x_{new} = x_{old} - y/d(y), \qquad (6)$$

and the process is repeated until the relative correction is less than a

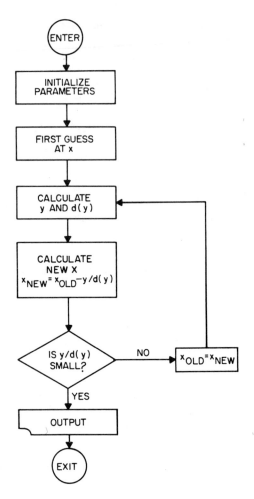

Fig. 3. Block diagram of Newton-Raphson procedure to calculate a root of an arbitrary function.

specified amount. A block diagram of this procedure is illustrated in Fig. 3. A FORTRAN progran using this basic method but set up for solving the cyclic chronopotentiometric equations for the reversible catalytic mechanism,

$$O + ne = R$$

$$\underset{k_2}{\overset{k_1}{\rightleftharpoons}}$$
, (7)

is shown in Fig. 4. When $k_1$ (RK1) is zero, the number of reversals (K) is 1, and the current program (RR) is 2, this program solves Eq. (3).

In some cases convergence is not obtained. Conditions for convergence are given by McCracken and Dorn [7]:

(1) The guess should be relatively close to the root.
(2) The second derivative of the function should not be too large.
(3) The first derivative of the function should not be too small.

It has been suggested [6] that a combination of the brute-force and Newton-Raphson methods might be advantageous. The former method would be used to calculate a close guess and the latter method would be used to compute the desired number of significant figures. In my experience of using the Newton-Raphson method to solve electrochemical equations, it was almost always possible to make a guess sufficiently close to the root so that the procedure converged. In the few instances where this was not the case it was a simple matter to use the brute-force method.

The equation solved in the previous example was relatively uncomplicated. In other situations, more complex procedures would have to be used. Professor Bard and I described in a recent paper [5] a computer program designed to solve the general equations of cyclic chronopotentiometry involving homogeneous kinetic complications. The basic flow pattern of Fig. 3 was used except that the function and its derivatives were generated with subprograms. The function includes terms in error function and square root multiplied by constants which were evaluated in another subprogram using a method suggested by Ashley and Reilley [9]. Notwithstanding these complications, the Newton-Raphson method successfully solved for the roots of the function and allowed tables of transition-time ratios to be prepared for this paper. A well-annotated listing of the program is given in Fig. 5. The input data consist of the number of reversals (K), the number of significant figures (NOSIG), and the current program (RR). The following reaction scheme is treated:

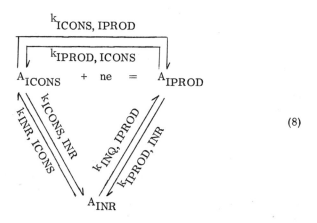

$$(8)$$

```
C        CYCLIC CHRONO CATALYTIC REVERSIBLE CASE T(1) IS RK1 + RK2 TIMES TAU
C        FRACT IS RATIO OF RED FORM TO OX FORM
         DIMENSION X(100),T(100),R(100),Z(100)
         READ (5,900) K,NOSIG,RR
         T(1)=1.0
         RK1=-0.5
         DO 500 JJ=1,6
         RK1=RK1+0.5
         RK2=0.0
         DO 500 KK=1,5
         RK2=RK2+0.5
         WRITE (6,903) RR,T(1),RK1,RK2
         RTK=RK1+RK2
         FRACT=RK1/RK2
         V=ERFUN(SQRT(RTK*T(1)))
         IF ((FRACT*V)-1.0)   65,500,500
C        GENERATION OF EQUATIONS
65       DO 200 N=2,K
         T(N)=0.5
         DO 199 II=1,100
10       DO 80 I=1,N
         SUM=0.0
         DO 60 J=I,N
60       SUM=SUM+T(J)
         X(I)=ERFUN(SQRT(RTK*SUM))
         Z(I)=1./1.772453809*EXP(-RTK*SUM)*1./SQRT(RTK*SUM)*RTK
80       CONTINUE
         Y=X(1)-V
         W=Z(1)
81       SIGN=1.0
         DO 95 L=2,N
         SIGN=-SIGN
         W=W+SIGN*RR*Z(L)
95       Y=Y+SIGN*RR*X(L)-SIGN*V-SIGN*FRACT*V
         T(N)=ABS(T(N)-Y/W)
         IF ((Y/W)**2/T(N)-10.**(-NOSIG))   200,200,199
199      CONTINUE
         WRITE (6,906) N,T(N),T(N-1)
906      FORMAT (17H DID NOT CONVERGE,5X,4HN = ,I3,5X,7HT(N) = ,E15.5,5X,9H
        1T(N-1) = ,E15.5)
         GO TO 500
200      CONTINUE
         DO 201 J=1,K
201      R(J)=T(J)/T(1)
         WRITE (6,901)
         WRITE (6,902) (N,T(N),R(N),N=1,K)
500      CONTINUE
900      FORMAT (2I10,4F10.5)
901      FORMAT (4(3X,1HN,9X,4H TAU,8X,5HRATIO)//)
```

Fig. 4. FORTRAN program listing to calculate cyclic chronopotentiometric transition time ratios when electrode process is perturbed by a reversible catalytic reaction. Procedure illustrated in Fig. 3 was used.

It is necessary to specify in the input data the subscripts of the species produced and consumed during the first transition and the subscript of the electroinactive substance. Input data also include the rate constants indicated in Eq. (8) (some of which may be zero). Subscripts on the rate constants indicate the path followed, i.e., $k_{12}$ refers to the path whereby $A_1$ reacts to form $A_2$. Output is of course a table of transition-time ratios for

```
902    FORMAT (4(1X,1I3,2E13.5)/)
903    FORMAT (2X,5HRR = ,1X,F10.5,1X,7HT(1) = ,1X,F10.5,1X,6HRK1 = ,F10.
      15,1X,6HRK2 = ,F10.5)
904    FORMAT (I10)
       STOP
       END
       FUNCTION ERFUN (X)
       XP=0.3275911
       A1=0.225836846
       A2=-0.252128668
       A3=1.259695130
       A4=-1.287822453
       A5=0.940646070
       Y=1./(1.+XP*X)
       ERFUN=1.-(A1*Y+A2*Y**2+A3*Y**3+A4*Y**4+A5*Y**5)*(2./1.772453809055
      12)*EXP(-X*X)
       RETURN
       END
          20         4      2.
```

Fig. 4 (cont.)

that particular mechanism and value of input parameters.

It should be pointed out that in some cases the differences may be subtle and it will not be possible to differentiate among various mechanisms. However, many possibilities usually can be eliminated and other evidence can be employed to help solve the puzzle. These points and others are elaborated further in the original article [5].

### B. Numerical Solution Necessary

Electrochemists have successfully produced analytical solutions for their peculiar boundary-value problems for several decades [10]. A chemical kinetic example chosen from current reversal chronopotentiometry was discussed above. The Laplace transform method [11], among others, has been widely used when the chemical reaction is first order. The method fails, except under restrictive conditions, when the chemical reaction is, for example, second order. In recent years finite-difference methods have been applied to situations where it is not possible to use Laplace transforms [12]. In these cases it is necessary to simulate the solution of Fick's law with a digital computer. The discussion below, using single potential step chronoamperometry [2] as an example, outlines the basic principles of this method.

The fundamental equation for simple electron transfer and linear diffusion is Fick's law,

$$\frac{\partial C_{\theta}(x,\ t)}{\partial t} = \frac{\partial^2 C_O(x,\ t)}{\partial x^2}. \tag{9}$$

Specification of initial and boundary conditions is necessary for a unique

```
C        THIS PROGRAM CALCULATES CYCLIC CHRONOPOTENTIOMETRIC RELATIVE
C        TRANSITION TIMES FOR VARIOUS HOMOGENEOUS KINETIC COMPLICATIONS
C        THE FOLLOWING EXAMPLE IS FOR THE SCHEME
C
C
C
C
C        A SUB ICONS + NE = A SUB IPROD
C
C
C
C                  A SUB INR
C        WHERE A3 IS THE CONCENTRATION OF SPECIES 3,ETC, AND K32 IS THE
C        RATE CONSTANT FOR THE PRODUCTION OF SPECIES 2 FROM SP. 3
C        ICONS IS SUBSCRIPT OF SPECIES WHICH IS CONSUMED DURING FIRST
C        TRANSITION TIME, IPROD IS SUBSCRIPT OF SPECIES WHICH IS PRODUCED
C        DURING FIRST TRAN TIME, INR IS SUBCRIPT OF NON ELECTROACTIVE SPECIES
         DIMENSION X(100),T(100),Z(100),XK(3,3,3),RK(3,3),A(3),C(3)
         READ (5,904) K,NOSIG,RR
C        DATA READ IN  K(NO. OF REVERSALS)
C        (2I10,F10.5) NOSIG(NO SIG FIGS IN ANS.)
C        RR      CURRENT PROGRAM( (IF+IR)/IF) 2. IF EQUAL CURRENTS
         T(1)=1.0
2        READ (5,909) ICONS,IPROD,INR
C        READ SUBSCRIPT OF SPECIES CONSUMED, PRODUCED AND NOT ELECTROACTIVE
C        ( 3I2)
909      FORMAT (3I2)
         IF (ICONS.EQ.0) STOP
C        BLANK CARD WHEN READING ICONS STOPS PROGRAM
         A(IPROD)=-1.
         A(ICONS)=1.
         A(INR)=0.
C        -1 MEANS PRODUCED DURING FIRST TRANSITION
C         1 MEANS CONSUMED DURING FIRST TRANSITION
C        0. MEANS NOT ELCTROACTIVE
         WRITE (6,999)
999      FORMAT (1H1,' NOREVERSALS  NO SIG FIGS CURR. PROG. ')
         WRITE (6,900) K,NOSIG,RR
         WRITE (6,910) ICONS,IPROD,INR
910      FORMAT (1X,'A(',I1,') IS CONSUMED DURING FIRST TRANSITION'/,1X,'A(
        1',I1,') IS PRODUCED DURING FIRST TRANSITION'/,1X,'A(',I1,') IS NOT
        1 ELECTROACTIVE')
1        READ (5,907) RK(INR,ICONS),RK(ICONS,INR),RK(IPROD,ICONS),RK(ICONS,
        1IPROD),RK(IPROD,INR),RK(INR,IPROD)
         IF (RK(INR,ICONS).LT.0.) GO TO 2
C        NEGATIVE VALUE FOR RK(INR, ICONS) TRANSFERS TO 2 AND READS ANOTHER
C        VALUE  FOR ICONS , ETC.
         WRITE (6,908)
908      FORMAT (///)
         WRITE (6,911) INR,ICONS,ICONS,INR,IPROD,ICONS,ICONS,IPROD,IPROD,IN
        1R,INR,IPROD
911      FORMAT (6(1X,'RK(',I1,',',I1,')  '))
         WRITE (6,907) RK(INR,ICONS),RK(ICONS,INR),RK(IPROD,ICONS),RK(ICONS
        1,IPROD),RK(IPROD,INR),RK(INR,IPROD)
907      FORMAT (6F10.5)
```

Fig. 5. FORTRAN program listing to calculate cyclic chronopotentiometric relative transition time ratios when electrode process is perturbed by general kinetic scheme. Procedure illustrated in Fig. 3 was used.

```
        FRACT=0.
        IF (RK(IPROD,ICONS).NE.0.0)FRACT=RK(ICONS,IPROD)/RK(IPROD,ICONS)
        IF ((FRACT*V).GE.1.) GO TO 500
        CALL ASHLEY(RK,XK,C)
C       SUBROUTINE ASHLEY CALCULATES SYSTEM CONSTANTS
63      V=AR(T(1),ICONS,XK,C,A)
C       CALC OF F(T(1)) FOR SPECIES A SUB ICONS WITH SYSTEM CONSTANTS
C       CALC. FROM ASHLEY SUBPROGRAM
C       GENERATION OF EQUATIONS
C       K RELATIVE TRAN. TIMES ARE CALC. IN THIS LOOP
65      DO 200 N=2,K
C       START SOLUTION BY NEWTON RAPH.
        T(N)=0.5
C       FIRST GUESS
        DO 199 II=1,100
C       100 MAX. ITERATIONS FOR EQUATIONS WHICH DO NOT CONVERGE
10      DO 80 I=1,N
        SUM=0.0
C       CALC OF SUMS OF T(N) S
        DO 60 J=I,N
60      SUM=SUM+T(J)
66      IF (MOD(N,2))  62,61,62
C       N  ODD
62      X(I)=AR(SUM,ICONS,XK,C,A)
C       AT TRAN TIME  A SUB ICONS IS ZERO
        Z(I)=DAR(SUM,ICONS,XK,C,A)
        GO TO 80
C       N EVEN
61      X(I)=AR(SUM,IPROD,XK,C,A)
C       AT TRAN TIME A SUB IPROD IS ZERO
        Z(I)=DAR(SUM,IPROD,XK,C,A)
80      CONTINUE
        Y=X(1)-V
        W=Z(1)
81      SIGN=1.0
        DO 95 L=2,N
        SIGN=-SIGN
        W=W+SIGN*RR*Z(L)
95      Y=Y+SIGN*RR*X(L)-SIGN*V+SIGN*FRACT*V
        T(N)=ABS(T(N)-Y/W)
        IF ((Y/W)**2/T(N)-10.**(-NOSIG))  200,200,199
199     CONTINUE
        WRITE (6,906) N
906     FORMAT (17H DID NOT CONVERGE,5X,4HN = ,I3)
        GO TO 500
C       RELATIVE ERROR SMALL ENOUGH
200     CONTINUE
C    3 EQUATION SOLVED PRINT ANSWER
        WRITE (6,901)
C       OUTPUT OF RESULTS
        WRITE (6,902) (N,T(N),N=1,K)
500     CONTINUE
900     FORMAT (2I10,6X,F10.5,///)
901     FORMAT (5(3X,1HN,9X,4H TAU)//)
```

Fig. 5 (cont.)

```
902     FORMAT (5(1X,I3,F13.5)/)
904     FORMAT (2I10,F10.5)
        GO TO 1
        END
        FUNCTION AR(X,IF,XK,C,A)
C       INVERSE TRANSFORM OF EQUATION 2 IN ASHLEY AND REILLEY
C       SYSTEM CONSTANTS ARE CALC. BY ASHLEY   SUBROUTINE
        DIMENSION C(3),XK(3,3,3),A(3)
        AR=0.0
        IF (C(3))  3,1,3
1       DO 2 IG=1,3
2       AR=(2./1.772453851*SQRT(X)*XK(IF,IG,1)+1./SQRT(C(2))*ERFUN(SQRT(C(
        12)*X))*XK(IF,IG,2)+2./1.772453851*SQRT(X)*XK(IF,IG,3))*A(IG)+AR
        RETURN
3       DO 4 IG=1,3
4       AR=(2./1.772453851*SQRT(X)*XK(IF,IG,1)+1./SQRT(C(2))*ERFUN(SQRT(C(
        12)*X))*XK(IF,IG,2)+1./SQRT(C(3))*ERFUN(SQRT(C(3)*X))*XK(IF,IG,3))*
        1A(IG)+AR
        RETURN
        END
        FUNCTION DAR(X,IF,XK,C,A)
C       DERIVATIVE OF INVERSE TRANSFORM OF EQUATION 2 IN ASHLEY REILLEY
        DIMENSION C(3),XK(3,3,3),A(3)
        DERFUN(A,X)=1./1.77245385/SQRT(A*X)*A*EXP(-A*X)
        DAR=0.0
        IF (C(3))  3,1,3
1       DO 2 IG=1,3
2       DAR=(1./1.772453851/SQRT(X)*XK(IF,IG,1)+1./SQRT(C(2))*DERFUN(C(2),
        1X)*XK(IF,IG,2)+1./1.772453851/SQRT(X)*XK(IF,IG,3))*A(IG)+DAR
        RETURN
3       DO 4 IG=1,3
4       DAR=(1./1.772453851/SQRT(X)*XK(IF,IG,1)+1./SQRT(C(2))*DERFUN(C(2),
        1X)*XK(IF,IG,2)+1./SQRT(C(3))*DERFUN(C(3),X)*XK(IF,IG,3))*A(IG)+DAR
        RETURN
        END
        SUBROUTINE ASHLEY(RK,XK,C)
C       SUBROUTINE TO CALCULATE SYSTEMS CONSTANTS FOR THREE COMPONENT
C       SYSTEM ( SEE ASHLEY AND REILLEYS PAPER IN J ELECTRUANAL CHEM)
        DIMENSION RK(3,3),XK(3,3,3),C(3),G(3),Q(3)
        S=(RK(1,2)+RK(2,1)+RK(1,3)+RK(3,1)+RK(2,3)+RK(3,2))/2.0
        IF (S)  98,98,99
C       S = 0.
98      C(1)=0.0
        C(2)=1.
        C(3)=1.
        DO 23 I=1,3
        DO 23 J=1,3
23      XK(I,J,1)=1.
        DO 24 K=2,3
        DO 24 I=1,3
        DO 24 J=1,3
24      XK(I,J,K)=0.
        RETURN
C       S NE 0.
```

Fig. 5 (cont.)

```
99      G(1)=RK(2,1)*RK(3,1)+RK(3,2)*RK(2,1)+RK(2,3)*RK(3,1)
        G(2)=RK(1,2)*RK(3,2)+RK(3,1)*RK(1,2)+RK(1,3)*RK(3,2)
        G(3)=RK(1,3)*RK(2,3)+RK(2,1)*RK(1,3)+RK(1,2)*RK(2,3)
        GG=G(1)+G(2)+G(3)
        IF (GG)   3,4,3
C       GG NE 0.
3       DO 7 I=1,3
7       Q(I)=G(I)/GG
        D=ABS(SQRT(S**2-GG))
        GO TO 5
C       GG = 0.
4       DO 8 I=1,3
        Q(I)=0.0
        DO 8 J=1,3
        IF (I-J)  21,8,21
21      IF (RK(J,I)+RK(I,J))   22,8,22
22      Q(I)=RK(J,I)/(RK(J,I)+RK(I,J))+Q(I)
8       CONTINUE
        D=S
5       CONTINUE
C       BOTH CASES =0. NEO. MERGE HERE
        C(1)=0.0
        C(3)=S-D
        C(2)=S+D
C       START CALC OF K SUB FFH
        DO 18 IF=1,3
        XK(IF,IF,1)=Q(IF)
        DO 12 I=1,3
        IF (IF-I)  13,12,13
13      XK(IF,I,1)=Q(IF)
12      CONTINUE
        IF (D)   9,10,9
9       XK(IF,IF,2)=0.0
        DO 19 I=1,3
        IF (IF-I) 1,19,1
1       XK(IF,IF,2)=XK(IF,IF,2)+1./(2.*D)*(RK(IF,I)-Q(I)*C(3))
19      CONTINUE
        XK(IF,IF,3)=0.0
        DO 6 I=1,3
        IF (IF-I)  2,6,2
2       XK(IF,IF,3)=XK(IF,IF,3)-1./(2.*D)*(RK(IF,I)-Q(I)*C(2))
6       CONTINUE
C       END CALC OF K SUB FFH
C       START CALC OF K SUB FRH
        DO 14 I=1,3
        IF (IF-I)   15,14,15
15      XK(IF,I,2)=-1./(2.*D)*(RK(I,IF)-Q(IF)*C(3))
        XK(IF,I,3)=1./(2.*D)*(RK(I,IF)-Q(IF)*C(2))
14      CONTINUE
        GO TO 18
10      XK(IF,IF,2)=0.0
        DO 11 I=1,3
11      XK(IF,IF,2)=XK(IF,IF,2)+1./2.*Q(I)
        XK(IF,IF,3)=XK(IF,IF,2)
```

Fig. 5 (cont.)

```
      DO 16 I=1,3
      IF (IF-I)  17,16,17
17    XK(IF,I,2)=-1./2.*Q(IF)
      XK(IF,I,3)=XK(IF,I,2)
16    CONTINUE
18    CONTINUE
C     END CALC OF K SUB FRH
      RETURN
      END
      FUNCTION ERFUN(X)
C     HASTINGS APPROXIMATION TO ERROR FUNCTION ( NOW AVAILABLE IN
C     FORTRAN IV FOR IBM 360 AS ERF )
      XP=0.3275911
      A1=0.225836846
      A2=-0.252128668
      A3=1.259695130
      A4=-1.287822453
      A5=0.940646070
      Y=1./(1.+XP*X)
      ERFUN=1.-(A1*Y+A2*Y**2+A3*Y**3+A4*Y**4+A5*Y**5)*(2./1.772453809055
     12)*EXP(-X*X)
      RETURN
      END
```

Fig. 5 (cont.)

solution. The initial and boundary conditions for single potential step chronoamperometry are

$$C_O(x, \ t < 0) = C_O^o, \quad C_O(0, \ t) = 0, \quad C_O(x \to \infty, \ t) = C_O^o. \tag{10}$$

A particularly useful solution involving current as a function of time, Cotrell's equation is [13]

$$i = \frac{nFAD^{1/2}C_O^o}{(\pi t)^{1/2}}. \tag{11}$$

The finite-difference method obviously does not give us Eq. (11). But it will allow us to calculate the current-time behavior and possibly deduce the consistency of the current-root-time product. When chemical kinetics is important, the finite-difference method will allow calculation of working curves even in cases where no simple analytical solution is known.

The finite-difference analog of Eq. (9) is

$$[C_O(x, t + \Delta t) - C_O(x, \ t)]/\Delta t$$
$$= D[C_O(x + \Delta x, \ t) - 2C_O(x, \ t) + C_O(x - \Delta x, \ t)]/(\Delta x)^2. \tag{12}$$

When the increments are small Eq. (12) has been shown to be a good approximation of Fick's law [12]. Stability criteria restrict the value of $D\Delta t/(\Delta x)^2$ to less than one-half [14]. The concentrations at $t + \Delta t$ can be calculated using Eq. (12) if we specify $D\Delta t/(\Delta x)^2$ and know the concentrations at t. The procedure has been dubbed the "forward difference method" because of the way the partial differentials in Fick's law have been approximated.

Current is calculated using the following boundary condition,

$$i = \frac{nFAD \, \partial C_0(0, \, t)}{\partial x} . \tag{13}$$

The finite difference analog of Eq. (13) is

$$i = \frac{nFAD[C_0(2, \, t) - C_0(0, \, t)]}{2\Delta x} \tag{14}$$

where the parenthetical subscripts 2 and 0 refer to the row of concentrations once removed from the surface on either side of the electrode. The concentration with the subscript zero, used only for convenience, is eliminated using the following equations,

$$C_0(1, \, t+\Delta t) - C_0(1, \, t) = D\Delta t/(\Delta x)^2 [C_0(2, \, t) - 2C_0(1, \, t) + C_0(0, \, t)] \tag{15}$$

$$C_0(1, \, t) = C_0(1, \, t + \Delta t) = 0. \tag{16}$$

The equation for current then becomes

$$i = nFAD \frac{C_0(2, \, t)\Delta x}{\Delta t} . \tag{17}$$

It is easiest to give $\Delta x$ and $\Delta t$ the value of 1 and satisfy the stability requirement by proper definition of D, i.e., $D \leq \frac{1}{2}$. The particular values picked for $C_0^0$, D, $\Delta x$, and $\Delta t$ need not be realistic as the problem is usually couched in dimensionless terms. However, if desired, a correspondence table could be set up.

A block diagram of the procedure for diffusion-controlled chronoamperometry is illustrated in Fig. 6 and the FORTRAN program in Fig. 7. The equations, with minor simplifications, are as described in the text. The concentration of R(Y) also is calculated even though it was not necessary for this simple example.

A more complicated FORTRAN example following this basic outline is shown in Fig. 8. This program simulates the solution of chronoamperometry and the ECE mechanism [15]. It also allows optional use of the equations involving a nonequilibrium nuance [16, 17]. The exact analytical solution was given by Alberts and Shain [15],

$$\frac{it^{1/2}}{(it)_\infty^{1/2}} = 1 - \frac{n_2 \exp(-kt)}{(n_1 + n_2)}. \tag{18}$$

The agreement between the exact answer and the simulation was quite good.

Again, much needless calculation is saved by the use of dimensionless parameters (current-root-time ratio and kt). It is important even when the exact solution is known, witness Eq. (8), to keep the number of working curves within manageable proportions.

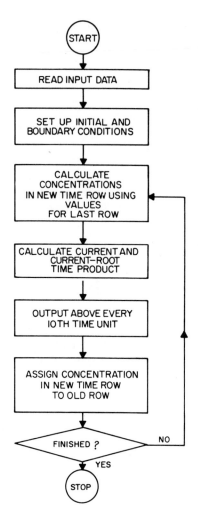

Fig. 6.  Block diagram of procedure used to simulate Fick's Law solution under chronoamperometric boundary conditions.

## III.  REGRESSION ANALYSIS

The previous discussion illustrated the determination of rate constants with working curves or tables. Occasionally graphical methods, plots of $y = ax + b$, are more convenient. The more reliable method of computing the slope (a) and intercept (b) of a straight line is by the method of least squares (linear regression on a single variable) [18]. Other applications of

```
C        THIS PROGRAM SIMULATES THE SOLUTION OF FICK'S LAW UNDER
C        CHRONOAMPEROMETRIC BOUNDARY CONDITIONS
         DIMENSION X(1000),Y(1000),XX(1000),YY(1000)
         READ (1,900) D,C
900      FORMAT (2E10.4)
         XITHF=SQRT(D)*C/1.77245
         WRITE (3,901) D,C,XITHF
901      FORMAT (4E15.5)
         B=(1.-2.0*D)
         DO 101 I=2,999
         X(I)=C
101      Y(I)=0.0
         X(1)=0.0
         DO 4 J=1,5000
         XM=J
         IK=5.*SQRT(D*XM)+3.
         DO 3 I=2,IK
         XX(I)=B*X(I)+D*(X(I+1)+X(I-1))
3        YY(I)=B*Y(I)+D*(Y(I+1)+Y(I-1))
         ZI=X(2)*D
         IF (MOD(J,10)) 6,5,6
5        XITHF=ZI*SQRT(XM)
         WRITE (3,901) ZI,XM,XITHF,Y(1)
6        YY(1)=B*Y(1)+D*(2.*Y(2)+2.*ZI/D)
         DO 61 K=2,IK
         X(K)=XX(K)
61       Y(K)=YY(K)
         Y(1)=YY(1)
4        CONTINUE
         STOP
         END
```

Fig. 7.  FORTRAN program illustrating procedure in Fig. 6.

linear regression, besides rate constant determination, suggest themselves after the rate constants are evaluated. For example, we may be interested in correlating substituent with rate (Hammet equation) [19] or in correlating ionic strength with rate (primary salt effect) [20]. These calculations, usually by digital computers, are simple in comparison with the data-gathering step.

Regression analysis is not limited to one independent variable. A commonly used procedure is multiple linear regression [18]. For example, the reaction whose rate is measured may be catalyzed by acids, bases, or both [21]. It is important to determine, for mechanistic purposes, whether this is true and which substances are the catalytic agents. The dependent variable in this case may be correlated with several other variables (concentrations of catalysts). Multiple linear regression, again using digital computers, is the proper way to treat these data.

Several equations which are used for data analysis are not linear. In some cases these equations can be linearized, e. g., Arrhenius' equation. However, we must be careful to weight the regression equations properly

```
C       THIS PROGRAM SIMULATES THE SOLUTION OF FICK'S LAW.
C       THE EQUATION IS PERTURBED BY AN ECE NUANCE UNDER
C       CHRONOAMPEROMETRIC BOUNDARY CONDITIONS
        DIMENSION X(1000),Y(1000),XX(1000),YY(1000),Z(1000),ZZ(1000),W(100
       10),WW(1000),ZI(3)
        READ (5,903) L
903     FORMAT (I10)
        DO 4 M=1,L
        WRITE (6,904)
904     FORMAT (1H1)
        READ (5,900) D,C,RK,RK1,RK2
900     FORMAT (5E10.4)
        XITHFO=SQRT(D)*C/1.77245
        WRITE (6,901) D,C,XITHFO,RK,RK1,RK2
901     FORMAT (7E15.5)
        B=(1.-2.0*D)
        DO 101 I=2,999
        X(I)=C
        Z(I)=0.
        W(I)=0.
101     Y(I)=0.0
        X(1)=0.
        Y(1)=C
        Z(1)=0.
        W(1)=0.
        DO 4 J=1,5000
        XM=J
        IK=5.*SQRT(D*XM)+3.
        DO 3 I=2,IK
        CC=RK1*Y(I)*Z(I)-RK2*X(I)*W(I)
        XX(I)=B*X(I)+D*(X(I+1)+X(I-1))+CC
        ZZ(I)=B*Z(I)+D*(Z(I+1)+Z(I-1))-CC+RK*Y(I)
        WW(I)=B*W(I)+D*(W(I+1)+W(I-1))+CC
3       YY(I)=B*Y(I)+D*(Y(I+1)+Y(I-1))-CC-RK*Y(I)
        CC=RK1*Y(1)*Z(1)-RK2*X(1)*W(1)
        ZI(1)=X(2)*D+CC/2.
        ZI(2)=Z(2)*D+(RK*Y(1)-CC)/2.
        ZI(3)=ZI(1)+ZI(2)
        IF (MOD(J,10))  6,5,6
5       XITHF=ZI(3)*SQRT(XM)
        XITHF=XITHF/XITHFO
        WRITE (6,901) ZI(3),XM,XITHF,Y(1)
6       YY(1)=B*Y(1)+2.*D*(Y(2)+ZI(1)/D)-CC-RK*Y(1)
        WW(1)=B*W(1)+2.*D*(W(2)+ZI(2)/D)+CC
        DO 61 K=2,IK
        X(K)=XX(K)
        Z(K)=ZZ(K)
        W(K)=WW(K)
61      Y(K)=YY(K)
        Y(1)=YY(1)
        W(1)=WW(1)
4       CONTINUE
        STOP
        END
```

Fig. 8.  FORTRAN program listing to calculate by simulation chronoamperometric current-root-time ratios when electrode process is perturbed by an
ECE reaction.

[22]. In other cases it is necessary to use the more complicated technique of nonlinear regression [23, 24]. For example, it may be necessary to measure the dissociation constants of the acids used in our buffer solutions. In a few situations well-separated dissociation constants allow equation simplification to be made. However, most of the time proper values can only be calculated using a nonlinear method.

## A. Least-Square Fit to a Straight Line

### 1. Discussion

Frequently data are acquired which can be correlated by a linear equation. A scientist may measure x-y data pairs and desire the slope (a) and intercept (b) of the following equation,

$$y = ax + b. \tag{19}$$

The best value of the constants is assumed to minimize the sum of the squares of the residuals $(y - ax_i - b)$. The equation for the sum,

$$S = \sum_{i=1}^{n} (y_i - ax_i - b)^2, \tag{20}$$

will be a minimum when the proper values of a and b are chosen. The name, least-square fit, follows naturally from the assumption above. The form of Eq. (20) implies no error in x and, conversely, all error in y. The derivations below will ascribe all of the uncertainty of measurement to y. The terms in the sum should be weighted if it is believed that they are not equally important. The present discussion will assume that all y values have equal weight, although it is not difficult to introduce unequal factors. It was felt that a discussion on these points would be more appropriate in the section on nonlinear regression.

The minimum in Eq. (20) is found by partial differentiation. Thus,

$$\left(\frac{\partial S}{\partial a}\right)_b = \frac{\partial[\Sigma(y - ax + b)^2]}{\partial a} = 0 \quad \text{and} \quad \left(\frac{\partial S}{\partial b}\right)_a = \frac{\partial[\Sigma(y - ax + b)^2]}{\partial b} = 0. \tag{21}$$

The sums are as before, over the number of data pairs. Two simultaneous equations result,

$$bn + a\Sigma x = \Sigma y \quad \text{and} \quad b\Sigma x + a\Sigma x^2 = \Sigma xy. \tag{22}$$

These equations may be solved by any convenient method, e.g., determinants. Thus,

$$a = [n\Sigma xy - \Sigma x\Sigma y]/d, \tag{23}$$

$$b = [\Sigma x^2\Sigma y - \Sigma x\Sigma xy]/d, \quad \text{and} \tag{24}$$

$$d = n\Sigma x^2 - (\Sigma x)^2. \tag{25}$$

The standard deviation of the slope and intercept can be calculated from the formulas,

$$s_a = (n/d)^{1/2} s_y, \tag{26}$$

$$s_b = (\Sigma x^2/d)^{1/2} s_y, \quad \text{and} \tag{27}$$

$$s_y = [\Sigma(y - ax - b)^2]^{1/2}/(n - 2). \tag{28}$$

Equation (26) is derived from the relationship for a, Eq. (23), and

$$s_a^2 = \sum_{j=1}^{n} (\partial a/\partial y_j)^2 s_y^2. \tag{29}$$

The derivative of Eq. (23) in Eq. (29) is

$$(\partial a/\partial y_j) = (nx_j - \Sigma x)/d. \tag{30}$$

The index j is explicitly shown to distinguish it from the usual index i, not shown. For our purposes the details are unimportant; however, if the reader is interested he can refer to Bevington [18]. The $s_b$ is calculated in much the same manner.

An important statistical parameter is called the linear correlation co-efficient (r) [18]. It is defined by the equation,

$$r = (aa')^{1/2} \tag{31}$$

where a is the normal slope in Eq. (19), and $a'$ is the slope in the equation,

$$x = b' + a'y. \tag{32}$$

If Eq. (23) and an analogous equation for $a'$ are substituted into Eq. (31) the following equation for r can be derived,

$$r = \frac{n\Sigma xy - \Sigma x \Sigma y}{[n\Sigma x^2 - (\Sigma x)^2]^{1/2}[n\Sigma y^2 - (\Sigma y)^2]^{1/2}} \tag{33}$$

where r can take on values from 0 to ±1, with the latter being a perfect correlation. However, the value of r by itself cannot be used to determine if the data are linearly correlated. For example, 10% of the time r will be greater than 0.9 for four uncorrelated data pairs, but for nine such pairs this would only happen less than one time in 1000. Table C3 in Ref. [18] gives information which can be used to evaluate a given correlation. A reasonable confidence level must be chosen along with the number of data pairs used.

A block diagram of a subprogram which can be used to calculate slope, intercept, standard deviation, and correlation coefficient is shown in Fig. 9. A listing of the corresponding FORTRAN function subroutine appears in Fig. 10. The subprogram is a simplified version of the one appearing in Bevington's text [18]. Most computation centers will have a similar one available for their users.

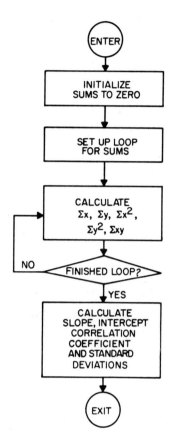

Fig. 9. Block diagram of subroutine procedure used to calculate least-square values of slope, intercept, standard deviations, and correlation coefficient for a straight line.

## 2. Applications

Least-square fit to a straight line has several applications in organic electrode mechanism problems.

a. Rate Constant Determination. It may be convenient to evaluate the rate constant at small values of kt. At longer times, electrode fouling or convection may become important. For example, Alberts and Shain [15] in a chronoamperometric study derived the following equation for the ECE mechanism (electron transfer-chemical reaction-electron transfer).

$$it^{1/2} = \frac{FAC_o^o D^{1/2} \{n_1 + n_2 [1 - \exp(-kt)]\}}{\pi^{1/2}} \tag{34}$$

```
      SUBROUTINE LINFIT(X,Y,NPTS,B,SIGMAB,A,SIGMAA,R)
C     LEAST SQUARE FIT TO Y = A*X + B
C     PROGRAM WAS ADAPTED FROM ONE IN TEXT BY BEVINGTON
      DIMENSION X(500),Y(500)
      SUM=NPTS
      SUMX=0.
      SUMY=0.
      SUMX2=0.
      SUMXY=0.
      SUMY2=0.
      DO 50 I=1,NPTS
      XI=X(I)
      YI=Y(I)
      SUMX=SUMX+XI
      SUMY=SUMY+YI
      SUMX2=SUMX2+XI*XI
      SUMXY=SUMXY+XI*YI
      SUMY2=SUMY2+YI*YI
50    CONTINUE
      DELTA=SUM*SUMX2-SUMX*SUMX
      B=(SUMX2*SUMY-SUMX*SUMXY)/DELTA
      A=(SUMXY*SUM-SUMX*SUMY)/DELTA
      C=NPTS-2
      VARNCE=(SUMY2+B*B*SUM+A*A*SUMX2-2.*(B*SUMY+A*SUMXY-B*A*SUMX))/C
      SIGMAB=SQRT(VARNCE*SUMX2/DELTA)
      SIGMAA=SQRT(VARNCE*SUM/DELTA)
      R=(SUM*SUMXY-SUMX*SUMY)/SQRT(DELTA*(SUM*SUMY2-SUMY*SUMY))
      RETURN
      END
```

Fig. 10.   FORTRAN program illustrating procedure in Fig. 9.

If the first two terms of the series expansion for exponential are substituted into Eq. (34), we have

$$it^{1/2} = \frac{FAC_0^{\,0} D^{1/2} (n_1 + n_2 kt)}{\pi^{1/2}} \tag{35}$$

This equation is, of course, only valid for $kt \ll 1$ and must be used carefully.

A plot of $it^{1/2}$ vs. t gives a slope, $FAC_0^{\,0} D^{1/2} n_2 k/\pi^{1/2}$, and an intercept, $FAC_0^{\,0} D^{1/2} n_1/\pi^{1/2}$. When $n_1 = n_2$, division of the slope by the intercept evaluates k.

b.  Hammett Plot.  Linear-free energy correlations have been found very useful in mechanistic studies [19]. Substitution of different functional groups, meta or para and sometimes ortho, to the reaction site frequently influences the rate of a reaction. The correlations are between log k and substituant constant $(\sigma)$. Information on whether the rate increases or decreases with electron-withdrawing ability $(+\sigma$ values) can lend support to a plausible mechanism. The reader is referred to Wells [19] for more details.

c.  Activation Parameters.  Arrhenius plots are commonly used to determine activation parameters. The simplified equation

$$k = aT \exp(-b/T) \tag{36}$$

is linearized using logarithms. Thus

$$\ln k = \ln (aT) - b/T \quad \text{or} \quad \ln(k/T) = \ln a - b/T, \tag{37}$$

from which a and b can be evaluated and $\Delta H^{\ddagger}$ and $\Delta S^{\ddagger}$ calculated. Unweighted linear correlations involving logarithms should not give the best answers. However, in my experience the calculated parameters are not that different. This could be caused by the narrow range of the correlation.

   d. Ionic Strength Effects.   The primary salt effect can be used to determine the charges on the ions which react to form the transition state complex [20]. If the reaction is

$$A^{Z_A} + B^{Z_B} = (AB) \rightarrow \text{products} \tag{38}$$

the following simplified equation illustrates the ionic strength dependence,

$$\log k = a + \frac{b(K)^{1/2}}{1 + (I)^{1/2}} \tag{39}$$

Least-square evaluation of $b(Z_A Z_B)$ gives the product of the charges on the ions involved.

## B.  Multiple-Linear Regression

### 1.  Discussion

   The preceding material on regression was limited to one independent variable. When the dependent variable can be correlated with more than one independent variable the least-square procedure is called multiple-linear regression [25]. The best value of the constants in the equation

$$y = C_1 x_1 + C_2 x_2 + \cdots + C_m x_m \tag{40}$$

can be evaluated using the least-square condition, see Eq. (21), for each independent variable $(x_i)$. The following m equations in m unknowns $(C_i)$ results,

$$C_1 \Sigma x_1^2 + C_2 \Sigma x_2 x_1 + C_3 \Sigma x_3 x_1 + \cdots + C_m \Sigma x_m x_1 = \Sigma y x_1$$

$$C_1 \Sigma x_1 x_2 + C_2 \Sigma x_2^2 + C_3 \Sigma x_3 x_2 + \cdots + C_m \Sigma x_m x_2 = \Sigma y x_2 \tag{41}$$

$$\vdots$$

$$C_1 \Sigma x_1 x_m + C_2 \Sigma x_2 x_m + C_3 \Sigma x_3 x_m \cdots + C_m \Sigma x_m^2 = \Sigma y x_m$$

or in matrix notation

$$\underline{A}\,\underline{C} = \underline{B} \tag{42}$$

where

$$A = \begin{vmatrix} \Sigma x_1{}^2 & \Sigma x_2 x_1 & \Sigma x_3 x_1 & \cdots & \Sigma x_m x_1 \\ \Sigma x_1 x_2 & \Sigma x_2{}^2 & \Sigma x_3 x_2 & \cdots & \Sigma x_m x_2 \\ & & \vdots & & \\ \Sigma x_1 x_m & \Sigma x_2 x_m & \Sigma x_3 x_m & \cdots & \Sigma x_m{}^2 \end{vmatrix}, \quad C = \begin{vmatrix} C_1 \\ C_2 \\ \vdots \\ C_m \end{vmatrix}, \quad B = \begin{vmatrix} \Sigma y x_1 \\ \Sigma y x_2 \\ \vdots \\ \Sigma y x_m \end{vmatrix}. \quad (43)$$

The value of each C and its standard deviation can be found if the inverse matrix of $A$ (i.e., $A^{-1}$) can be evaluated [26]. Thus,

$$A^{-1} A C = A^{-1} B \quad \text{and} \quad C = A^{-1} B \tag{44}$$

and if the elements of $A^{-1}$ are called $a_{ij}$ it is possible to write the standard deviation of $C_i$ from the elements $a_{ii}$ with the equation

$$s_{c_i} = (a_{ii})^{1/2} s_y \tag{45}$$

where

$$s_y = [\Sigma(y - c_1 x_1 - c_2 x_2 - \cdots - c_m x_m)^2 / (n - m - 1)]^{1/2}. \tag{46}$$

The value of the constant $x_0$ implied in Eq. (40) can be evaluated by setting all of the values of a given variable equal to one.

A simplified approach to multiple-linear regression is discussed by Pennington [27]. Equation (40) can be written in matrix form,

$$XC = Y \tag{47}$$

where

$$X = \begin{vmatrix} x_{11} & x_{12} & \cdots & x_{2m} \\ x_{21} & x_{22} & \cdots & x_{2m} \\ & & \vdots & \\ x_{m1} & x_{n2} & \cdots & x_{nm} \end{vmatrix}, \quad Y = \begin{vmatrix} y_1 \\ y_2 \\ \vdots \\ y_n \end{vmatrix}, \tag{48}$$

and $C$ is defined earlier [Eq. (43)]. Since there are more equations than unknowns (m > n) the system is overdetermined. The least-square condition is obtained by premultiplication of the matrix $X$ by its transpose ($\tilde{X}$). Thus,

$$\tilde{X} X C = \tilde{X} Y. \tag{49}$$

The matrices $\tilde{X} X$ and $\tilde{X} Y$ are identical to $A$ and $B$ respectively in Eq. (42). Regression coefficients are then obtained in the same manner as before.

The implementation of either method is not difficult. A block diagram of a procedure based on the former method is shown in Fig. 11. Output is the coefficient matrix, $C$, in Eqs. (42) and (47). The latter method is most easily programmed using a language such as BASIC [28] which includes matrix multiplication, inversion, and transposition as an integral part of the

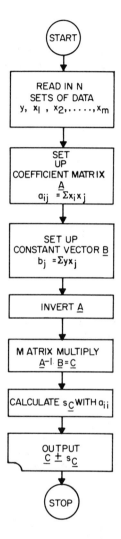

Fig. 11. Block diagram of procedure for multiple linear regression. Program not shown.

language. Notwithstanding the relative ease of implementation, chemists and others have taken ligitimate advantage of the multiple-linear regression programs offered by many computation centers as part of their programming services. It is certainly more convenient to use one of these programs than to attempt to write and debug one of your own.

## 2. Application to Acid-Base Catalysis

An important use of multiple-linear regression is in the interpretation of experiments in acid-base catalysis. The observed rate constant can be composed of several factors [21],

$$k_{obs} = k_o + k_{H^+}[H^+] + k_{OH^-}[OH^-] + \Sigma k[A] + \Sigma k[B] \tag{50}$$

where contributions from all acids A and bases B are initially taken into account. The following parameters need to be known before the hypothesis represented by Eq. (50) can be tested: hydrogen ion concentration, concentration autoprotolysis constant, and analytical concentration and dissociation constants of each acid. For example, the phosphate concentration can be calculated [29] at a given hydrogen ion concentration by

$$[PO_4^{-3}] = \frac{K_{a_1} K_{a_2} K_{a_3} C_{H_3PO_4}}{[H^+]^3 + K_{a_1}[H^+]^2 + K_{a_1} K_{a_2}[H^+] + K_{a_1} K_{a_2} K_{a_3}} \tag{51}$$

$$= \alpha_o C_{H_3PO_4}$$

A first pass through the regression procedure, using concentrations such as those calculated in equations like Eq. (51), allows calculation of the constants in Eq. (50). Variables that do not contribute significantly to the regression can be rejected by the t test; t can be calculated for the j-th coefficient by the equation

$$t = C_j / s_{C_j} . \tag{52}$$

If all values exceed those in a table like A8 in Ref. [25] (for n - m - 1 degrees of freedom and a given level of significance), the null hypothesis may be rejected and each coefficient retained in our equation. If not, the regression procedure is repeated with rejected variables excluded from the analysis.

Needless to say a great deal of preliminary work has not been discussed. First, equilibrium constants such as dissociation constants of acids must be known or measured if not available. Data analysis for that step is discussed below. Second, the input data to a multiple-linear regression program must be properly prepared. The concentrations of all the acids and bases are calculated using equations like Eq. (51) and inputted as independent variables. A computer program in PL/1 [30] was written which created a file on disk, called output, which could then be accessed by the multiple-linear regression program [31] supplied by the computer center. The former program is listed in Fig. 12. The job control language statements necessary to create and assess the file output would be different at each computation center and is not shown. The two-step procedure described here was used specifically to save intermediate card handling and has worked very well. A procedure similar to

```
DATA: PROCEDURE OPTIONS(MAIN);
ON ENDFILE GO TO LAST ;
DECLARE X(20),XK(5),TERM(6),ALFA(6),II(20) ;
GET LIST(N,M,NS);
  /* N IS THE NUMBER OF OBSERVATIONS, M IS THE NUMBER OF VARIABLES,
  NS SPECIFIES HOW MANY TIMES THE SELECTION OF A DEPENDENT VARIABLE
  AND A SET OF INDEPENDENT VARIABLES IS TO BE PERFORMED ON THE SAME
  SET OF ORIGINAL VARIABLES */
PUT FILE(OUTPUT) SKIP;
PUT FILE(OUTPUT) EDIT('N=',N,',M=',M,',NS=',NS,';')(        A,F(2),A,
F(2),A,F(2),A);
PUT              EDIT('N=',N,',M=',M,',NS=',NS,';')(        A,F(2),A,
F(2),A,F(2),A);
  /* M IS THE NUMBER WE HAVE ASSIGNED TO THE MOST PROTONATED FORM
  OF A GIVEN ACID, N IS THE NUMBER OF PROTONS ON THAT ACID AND XK IS
  IS THE ARRAY OF DISSOCIATION CONSTANTS OF THAT ACID */
START: GET LIST(M,N,(XK(I) DO I=1 TO N));
  /* ZEROS STOP INPUT */
IF N=0 THEN GO TO OUT;
X=0.0;
XKW=6.61E-15;
  /* RK IS THE RATE CONSTANT, PH IS -LOG H SUB C AND CA IS THE
  ANALYTICAL CONCENTRAION OF THE ACID SYSTEM USED AS A BUFFER */
CONT: GET LIST(RK,PH,CA);
  /* ZEROS STOP INPUT */
IF RK=0.0 THEN GO TO START;
X(1)=10.**(-PH);
X(2)=XKW*10.**(PH);
TERM(1)=X(1)**N;
DENOM=TERM(1);
DO J=2 TO N+1;
TERM(J)=TERM(J-1)*XK(J-1)/X(1);
DENOM=DENOM+TERM(J);
END;
DO J=1 TO N+1;
ALFA(J)=TERM(J)/DENOM;
X(M-1+J)=ALFA(J)*CA;
END;
PUT FILE(OUTPUT) EDIT(RK)(SKIP ,E(15,5)) ;
PUT              EDIT(RK)(SKIP ,E(15,5)) ;
PUT FILE(OUTPUT) EDIT ((X(I) DO I=1 TO 12))(SKIP,4(E(15,5)));
GO TO CONT;
OUT: DO JJ = 1 TO NS;
  /* NRESI IS AN OPTION CODE FOR A TABLE OF RESIDUALS (1-YES, 0-NO),
  NDEP IS THE  NUMBER ASSIGNED TO THE DEPENDENT VARIABLE (1), AND
  K IS THE NUMBER OF INDEPENDENT VARIABLES */
GET LIST(NRESI,NDEP,K);
PUT FILE(OUTPUT) EDIT('NRESI=',NRESI,',NDEP=',NDEP,',K=',K,';')(SKIP ,
A,F(1),A,F(1),A,F(2),A);
PUT              EDIT('NRESI=',NRESI,',NDEP=',NDEP,',K=',K,';')(SKIP ,
A,F(1),A,F(1),A,F(2),A);
GET LIST((II(J) DO J = 1 TO K));
PUT FILE(OUTPUT) EDIT((II(J) DO J = 1 TO K))(SKIP,11(F(3)));
PUT              EDIT((II(J) DO J = 1 TO K))(SKIP,11(F(3)));


END OUT;
PUT FILE(OUTPUT) SKIP  ;
PUT              SKIP  ;
LAST: END DATA;
```

Fig. 12.  PL/1 program illustrating acid-base catalysis data treatment before using multiple linear regression program.

this, without disk storage, was used in an earlier paper by Blount and myself [32].

## C. Nonlinear Regression

### 1. Discussion

Data cannot always be analyzed with linear relationships. It is not possible to manipulate many useful functions so that only first-power terms in the independent variable(s) remain. However, the least-square condition can still be applied if proper weight is given to each residual. Thus,

$$S = \Sigma w f^2 \tag{52}$$

where

$$w = \left( f_{x_1}^2 \sigma_{x_1}^2 + f_{x_2}^2 \sigma_{x_2}^2 + \cdots + f_{x_m}^2 \sigma_{x_m}^2 \right)^{-1} \tag{53}$$

and

$$f = F(x_1, x_2, \ldots, x_m, C_1, C_2, \ldots, C_n). \tag{54}$$

The function in Eq. (54) is a general function of m independent variables and n calculable parameters. The weighting function w contains partial derivatives ($f_{x_1}$) and standard deviations ($\sigma_{x_1}$) for each variable (in this case $x_1$).

The least-square procedure discussed here [18] differs from those discussed before in that it requires partial derivatives, standard deviations, and initial guesses at the sought-for parameters. It is true that the derivatives can be approximated by finite difference equations and the standard deviation can be estimated from the weighted sum, nevertheless, convergence still depends on an intuitive estimation of the approximate value of the parameters. The concept is the same as in the Newton-Raphson method; a guess gives a correction which in turn gives a new guess. The correction $\Delta C_i$ is calculated from the following set of simultaneous equations in matrix form,

$$\underline{A}\underline{\Delta C} = \underline{B} \tag{55}$$

where

$$A = \begin{vmatrix} \Sigma w f_{C_1} f_{C_1} & \Sigma w f_{C_2} f_{C_1} & \cdots & \Sigma w f_{C_n} f_{C_1} \\ \Sigma w f_{C_1} f_{C_2} & \Sigma w f_{C_2} f_{C_2} & \cdots & \Sigma w f_{C_n} f_{C_2} \\ & \vdots & & \\ \Sigma w f_{C_1} f_{C_n} & \Sigma w f_{C_2} f_{C_n} & \cdots & \Sigma w f_{C_n} f_{C_n} \end{vmatrix}, \quad \underline{\Delta C} = \begin{vmatrix} \Delta C_1 \\ \Delta C_2 \\ \vdots \\ \Delta C_n \end{vmatrix}, \quad B = \begin{vmatrix} \Sigma w f_{C_1} f \\ \Sigma w f_{C_2} f \\ \vdots \\ \Sigma w f_{C_n} f \end{vmatrix}.$$

$$\tag{56}$$

As before the solution is most easily accomplished by matrix inversion. Thus,

$$\underline{A}^{-1}\underline{A}\Delta\underline{C} = \underline{A}^{-1}\underline{B} \quad \text{and} \quad \Delta\underline{C} = \underline{A}^{-1}\underline{B}. \tag{57}$$

The old value of $C_i$ may be corrected by

$$(C_i)_{new} = (C_i)_{old} - \Delta C_i. \tag{58}$$

Iteration is continued until each element of $\Delta\underline{C}$ is as small as desired. The standard deviation of each parameter may be computed from the diagonal terms, $a_{ii}$, in the inverse matrix of $\underline{A}$, if estimates of $\sigma$ were used in the weighting function. If not, and each $\bar{\sigma}$ is assumed to be 1, the standard deviations can be approximated by [18]

$$s_{C_i} = a_{ii}^{1/2}(\Sigma wf^2)^{1/2} (N - n - 1)^{-1/2} \tag{59}$$

where N is the number of observations.

The regression procedure will not always converge on physically reasonable values as there may be other local minimums on the hypersurface in n-dimensional space. In that case, more sophisticated methods must be used [18, 33]. Justification for their use was not apparent in the applications discussed below.

2. Applications

a. Activation Parameters. The entropy and enthalpy of activation can be determined (most accurately) by a nonlinear regression procedure [23]. Equation (36) is rearranged to the necessary form,

$$f = -x + ay \exp(-b/y), \tag{60}$$

where x and y are k and T, and a and b are functions of the activation parameters. In this case it is comparatively simple to determine the derivatives analytically. Thus,

$$f_x = -1 \quad f_y = (ab/y) \exp(-b/y) + a \exp(-b/y)$$
$$f_z = y \exp(-b/y) \quad f_b = -a \exp(-b/y). \tag{61}$$

The weighting factors can be calculated from $f_x$ and $f_y$ (assuming $\sigma_x$ and $\sigma_y$ are 1) by

$$w = (f_x^2 + f_y^2)^{-1}. \tag{62}$$

The previous discussion on nonlinear regression gave the general equation, in matrix form, involving the correction factors. This now becomes, for the present system (in slightly different notation),

$$(\Sigma wf_a f_a)\Delta a + (\Sigma wf_a f_b)\Delta b = \Sigma wf_a f$$
$$(\Sigma wf_a f_b)\Delta a + (\Sigma wf_b f_b)\Delta b = \Sigma wf_b f. \tag{63}$$

In this case a solution using the method of determinants is probably more appropriate [34]. As before, the initial and subsequent guesses are corrected with the equations [24],

$$a_{new} = a_{old} - \Delta a \qquad b_{new} = b_{old} - \Delta b \tag{64}$$

where

$$\Delta a = \frac{\begin{vmatrix} \Sigma wf_a f & \Sigma wf_z f_b \\ \Sigma wf_b f & \Sigma wf_b f_b \end{vmatrix}}{denom}, \qquad \Delta b = \frac{\begin{vmatrix} \Sigma wf_a f_a & \Sigma wf_a f \\ \Sigma wf_a f_b & \Sigma wf_b f \end{vmatrix}}{denom}, \qquad and$$

$$denom = \begin{vmatrix} \Sigma wf_a f_a & \Sigma wf_a f_b \\ \Sigma wf_a f_b & \Sigma wf_b f_b \end{vmatrix}. \tag{65}$$

The standard deviations of a and b are calculated with the equations,

$$s_a = (\Sigma wf_b f_b / denom)^{1/2} s_f, \qquad s_b = (\Sigma wf_a f_a / denom)^{1/2} s_f, \qquad and$$
$$s_f = (\Sigma wf^2)^{1/2} / (n - 3)^{1/2}. \tag{66}$$

The activation parameters can be calculated from the values of a and b with

$$\Delta H^{\ddagger} = Rb \qquad and \qquad \Delta S^{\ddagger} = R \ln(ah/k_{Boltz}). \tag{67}$$

A block diagram of a procedure which evaluates activation parameters using the method above is shown in Fig. 13. The corresponding PL/1 program is illustrated in Fig. 14. This program was originally written in a conversational language, CPS [35], and was converted to PL/1 using a program supplied by the computer center [36]. Batch operation through PL/1 was found to be more efficient.

b. Dissociation Constants of Acids. The evaluation of unknown dissociation constants with nonlinear regression is essential to the success of a catalysis study. Again the constants are needed to calculate the concentrations of suspected catalysts.

The experimental procedure is comprised of three parts:

(1) Calibration of the response ($pH_a$ vs. mV) of a pH-sensitive electrode with NBS buffers. An ordinary linear regression can be used.

(2) Titration of a strong acid with a strong base in order to evaluate the concentration autoprolysis constant of water. A procedure to account for the junction potential in strong acid and base solution, similar to the one suggested by Liberti and Light [37], has been used. The following equations are used to extrapolate to zero hydrogen and hydroxide concentration,

$$pH_C - pH_a = a + b[H^+] \tag{68}$$

and

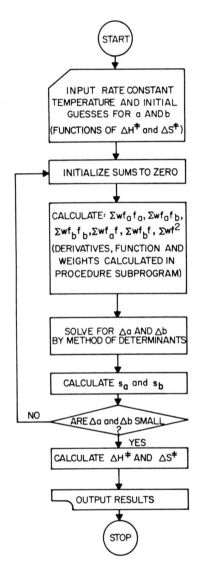

Fig. 13. Block diagram of procedure used to evaluate activation parameters by nonlinear regression.

```
MAIN: PROCEDURE OPTIONS(MAIN);
       DECLARE (X(20),Y(20)) FLOAT DEC(14);
       ON ENDFILE(SYSIN) STOP;
       DECLARE (A,B,FAFA,FBFB,FAFB,FAF,FBF,FF,DENOM,DELA,DELB,
       SA,SB,DELTAH,DELTAS) FLOAT DEC(14);
       DECLARE(#SETS,J,N,I) FIXED;
       GET LIST (#SETS) COPY;
       DO J = 1 TO #SETS;
       GET LIST(N) COPY;
       ON ERROR BEGIN;PUT LIST('ERROR IN DATA');GO TO NEWDATA;END;
LOOP1: DO I=1 TO N;
       /* INPUT OF RATE CONSTANT(X) AND ABS TEMP(Y) */
       GET LIST(X(I),Y(I)) COPY;
       END LOOP1;
       /*  INPUT OF FIRST GUESS OF A AND B IN EQUATION K = */
       /*  A T EXP(-B/T) */
       GET LIST(A,B) COPY;
LOOP:  FAFA,FBFB,FAFB,FAF,FBF,FF=0;
LOOP2: DO I=1 TO N;
       FAFA=FAFA+FA(Y(I))**2*W(X(I),Y(I));
       FAFB=FAFB+FA(Y(I))*FB(Y(I))*W(X(I),Y(I));
       FBFB=FBFB+FB(Y(I))**2*W(X(I),Y(I));
       FAF=FAF+FA(Y(I))*F(X(I),Y(I))*W(X(I),Y(I));
       FBF=FBF+FB(Y(I))*F(X(I),Y(I))*W(X(I),Y(I));
       FF=FF+F(X(I),Y(I))**2*W(X(I),Y(I));
       END LOOP2;
       DENOM=FAFA*FBFB-FAFB*FAFB;
       DELA=(FAF*FBFB-FAFB*FBF)/DENOM;
       DELB=(FAFA*FBF-FAF*FAFB)/DENOM;
       A=A-DELA;
       SA=SQRT(FBFB/DENOM)*SQRT(FF/(N-3));
       B=B-DELB;
       SB=SQRT(FAFA/DENOM)*SQRT(FF/(N-3));
       IF ABS(DELA/A)>.0001&ABS(DELB/B)>.0001
THEN   GO TO LOOP;
       PUT SKIP;
       PUT LIST(' A AND B IN K = A T EXP(-B/T) ARE: ');
       PUT SKIP;
       PUT LIST('A = ',A,' B = ',B);
       PUT SKIP;
       PUT LIST(' STD DEV A = ',SA,' STD DEV B = ',SB);
       /* IN K CALS PER MOLE */
       DELTAH=1.987*B*10**-3 ;
       /* IN ENTROPY UNITS */
       DELTAS=1.987*LOG(A/.20836E11) ;
       PUT SKIP;
       PUT LIST('DELTA H (KCALS/MOLE) = ',DELTAH);
       PUT SKIP;
       PUT LIST('DELTA S(E.U.) = ',DELTAS);
NEWDATA: END;
       STOP ;
       F:PROC(XX,YY);
       RETURN(-XX+A*YY*EXP(-B/YY));
       END;
```

Fig. 14.  PL/1 program illustrating procedure in Fig. 13.

```
FX:PROC(XX);
RETURN(-1);
END:
FY:PROC(YY);
RETURN(A*B/YY*EXP(-B/YY)+A*EXP(-B/YY));
END;
FA:PROC(YY);
RETURN(YY*EXP(-B/YY));
END;
FB:PROC(YY);
RETURN(-A*EXP(-B/YY));
END;
W:PROC(XX,YY);
RETURN(1/((FX(XX))**2+(FY(YY))**2));
END;
/**/END;
```

Fig. 14 (cont.)

$$pH_a - pOH_C + a = p(K_w)_C + b'[OH^-] \tag{69}$$

The subscripts a and C refer to activity and concentration, respectively.

The intercept, a, of Eq. (68) is used in Eq. (69) to establish the autoprolysis constant by intercept. Similarly, ordinary regressions are used.

(3) Titration of a weak acid with a strong base to determine the dissociation constant(s). Data used in the calculation are: acid and base concentrations, initial volume, autoprolysis constant of water, pairs of $pH_C$-base volume points and established standard errors of these numbers.

The data in part 3 can be analyzed by modifying the general procedure for nonlinear regression. The general procedure, shown in Fig. 15, accepts an arbitrary number of variables and parameters. Matrix inversion and multiplication are used to calculate the corrections to the parameters. Finite difference approximation of derivatives was used in place of explicit representation. Thus,

$$\frac{\partial f}{\partial x} = \frac{f(x + \Delta x) - f(x - \Delta x)}{2\Delta x} \tag{70}$$

The PL/1 program, shown in Fig. 16, illustrates the determination of the dissociation constant of a triprotic acid. The function, f, consists of the charge balance and fraction dissociated relations which embody all the independent information known about the system [38]. For phosphoric acid the function is

$$f = [Na^+] + [H^+] - [H_2PO_4^-] - 2[HPO_4^{-2}] - 3[PO_4^{-3}] - [OH^-]$$

or

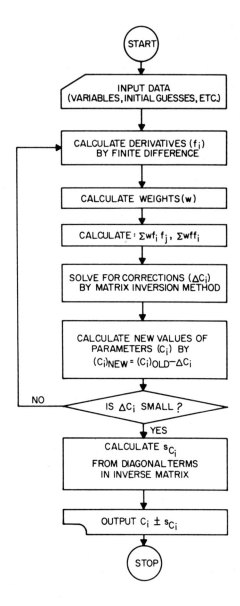

Fig. 15. Block diagram of general procedure used to evaluate parameters by nonlinear regression.

```
MAIN: PROCEDURE OPTIONS(MAIN);
      DECLARE (DELV(6),V(15),SV(15),PH(75),VB(75),A(4,4),B(4),
      D(15),RELV(15)) FLOAT DEC(14);
      DECLARE (#SETS,L,M,O,P,I,CNTR,J,II,K) FIXED;
      ON ENDFILE (SYSIN) BEGIN; PUT LIST('NOT ENOUGH DATA');STOP;8HD;
      DECLARE (DELX,RSDL,FF,W,T,K1,K2,K3,CA,CB,VA,KW,PHC,VOLB,H,
      OH,NA,DENOM,DF,H2PO4,HPO4,PO4) FLOAT DEC(14);
      GET LIST (#SETS) COPY;
      /* # OF TITRATION CURVES TREATED WITH ONE XEQ */
      D=0;
      DELV=0;
      DO L=1 TO #SETS;
      GET LIST (M,O,N) COPY;
      /* M IS # OF UNKNOWNS, K1 K2 ETC.; O IS # OF OTHER KNOWN*/
      /* VARIABLES; N IS # OF DATA PAIRS IN TITRATION CURVE*/
      DELX=.001;
      P=M+O;
      /* P IS SUM OF KNOWN AND UNKNOWN VARIABLES */
      DO I=1 TO P-2;
      GET LIST (V(I),SV(I)) COPY;
      /* V(1) TO V(M) ARE FIRST GUESSES AT UNKNOWN VARIABLES*/
      /*V(M+1) TO V(M+O-2) ARE VALUES OF KNOWN OTHER VARIABLES*/
      /* EXCEPT PHC AND VB WHICH ARE READ IN IN DO LOOP BELOW*/
      /* IT IS CONVENIENT TO DEFINE THEIR MEANINGS IN THE */
      /* F PROCEDURE BELOW */
      /* SV'S ARE STANDARD DEVIATION OF VARIABLES */
      END ;
      ON ERROR BEGIN; PUT LIST (' ERROR IN DATA, TRY ANOTHER GUESS');
      PUT LIST (' IF ALL ELSE CORRECT'); GO TO NEWDATA; END;
      Q=1;
      DO I=1 TO N;
      GET LIST (PH(I),VB(I)) COPY;
      /* SV(P-1) AND SV(P) ARE STANDARD DEVIATION OF PHC AND VB */
      END ;
      GET LIST (SV(P-1),SV(P)) COPY;
LOOP: A=0;
      B=0;
      RSDL=0;
      DO J=1 TO N;
      V(P-1)=PH(J);
      V(P)=VB(J);
      FF=F;
      DO I=1 TO P;
      CALL DERIV;
      END ;
      W=0;
      DO I=M+1 TO P;
      W=W+(SV(I)*D(I))**2;
      END ;
      W=1/W;
      RSDL=RSDL+FF**2*W;
      DO I=1 TO M;
      DO II=1 TO M;
      A(I,II)=A(I,II)+D(I)*D(II)*W;
```

Fig. 16.  PL/1 program listing to calculate the dissociation constants of a triprotic acid by nonlinear regression. Procedure illustrated in Fig. 15 was used.

```
            END ;
            B(I)=B(I)+D(I)*F*W;
            END ;
            END ;
            PUT SKIP;
            PUT LIST (' SUM OF WEIGHTED SQUARES OF RESIDUALS IS',RSDL);
            CALL SMLEQ;
            DO I=1 TO M;
            V(I)=V(I)-DELV(I);
            RELV(I)=DELV(I)/V(I);
            PUT SKIP;
            PUT LIST('V(',I,') = ',V(I));
            END ;
            DO I=1 TO M;
            IF ABS(RELV(I))>.0001
     THEN   GO TO LOOP;
            END ;
     OUT:   DO I=1 TO M;
            PUT SKIP;
            SV(I)=SQRT(A(I,I));
            PUT LIST('STD DEV SV(',I,') = ',SV(I));
            END ;
 NEWDATA: END;
            STOP ;
  DERIV: PROCEDURE ;
            V(I)=(1-DELX)*V(I);
            D(I)=-F;
            V(I)=(1+DELX)*V(I)/(1-DELX);
            D(I)=F+D(I);
            V(I)=V(I)/(1+DELX);
            D(I)=D(I)/(DELX*V(I)*2);
            RETURN ;
            END DERIV;
  SMLEQ: PROCEDURE ;
            DO I=1 TO M;
            T=A(I,I);
            A(I,I)=1;
            DO J=1 TO M;
            A(I,J)=A(I,J)/T;
            END ;
            DO K=1 TO M;
            IF K=I
     THEN   GO TO ELP3;
            T=A(K,I);
            A(K,I)=0;
            DO J=1 TO M;
            A(K,J)=A(K,J)-T*A(I,J);
            END ;
  ELP3:   END ;
            END ;
            DELV=0;
            DO J=1 TO M;
            DO I=1 TO M;
            DELV(J)=DELV(J)+A(J,I)*B(I);
```

Fig. 16 (cont.)

```
          END ;
          END ;
          RETURN ;
          END SMLEQ;
F:        PROCEDURE ;
          /* THE DEFINITIONS FOLLOW OF ALL VARIABLES */
          /* THE CHARGE BALANCE IS USUALLY  IN THE RETURN STATEMENT */
          K1=V(1);
          K2=V(2);
          K3=V(3);
          CA=V(4);
          CB=V(5);
          VA=V(6);
          KW=V(7);
          PHC=V(8);
          VOLB=V(9);
          H=10**-PHC;
          OH=KW/H;
          NA=CB*VOLB/(VOLB+VA);
          DENOM=H**3+K1*H**2+K1*K2*H+K1*K2*K3;
          DF=VA/(VA+VOLB);
          H2PO4=CA*DF*K1*H**2/DENOM;
          HPO4=CA*DF*K1*K2*H/DENOM;
          PO4=CA*DF*K1*K2*K3/DENOM;
          RETURN (H2PO4+2*HPO4+3*PO4+OH-H-NA);
          END F;
          /**/END;
```

Fig. 16 (cont.)

$$f = \frac{C_b V_b}{V_b + V_a} + 10^{-pH_C} - \frac{C_a V_a}{V_b + V_a} \times \frac{[H^+]^2 K_{a_1} + 2[H^+]K_{a_1} K_{a_2} + 3K_{a_1} K_{a_2} K_{a_3}}{D}$$

$$- (K_W)_C 10^{+pH_C}$$

where

$$D = [H^+]^3 + [H^+]^2 K_{a_1} + [H^+]K_{a_1} K_{a_2} + K_{a_1} K_{a_2} K_{a_3} .$$

This equation, or one like it for other systems, appears in a procedure subprogram at the end of the listing. Fewer than ten iterations for convergence is not uncommon although this is quite dependent on the initial guesses. Conversion from a conversational language to PL/1 was necessitated in this case when the connect time became inconveniently long. The program mentioned before was used [36].

## IV.  GENERAL COMMENTS

The examples of computer applications discussed here are representative of the ones the author and his students have used in the solution of organic electrode reaction mechanisms. Use of the computer saved time and allowed more exact answers to be calculated for a given effort. However, if for any reason computer use increases the total time on a given problem, other methods should be adopted. Problems like these were being solved long before the widespread use of high-speed digital computers.

The important subject of data acquisition and control [39] has been ignored in this chapter. The author's experience in this area is very recent and limited. This subject is covered in another chapter in the series. I have made the assumption that the data have been acquired by some means, either conventional operational amplifier-oscilloscope technology or computerized data acquisition. I am cognizant of the advantages of the latter method and view the increasing trend toward it as a good thing. However, in any multistep process like research, the stow step is rate-determining. Experimental design and interpretation will in many cases still limit the amount of work accomplished. Moreover, while the price of computers has declined rapidly in recent years, the total cost still is not in the price range of every laboratory. Perhaps some day!

## REFERENCES

[1].   H. Strehlow, in Technique of Organic Chemistry (A. Weissberger, ed.), Vol. VIII, Part II, Wiley-Interscience, New York, 1963, p. 799.

[2].   P. Delahay, in Treatise on Analytical Chemistry (I. M. Kolthoff and P. J. Elving, eds.), Part I, Vol. 4, Wiley-Interscience, New York, 1963, p. 2233.

[3].   A. C. Testa and W. H. Reinmuth, Anal. Chem., 32:1512 (1960).

[4].   H. B. Herman and A. J. Bard, Anal. Chem., 35:1121 (1963).

[5].   H. B. Herman and A. J. Bard, J. Electrochem. Soc., 115:1028 (1968).

[6].   D. Secrest, J. Chem. Ed., 42:625 (1965).

[7].   D. D. McCracken and W. S. Dorn, Numerical Methods and FORTRAN Programming, Wiley, New York, 1964, p. 133.

[8].   D. D. McCracken, A Guide to FORTRAN IV Programming, Wiley, New York, 1965, p. 31.

[9].   J. W. Ashley and C. N. Reilley, J. Electroanal. Chem., 7:253 (1964).

[10].   P. Delahay, New Instrumental Methods in Electrochemistry, Wiley-Interscience, New York, 1954.

[11].   R. V. Churchill, Operational Mathematics, McGraw-Hill, New York, 1958.

[12].   S. W. Feldberg, in Electroanalytical Chemistry (A. J. Bard, ed.), Vol. 3, Dekker, New York, 1969, p. 199.

[13].   C. N. Reilley, in Treatise on Analytical Chemistry (I. M. Kolthoff and P. J. Elving, eds.), Part I, Vol. 4, Wiley-Interscience, New York, 1963, p. 2109.

[14]. G. D. Smith, Numerical Solutions of Partial Differential Equations, Oxford, New York, 1965, p. 16.

[15]. G. S. Alberts and I. Shain, Anal. Chem., 35:1859 (1963).

[16]. M. D. Hawley and S. W. Feldberg, J. Phys. Chem., 70:3459 (1966).

[17]. H. N. Blount and H. B. Herman, J. Electrochem. Soc., 117:504 (1970).

[18]. P. R. Bevington, Data Reduction and Error Analysis for the Physical Sciences, McGraw-Hill, New York, 1969.

[19]. P. R. Wells, Linear Free Energy Relationships, Academic Press, New York, 1968, Chap. 3.

[20]. G. M. Harris, Chemical Kinetics, Heath, Boston, 1966, p. 101.

[21]. R. P. Bell, The Proton in Chemistry, Cornell Univ. Press, Ithaca, 1959, Chap. IX.

[22]. W. E. Deming, The Statistical Adjustment of Data, Wiley, New York, 1943, Chap. 10.

[23]. W. E. Wentworth, J. Chem. Educ., 42:96, 162 (1965).

[24]. P. J. Lingane, Anal. Chem., 39:485 (1967).

[25]. A. M. Neville and J. B. Kennedy, Basic Statistical Methods, International Textbook, Scranton, 1964, Chap. 17.

[26]. J. A. N. Lee, Numerical Analysis for Computers, Reinhold, New York, 1966, p. 165.

[27]. R. H. Pennington, Introductory Computer Methods and Numerical Analysis. Macmillan, New York, 1970, p. 410.

[28]. J. G. Kemeny and T. E. Kurtz, Basic Programming, Wiley, New York, 1967.

[29]. J. N. Butler, Ionic Equilibrium, Addison-Wesley, Reading, Mass., 1964, p. 210.

[30]. R. C. Sprowls, Introduction to PL/1 Programming, Harper and Row, New York, 1969.

[31]. North Carolina Educational Computing Service Memorandum LP-05-2, Research Triangle Park, North Carolina, 1968.

[32]. H. N. Blount and H. B. Herman, J. Phys. Chem., 72:3006 (1968).

[33]. L. G. Sillen, Acta Chem. Scand., 16:159 (1962).

[34]. A. G. Worthing and J. Geffner, Treatment of Experimental Data, Wiley, New York, 1943, p. 309.

[35]. Triangle Universities Computation Center Memorandum LS-55-1, Research Triangle Park, North Carolina, 1969.

[36]. Triangle Universities Computation Center Memorandum LS-68, Research Triangle Park, North Carolina, 1969.

[37]. A. Liberti and T. S. Light, J. Chem. Ed., 39:236 (1962).

[38]. L. G. Sillen, in Treatise on Analytical Chemistry (I. M. Kolthoff and P. J. Elving, eds.), Part I, Vol. 1, Wiley-Interscience, 1959, p. 277.

[39]. G. Lauer and R. A. Osteryoung, Anal. Chem., 40(10):30A (1968).

Chapter 4

COMPUTER ANALYSIS OF DATA OBTAINED BY

ELECTROCHEMICAL TRANSIENT PERTURBATION TECHNIQUES

Robert F. Martin

Department of Chemistry
Vassar College
Poughkeepsie, New York 12603

and

Donald G. Davis

Department of Chemistry
Louisiana State University in New Orleans
New Orleans, Louisiana 70122

I. INTRODUCTION

  Small-amplitude perturbation techniques have been used by a number of workers for the measurement of kinetic parameters. Especially when fast reactions are considered, inadequate correction for mass transfer has often in the past led to low values for the apparent exchange current. We do not propose to review the field of the measurement of kinetic data but rather to show how the use of a computer can make complex calculations possible and that in some cases true results can be obtained only in this way.

The techniques of coulostatics [1-4] and single-pulse galvanostatics [5] appear to have some specific advantages over other techniques for measurement of fast electron transfer reactions. Thus we have concentrated on these two techniques and developed programs to be used with them. The periodic small-amplitude methods, faradaic rectification, and faradaic impedance suffer from the fact that at high frequencies (needed to measure fast reactions), ohmic resistance dominates the cell impedance [6].

The transient methods, which include coulostatics, galvanostatics, and potentiostatics, all suffer from the fact that in order to take into account the rate of mass transfer on the total rates of heterogeneous reactions, equations containing a term of the form $\exp(z)^2 \operatorname{erfc}(z)$ must be solved for real and complex arguments. This task is difficult to do without the aid of a computer. Thus, in order to apply coulostatics or galvanostatics to fast reactions a computer program was developed to solve this problem of data treatment. Potentiostatics was not treated because of the experimental difficulty of obtaining instantaneous potential steps. The problem of including mass transfer in the treatment of potentiostatic data has been considered in an approximate way [7]. It would be a relatively simple task to redo the programs developed here for the potentiostatic case.

## II. STATEMENT OF PROBLEM

The general equation which, with some modification, can be used for double- or single-pulse galvanostatics or coulostatics was derived by Matruda et al. [8], is

$$
\eta = \frac{I_1}{C_{dl}(\beta - \gamma)} \left\{ \frac{\gamma}{\beta^2} \left[ \exp(\beta^2 t) \operatorname{erfc}(\beta t^{1/2}) + 2\beta\left(\frac{t}{\pi}\right)^{1/2} - 1 \right] \right.
$$
$$
\left. - \frac{\beta}{\gamma^2} \left[ \exp(\gamma^2 t) \operatorname{erfc}(\gamma t^{1/2}) + 2\gamma\left(\frac{t}{\pi}\right)^{1/2} - 1 \right] \right\}
$$
$$
- \frac{I_1 - I_2}{C_{dl}(\beta - \gamma)} \left\{ \frac{\gamma}{\beta^2} \left[ \exp(\beta^2) \operatorname{erfc}(\beta^{1/2}) + 2\beta\left(\frac{T}{\pi}\right)^{1/2} - 1 \right] \right.
$$
$$
\left. - \frac{\beta}{\gamma^2} \left[ \exp(\gamma^2) \operatorname{erfc}(\gamma^{1/2}) + 2\gamma\left(\frac{T}{\pi}\right)^{1/2} - 1 \right] \right\} \tag{1}
$$

in which

$$
\beta, \gamma = \frac{I_a}{2nF}\left(\frac{1}{C_O{}^O D_O^{1/2}} + \frac{1}{C_R{}^O D_R^{1/2}}\right) \pm \left[\frac{(I_a{}^O)^2}{4n^2 F^2}\left(\frac{1}{C_O{}^O D_O^{1/2}} + \frac{1}{C_R{}^O D_R^{1/2}}\right)^2 - \frac{nFI_a^O}{RTC_{dl}}\right]^{1/2} \tag{2}
$$

where the plus sign is associated with $\beta$ and the negative sign with $\gamma$. The time, t, is reckoned from the initiation of the current pulse. The time for charge injection is $t_1$, and $\tau = t - t_1$. Here $I_a^o$ is the apparent exchange current $C_{dl}$ is the double-layer capacity, and the other symbols have their usual significance. The current $I_1$ and $I_2$ are for the first and second pulse for the double-pulse galvanostatic method. If $I_2$ is set equal to zero, an equation appropriate for coulostatics results which includes a correction for the occurrence of faradaic reaction during the transfer of charge. The occurrence of faradaic reaction during transfer can be neglected if desirable and the equation simplified [4]. We have included faradaic reaction during transfer in our program and have found for the data of Weir and Enke [9] that exchange current densities were about 1% too low and the double-layer capacities about 6% too high if the occurrence of faradaic reaction during injection of charge was neglected. As we shall see this is a minor error compared to that caused by the neglect of mass transfer in many cases. Equation (1), however, does neglect the coupling of double-layer charging and faradaic charge transfer caused by specific adsorption.

### III. APPROXIMATE METHODS OF DATA ANALYSIS

In much of the previous work the difficulty in evaluating $\exp(z)^2 \, \text{erfc}(z)$ for both real and complex arguments have lead workers to seek approximate solutions to Eq. (1). Normally for the galvanostatic method this problem is solved by expansion in series of the exponential for long and short times [5, 10]. It is evident from the literature (see Table 1) that results vary from worker to worker and are lower than those values generated by a more complete data analysis. The expansion technique is seriously wanting in that at short times the reaction is primarily controlled by the double-layer and at long times by diffusion. Thus much information is lost.

Like the galvanostatic case, approximations for coulostatic data treatment have also been devised [2, 4]. The most straightforward approach is to neglect the effect of mass transport. In this case a very simple equation results:

$$\ln \eta = \ln \eta_{t=0} - \left[\frac{-I_a^o nF}{C_{dl}RT}\right]\tau \tag{3}$$

where $\eta$ is the overpotential at time $\tau$ after charge injection, and the other symbols have their usual meaning. Delahay [4] indicated that if

$$\frac{nF}{RT}\frac{1}{C_{dl}I_a^o} \gg \left[\frac{1}{nF}\left(\frac{1}{C_o^o D_o^{1/2}} + \frac{1}{C_R^o D_o^{1/2}}\right)\right]^2 \tag{4}$$

then pure control by the charge transfer reaction would result. Unfortunately, when relation (4) is applied in practice, serious errors can result because $I_a^o$ is first estimated from Eq. (3). It is very likely that the use of this false $I_a^o$ will produce a value that will appear to satisfy (4) and thus apparently validate (3). A similar situation existing in the potential step

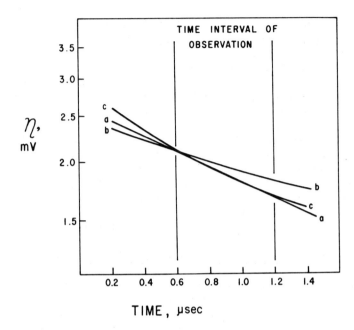

Fig. 1. Coulostatic curves from Weir and Enke [9]. Curve a, mass trans-
fer neglected — Eq. (3); curve b, $C_{dl}$ and $I_a^o$ values from Eq. (3) used in
Eq. (1); curve c, computer program for Eq. (1) using data of Weir and
Enke [9].

technique has been pointed out [7].

An examination of Fig. 1 will throw more light on this problem. The
curves in Fig. 1 are plotted from the Hg(I) reduction data of Weir and Enke
[9]. Curve a is plotted using Eq. (3) — mass transfer is not taken into
account. A reasonably good straight line results as would be expected.
However, if the $I_a^o$ and $C_{dl}$ values from the simple data treatment are
plugged into Eq. (1), then curve b results. Curves a and b should be identical or nearly so if one were actually justified in neglecting mass transfer.

The time domain during which only kinetics governs relaxation is
defined as the interval over which a plot of ln $\eta$ vs. t is linear and has the
correct slope and intercept. For the curve designated 0.05 in Fig. 2, the
termination of the region controlled by kinetics only is identified by the
small vertical line at about 2 $\mu$sec. The vertical lines have the same significance for the curves for $k_a^o$ = 0.1 and 0.3. Beyond the indicated times
both kinetics and diffusion control play important roles. We have found
during a coulostatic determination of the heterogeneous rate constant of the
dicyanide hemichrome-hemochrome couple [11] that substantial errors are

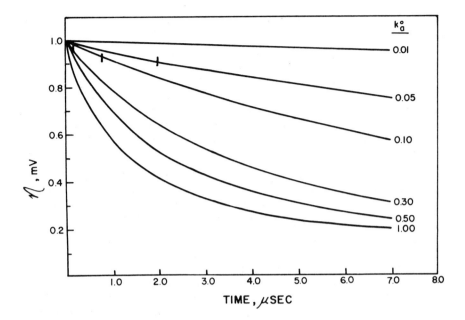

Fig. 2. Computer-calculated coulostatic curves for various values of $k_a^o$ under standard conditions. The following values were assumed: n = 1; $C_O = C_R = 5 \times 10^{-6}$ moles cm$^{-3}$; $D_O = D_R = 7.0 \times 10^{-6}$ cm$^2$ sec$^{-1}$; $C_{dl} = 2.0 \times 10^{-5}$ farads cm$^{-2}$; A = 8.0 $\times 10^{-2}$ cm$^2$. Vertical lines indicate start of the time region where diffusion control becomes important.

obtained if it is assumed that the region of pure kinetic control extends into the region where both kinetic and mass transfer control are important. Unfortunately, errors of this type are extremely easy to make since plots of ln $\eta$ vs. t often appear linear when they are not — especially if a small time interval not too close to zero time is selected.

## IV. MORE APPROPRIATE METHODS OF DATA ANALYSIS

Since it appeared clear that the effect of mass transfer must be considered for all but slow reactions, various methods of data analysis have been devised. Kooijman and Sluyters [12] have proposed the use of numerical tables and Kooijman [13] has published a set of nomograms from which calculations can be made. Improved results can be obtained from these techniques but extrapolating between numbers in a table or reading numbers from a graph introduces unwanted errors and does not allow the experimenter to know whether or not he recorded his data in the most appropriate time domain.

## V.  DATA ANALYSIS BY COMPUTER

To use Eq. (1) for the analysis of relaxation data two main problems must be overcome.  First, ways must be found of evaluating $\exp(z)^2 \, \mathrm{erfc}(z)$ for real and complex arguments.  Second, the fact that there are two unknowns in Eq. (1) dictates the use of some kind of curve-fitting technique. In this work a nonlinear least-squares method was employed.  The first problem was solved by the use of several different series expansions or rational approximations, the choice depending on the value of the terms in the complex argument [14].

The computer program developed here is set forth in the appendix at the end of this chapter.  Results obtained using the program have been reported [11, 14, 23].  While our work was in progress others have also developed computer programs to handle problems of this sort.

Kooijman [15] has developed a curve-fitting computer program for the analysis of galvanostatic and coulostatic data.  Like the program reproduced here, it contains a subroutine which evaluates $\exp(z)^2 \, \mathrm{erfc}(z)$ for real and complex arguments [16].  The basic equation used here, unlike Eq. (1), includes the effects of the coupling of double-layer charging with faradaic charge transfer.  The complexity of the resulting equation dictates some simplifying assumptions.  It is not known at present if these are justified or not but the analysis of galvanostatic data gives reasonable results in the light of other work.  Coulostatic data could be treated only if the charge injection is accomplished in a very short time — which is not the case experimentally.

More recently two more programs have been developed [17, 18].  That of Copa [17] does not make use of Eq. (1) but rather uses a newly developed equation which includes coupling of double-layer charging and faradaic charge transfer.  It appears that this may be the most complete approach yet available but sufficient results have not been completed for certainty on this point.

## VI.  COMPARISON OF RESULTS

The data of Weir and Enke [9] appear to be subject to error due to neglecting the contribution of mass transport to their data [15].  When our program was used to process their data, line c in Fig. 1 was the result. Interestingly, line a and line c are almost superimposed during the marked time interval during which data were actually taken.  However, the complete treatment yielded exchange current densities larger than those calculated by Eq. (3) by almost a factor of two.  This clearly indicates how easy it is to misinterpret data in situations of this type and strongly argues for the computer analysis of data by the most complete equation known.

TABLE 1

Comparison of Results of Kinetic Studies of the Electrochemical Reduction of 1 mM Mercury (1)[a] at 25°C

| Technique | Data treatment | Apparent exchange current density, $I_a^o$, A/cm$^2$ | Reference |
|---|---|---|---|
| Faradaic impedance | | 0.16 | 20 |
| Potentiostatic | | 0.07 | 20 |
| Galvanostatic, double pulse | | 0.13 | 21 |
| Galvanostatic, double pulse | | 0.25 | 8 |
| Faradaic rectification | | 13 | 22 |
| Faradaic impedance | Complex plane | $\geq 0.46$ | 6 |
| Galvanostatic | | 0.38 | 10 |
| Current impulse | | 0.30 | 9 |
| Current impulse | Computer | 0.56 | 9, 23 |
| Galvanostatic | Numerical tables | Between 0.6 and 2.0[b] | 12 |
| Galvanostatic | Computer (adsorption included) | $0.8 \pm 0.3$ | 16 |

[a]If necessary results were extrapolated to 1 mM from other concentrations.

[b]Range due to the uncertainty in measurement of $C_{dl}$.

Table 1 lists most of the kinetic results which have appeared in the literature for the reduction of Hg(I) at a mercury electrode. The main point to notice is that the data which are submitted to proper treatment yield values of $I_a^o$ substantially higher than otherwise found. The recalculation of Weir and Enke's current impulse data by our computer program emphasizes this point. Clearly if proper results are to be gathered by galvanostatic or coulostatic experiments complete equations such as Eq. (1) must be used even though their complexity dictates the use of a computer. More work is needed on a variety of systems to allow conclusions to be drawn as to which experimental method and which particular data treatment method is superior.

## ACKNOWLEDGMENTS

The authors wish to thank Jon Stewart of the LSUNO Computer Center for providing the nonlinear least-squares fitting program and the Computer Center of the California Institute of Technology for the subroutine for evaluating the function $\exp(z)^2 \, \mathrm{erfc}(z)$ for both real and complex arguments.

## APPENDIX

Below is a listing of our computer program in its present form. The symbols are explained in Table 2. More information concerning this is available [14].

### TABLE 2

### Coding of Variables and Constants

| Program symbol | Represents | Units |
|---|---|---|
| AREA | $A$, electrode area | $cm^2$ |
| CAP | $C_{dl}$, differential double-layer capacitance | $F/cm^2$ |
| CO | $C_O$, bulk concentration of oxidant | $moles/cm^3$ |
| CR | $C_R$, bulk concentration of reductant | $moles/cm^3$ |
| DC | $dC_{dl}$, small percentage (1%) of $C_{dl}$ | $F/cm^2$ |
| DCAP | $\Delta C_{dl}$, correction to be applied to previous $C_{dl}$ | $F/cm^2$ |
| DEXCH | $\Delta I_a^o$, correction to be applied to previous $I_a^o$ | $A/cm^2$ |
| DNDC | $\left[ \dfrac{\partial \eta_i^{APP}}{\partial C_{dl}} \right] I_a^o$ | $V\text{-}cm^2$ $F^{-1}$ |
| DNDX | $\left[ \dfrac{\partial \eta_i^{APP}}{\partial I_a^o} \right] C_{dl}$ | $V\text{-}cm^2$ $A^{-1}$ |
| DO | $D_O$, diffusion coefficient of oxidant | $cm^2/sec$ |
| DR | $D_R$, diffusion coefficient of reductant | $cm^2/sec$ |
| DX | $dI_a^o$, small percentage (1%) of initial $I_a^o$ | $A/cm^2$ |

TABLE 2 (cont.)

| Program symbol | Represents | Units |
|---|---|---|
| ELN | $n$, number of electrons transferred | Equiv/mole |
| ER | $$\text{error} = \frac{\Sigma \lvert \eta_i^{\,c} - \eta_i^{\,o} \rvert}{\Sigma \lvert \eta_i^{\,o} \rvert}$$ | dimensionless |
| ERROR | error, see ER | dimensionless |
| ETACAL(I) | $\eta_i^{\,c}$, the i-th calculated overpotential | V |
| ETACl(I) | $\eta_i^{\,c}$, see ETACAL(I) | V |
| ETAO(I) | $\eta_i^{\,o}$, the i-th observed overpotential | V |
| ETAOB | $\eta_i^{\,o}$, see ETAO(I); appears in output | V |
| ETATO | $\eta_{t=0}$, overpotential at $t = 0$ | V |
| EXCH | $I_a^{\,o}$, apparent exchange current density | $A/cm^2$ |
| F | $F$, Faraday constant | C/equiv |
| FIXCAP | $C_1$, capacitance of calibrated capacitor | F |
| GAMMA(N) | $\Gamma_n$, the n-th gamma function | dimensionless |
| R | $R$, Gas constant | V-C/deg |
| RTCONT | $k_a^{\,o}$, apparent heterogeneous rate constant | cm/sec |
| T(I) | $t_i$, the i-th time | sec |
| TIME | $t_i$, see T(I); appears in output | sec |
| TP | $T$, absolute temperature | deg |
| VOLT | $B$, battery voltage | V |
| X(I) | $t_i$, see T(I); | sec |

COULOSTATIC DATA ANALYSIS INCLUDING CHARGE AND MASS TRANSFER DURING
AND AFTER CHARGE INJECTION

```
      DIMENSION T(7),TB(3),ETAO(7),X(10),B(12,12),A(10),LOGIC(4)
      COMMON CO,CR,ELN,F,R,TP,CRNT,T1,AREA
    5 READ(2,100)CRNT,T1,EXCH,CAP,CO,CR,G
      READ(2,100)(TB(M),M=1,3),DO,DR,AREA,TP,ELN
      READ(2,100)(T(I),I=1,7)
      READ(2,100)(ETAO(I),I=1,7)
      WRITE(5,700)CO,CR,DO,DR,ELN
      WRITE(5,701)CAP,CRNT,T1,AREA,TP
      BEWARE=1.0
      F=9.6487E+04
      R=8.3143
      DX=0.02*EXCH
      DC=0.02*CAP
   15 LOGIC(1)=1
      CALL NORMAL (X,2,Y,1.,B,A,LOGIC)
      DO 20 I=1,7
      TX=T(I)
      TAU=T(I)-T1
      EX2=EXCH+DX
      EX1=EXCH-DX
      CAP2=CAP+DC
      CAP1=CAP-DC
      ETEX2=ETAC1(TX,EX2,CAP,DO,DR) - ETAC1(TAU,EX2,CAP,DO,DR)
      ETEX1=ETAC1(TX,EX1,CAP,DO,DR) - ETAC1(TAU,EX1,CAP,DO,DR)
      ETCP2=ETAC1(TX,EXCH,CAP2,DO,DR) - ETAC1(TAU,EXCH,CAP2,DO,DR)
      ETCP1=ETAC1(TX,EXCH,CAP1,DO,DR) - ETAC1(TAU,EXCH,CAP1,DO,DR)
      ETXCP=ETAC1(TX,EXCH,CAP,DO,DR) - ETAC1(TAU,EXCH,CAP,DO,DR)
      X(1)=(ETEX2-ETEX1)/(2.0*DX)
      X(2)=(ETCP2-ETCP1)/(2.0*DC)
      Y=ETAO(I) - ETXCP
   20 CALL NORMAL (X,2,Y,1.,B,A,LOGIC)
      LOGIC(1)=3
      CALL NORMAL (X,2,Y,1.,B,A,LOGIC)
      WRITE(5,200)EXCH,A(1),CAP,A(2)
      BEWARE=BEWARE+1.0
      IF(BEWARE-8.0)30,30,5
   30 EXCH=EXCH+A(1)
      CAP=CAP+A(2)
      EXCH01=0.01*EXCH
      CAP01=0.02*CAP
      IF(ABS(A(1))-EXCH01)40,40,15
   40 IF(ABS(A(2))-CAP01)45,45,15
   45 IF(CO-CR)47,46,47
   46 RTCONT=EXCH/(ELN*F*CO)
      WRITE(5,201)EXCH,CAP,RTCONT
      GO TO 48
   47 WRITE(5,202)EXCH,CAP
   48 CONTINUE
      ETA1=ETAC1(T1,EXCH,CAP,DO,DR)
      WRITE(5,205)ETA1
      DO 49 M=1,3
      TIME=TB(M)
      TAU=TB(M)-T1
```

```
      ETACAL=ETAC1(TIME,EXCH,CAP,DO,DR) - ETAC1(TAU,EXCH,CAP,DO,DR)
   49 WRITE(5,206)TIME,ETACAL
      CONTINUE
      DX=0.02*EXCH
      EX2=EXCH+DX
      EX1=EXCH-DX
      DO 50 I=1,7
      TIME=T(I)
      TAU=T(I)-T1
      ETACAL=ETAC1(TIME,EXCH,CAP,DO,DR) - ETAC1(TAU,EXCH,CAP,DO,DR)
      ETAOB=ETAO(I)
      RESID=ETAOB-ETACAL
      ETEX2=ETAC1(TIME,EX2,CAP,DO,DR) - ETAC1(TAU,EX2,CAP,DO,DR)
      ETEX1=ETAC1(TIME,EX1,CAP,DO,DR) - ETAC1(TAU,EX1,CAP,DO,DR)
      PARDEV=(ETEX2-ETEX1)/(2.0*DX)
   50 WRITE(5,203)TIME,ETAOB,ETACAL,PARDEV
      IF(G-1.0)60,60,55
   55 WRITE(5,204)
      GO TO 5
   60 WRITE(5,207)
      STOP
  100 FORMAT(8(F8.2,2X))
  200 FORMAT (7H EXCH=,E10.4,3X,4HDEX=,E10.4,3X,5HCAP =,E10.4,3X,
     14HDCP=,E10.4)
  201 FORMAT (/7H EXCH=,E10.4,3X,5HCAP =,E10.4,3X,8HRTCONT =,E10.4)
  202 FORMAT (/7H EXCH=,E10.4,3X,5HCAP =,E10.4)
  203 FORMAT(/6H TIME=,1PE10.3,2X,6HETAOB=,E10.3,2X,7HETACAL=,E10.3,
     12X,7HPARDEV=,E10.3)
  204 FORMAT (///9HNEXT JOB)
  205 FORMAT(/6H ETA1=,1PE10.3)
  206 FORMAT(/6H TIME=,1PE10.3,3X,7HETACAL=,E10.3)
  207 FORMAT(///19H ALL JOBS COMPLETED)
     12X,4HELN=,E9.2)
  701 FORMAT (5H CAP=,1PE9.2,2X,5HCRNT=,E9.2,2X,3HT1=,E9.2,
     12X,5HAREA=,E9.2,2X,3HTP=,E9.2)
      END



          FUNCTION ETAC1. EVALUATES OVERPOTENTIALS

      FUNCTION ETAC1(T,EXCH,CAP,DO,DR)
      COMMON CO,CR,ELN,F,R,TP,CRNT,T1,AREA
      COMPLEX CERFC,CEXP,ACMLX,BCMLX,ACMLXT,ERFCAT,AXSQT,EXAXT
      COMPLEX PRODAX,PRODBX,BRKT1,BRKT2,RAT1,RAT2,PRTH,BCMLXT
      DIFF=1.0/(CO*SQRT(DO))+1.0/(CR*SQRT(DR))
      P=(EXCH*DIFF)/(2.0*ELN*F)
      Q=(ELN*F*EXCH)/(R*TP*CAP)
      IF((P**2)-Q)500,600,600
  500 ACMLX=CMPLX(P,SQRT(Q-(P**2)))
      BCMLX=CONJG(ACMLX)
  510 ACMLXT=ACMLX*SQRT(T)
      BCMLXT=BCMLX*SQRT(T)
      ERFCAT=CERFC(ACMLXT)
      AXSQT=ACMLXT**2
      EXAXT=CEXP(AXSQT)
```

```
      PRODAX=EXAXT*ERFCAT
      PRODBX=CONJG(PRODAX)
      BRKT1=PRODAX+2.0*ACMLXT/1.772454-1.0
      BRKT2=PRODBX+2.0*BCMLXT/1.772454-1.0
      RAT1=BCMLX/(ACMLX**2)
      RAT2=ACMLX/(BCMLX**2)
      PRTH=RAT1*BRKT1-RAT2*BRKT2
      ALLAB  =AIMAG(PRTH)/(2.0*AIMAG(ACMLX))
      GO TO 700
  600 A=P+SQRT((P**2)-Q)
      B=P-SQRT((P**2)-Q)
  610 EXAT=EXP((A**2)*T)
      EXBT=EXP((B**2)*T)
      ACMLX=CMPLX(A*SQRT(T),0.0)
      BCMLX=CMPLX(B*SQRT(T),0.0)
      ACMLXT=CERFC(ACMLX)
      ERFCAT=CERFC(BCMLX)
      ERFCA=REAL(ACMLXT)
      ERFCB=REAL(ERFCAT)
      PRDA=EXAT*ERFCA
      PRDB=EXBT*ERFCB
      RT1=B/(A**2)
      RT2=A/(B**2)
      BR1=PRDA+2.0*A*SQRT(T/3.141593)-1.0
      BR2=PRDB+2.0*B*SQRT(T/3.141593)-1.0
      PT=RT1*BR1-RT2*BR2
      ALLAB  =PT*(1.0/(A-B))
  700 ETAC1=(-CRNT*ALLAB)/(AREA*CAP)
      RETURN
      END




SUBROUTINE NORMAL. SETS UP AND SOLVES LEAST SQUARES NORMAL EQUATIONS

      SUBROUTINE NORMAL(X,N,Y,W,B,A,LOGIC)
      DIMENSION X(10),B(12,12),A(10),LOGIC(4)
      L=LOGIC(1)
      GO TO(1,2,30),L
    1 NM1=N-1
      N1=N+1
      N2=N+2
      LOGIC(1)=2
      KNT=0
      DO 11 I=1,N2
      DO 11 J=1,N2
   11 B(I,J)=0.
      RETURN
    2 KNT=KNT+1
      DO 23 I=1,N
      T=X(I)
      WT=W*T
      B(I,N1)=B(I,N1)+WT*Y
      B(I,N2)=B(I,N2)+T*Y
      B(N2,I)=B(N2,I)+T
      B(N1,I)=B(N1,I)+T*T
```

```
     DO 23 J=I,N
 23  B(I,J)=B(I,J)+WT*X(J)
     T=Y*Y
     B(N1,N1)=B(N1,N1)+W*T
     B(N1,N2)=B(N1,N2)+T
     B(N2,N2)=B(N2,N2)+Y
     RETURN
 30  LOGIC(1)=KNT
     FN=KNT
     FNMM=KNT-N
     DO 31 I=1,NM1
     J1=I+1
     DO 31 J=J1,N
 31  B(J,I)=B(I,J)
 33  CALL MATINV(B,N)
     DO 41 I=1,N
     T=0.
     DO 40 J=1,N
 40  T=T+B(I,J)*B(J,N1)
 41  A(I)=T
     RETURN
     END


     SUBROUTINE MATINV. INVERTS MATRICES FOR NORMAL

     SUBROUTINE MATINV(A,N)
     DIMENSION A(12,12)
     NM1=N-1
     NP2=N+2
     DO 10 IP=2,N
     IRLIM=IP-1
     DO 20 IR=1,IRLIM
     ISLIM=IR-1
     SUM=A(IP,IR)
     IF(ISLIM)23,23,21
 21  DO 22 IS=1,ISLIM
 22  SUM=SUM-A(IP,IS)*A(IS,IR)
 23  A(IP,IR)=SUM/A(IR,IR)
 20  CONTINUE
     DO 10 IR=IP,N
     ISLIM = IP-1
     SUM=A(IP,IR)
     DO 30 IS=1,ISLIM
 30  SUM=SUM-A(IP,IS)*A(IS,IR)
     A(IP,IR)=SUM
 10  CONTINUE
     DO 40 J=1,NM1
     I1=J+1
     DO 40 I=I1,N
     SUM=-A(I,J)
     K1=J+1
     KLIM=I-1
     IF(KLIM-K1)51,49,49
 49  DO 50 K=K1,KLIM
```

```
50 SUM=SUM-A(I,K)*A(K,J)
51 A(I,J)=SUM
40 CONTINUE
   DO 60 I=1,N
60 A(I,I)=1./A(I,I)
   DO 70 JX=2,N
   J=NP2-JX
   IXLIM=J-1
   DO 70 IX=1,IXLIM
   I=J-IX
   K1=I+1
   KLIM=J
   SUM=0.
   DO 61 K=K1,KLIM
61 SUM=SUM+A(I,K)*A(K,J)
   A(I,J)=-A(I,I)*SUM
70 CONTINUE
   DO 80 I=1,N
   DO 80 J=1,N
   SUM=0.
   DO 90 M=1,N
   IF(M-I)90,91,91
91 IF(M-J)90,92,93
92 SUM=SUM+A(I,M)
   GO TO 90
93 SUM=SUM+A(I,M)*A(M,J)
90 CONTINUE
   A(I,J)=SUM
80 CONTINUE
95 RETURN
   END
```

# REFERENCES

[1]. P. Delahay, Anal. Chem., 34:1161 (1962).

[2]. W. H. Reinmuth and D. E. Wilson, Anal. Chem., 34:1159 (1962).

[3]. P. Delahay and A. Aramata, J. Phys. Chem., 66:2208 (1962).

[4]. P. Delahay, J. Phys. Chem., 66:2204 (1962).

[5]. D. J. Kooijman and J. H. Sluyters, J. Electroanal. Chem., 13:152 (1967).

[6]. M. Sluyters-Rehbach and J. H. Sluyters, Rec. Trav. Chim., 83:967, 983 (1964).

[7]. K. B. Oldham and R. A. Osteryoung, J. Electroanal. Chem., 11:397 (1966).

[8]. M. Matruda, S. Oka and P. Delahay, J. Am. Chem. Soc., 91:5077 (1969).

[9]. W. K. Weir and C. G. Enke, J. Phys. Chem., 71:275, 280 (1967).

[10]. R. L. Birke and D. K. Roe, Anal. Chem., 37:450 (1965).

[11]. R. F. Martin and D. G. Davis, Biochem., 7:3906 (1968).

[12]. D. J. Kooijman and H. Sluyters, Electrochim. Acta, 12:1579 (1967).

[13]. D. J. Kooijman, J. Electroanal. Chem., 18:81 (1968).

[14]. R. F. Martin, Ph. D. Thesis, Louisiana State University of New Orleans, La., 1967.

[15]. D. J. Kooijman, J. Electroanal. Chem., 19:365 (1968).

[16]. D. J. Kooijman, J. Electroanal. Chem., 19:445 (1968).

[17]. William Copa, Ph. D. Thesis, University of Minnesota, 1969.

[18]. Peter Daum, Ph. D. Thesis, Michigan State University, 1969.

[19]. F. C. Anson, in Ann. Rev. Phys. Chem. (H. Eyring, ed.), Annual Reviews, 1968, pp. 85, 86.

[20]. H. Gerischer and K. Stanbach, Z. Physik. Chem. (Frankfurt), 6:118 (1958).

[21]. H. Gerischer and M. Krause, Z. Physik. Chem. (Frankfurt), 14:184 (1958).

[22]. H. Imai and P. Delahay, J. Phys. Chem., 66:1108 (1962).

[23]. R. F. Martin and D. G. Davis, Southwest Meeting, American Chemical Society, Austin, December 1968.

Chapter 5

# NUMERICAL SOLUTION OF INTEGRAL EQUATIONS

Richard S. Nicholson

National Science Foundation
Washington, D.C. 20550

and

Michael L. Olmstead

Bell Telephone Laboratories
Murray Hill, New Jersey

## I. INTRODUCTION

The usefulness of many electrochemical techniques rests in part on the fact that over the past few decades it has been possible to develop mathematical models for these techniques. In most cases these models are based on the partial differential equations that describe diffusion. It is amply established by now that these diffusion equations accurately describe electrochemical processes. In fact, for the correct model these "theories" and experiment apparently agree within the uncertainties associated with experimental data. In many cases solutions of these mathematical models can be obtained rigorously. This has been the case for most of the published polarographic theory, although usually the theory is based on approximate models (the expanding-plane approximation, for example). In other cases, however, the models cannot be solved mathematically. In these cases the use of digital computers has been a necessary, if not always elegant, approach. Indeed, without computers there are some electrochemical problems which simply could never be solved. The point in these problems at which the computing machine is introduced varies, ranging from direct finite difference to evaluating infinite series solutions of the differential equations. In a few rare cases totally finite difference methods are the only choice possible. Oftentimes, however, it is possible to perform some relatively simple mathematical transformations prior to turning the problem over to the computer. If this were done only to lend an aura of sophistication, it would be difficult to justify to those interested only in answers. In fact, however, these transformations are so simple, and reduce the complexity of the problem so significantly, that it seems difficult to justify other approaches. Thus, most linear partial differential equations can be transformed very easily to corresponding integral forms. These integral expressions can be combined algebraically with boundary conditions to generate integral equations, that is, equations where the unknown function appears as the argument of the integration operator. For example, by essentially algebraic manipulation a boundary value problem consisting of two (linear) partial differential equations with two initial and four boundary conditions (not necessarily linear) can be reduced to a single integral equation involving a single unknown. Of course, if the original system of differential equation cannot be solved analytically, then necessarily no analytical solution exists for the integral equation. Nevertheless, as we hope to illustrate in this chapter, integral equations (at least those of interest to the electrochemist) can be solved numerically by procedures that are very straightforward and readily adaptable to modern computing machines and the FORTRAN language.

Of course, unless one can derive integral equations, there is little point in knowing how to solve them numerically. It is, therefore, with some reservations that we are restricting the treatment in this chapter to the numerical solution of one particular class of integral equations. Nevertheless, this point of view is in keeping with the objectives of this series.

Moreover, we feel it is justified by the fact that there are several excellent sources in the electrochemical literature, and elsewhere, dealing with the deviation of integral equations, but there is no computer-oriented source describing the numerical solution of integral equations. Thus, this chapter deals solely with the solution of integral equations, and in fact the illustrative examples are not even of electrochemical origin. As references to the derivation of integral equations we recommend the book by Churchill [1], the review article by Reinmuth [2], as well as the many literature examples (e.g., [3-16]).

In a sense our objective in writing this chapter is to illustrate as simply as possible how an integral equation may be written in FORTRAN. Thus, we present no existence theorems, proofs of convergence, or detailed mathematics of error analysis. Instead, we have attempted to keep the mathematics as simple as is consistent with adequate clarity. Although there are many different numerical methods for integral equations, our treatment is restricted to two methods which in our experience have proved most useful and reliable. We intend that at best our only original contribution should be one of presentation, and our only objective is to provide a ready reference for those electrochemists who may have occasion to solve integral equations.

## II. GENERAL FORM OF INTEGRAL EQUATIONS

Our aim is to describe a general and systematic procedure for numerical solution of integral equations. Since integral equations appear in a variety of forms, there are two distinctly different approaches that could be adopted in developing a general treatment. We could proceed by writing a single, very general (and necessarily complicated) integral equation, and then develop a numerical method for solving it. Specific integral equations would be treated as limiting cases of this general equation. Alternatively, we could focus at the outset only on the terms in integral equations which contain the unknown function. The former approach is the one usually presented in the literature. Nevertheless, in our experience the latter approach is much less complicated, more easily understood conceptually, and more readily applied to a given integral equation. Indeed, with this approach integral equations can be solved numerically in a nearly cookbook fashion. Hence, we will adopt the latter approach, and proceed by defining the following terms which are common to integral equations:

$$f(x); \quad \int_0^x f(z)g(x - z)dz. \qquad \text{(II-1)}$$

There $f(x)$ is the unknown function — i.e., it is the sought-for solution of a given integral equation -- and $g(x - z)$ is called the kernel function. This kernel is of the convolution form since we have explicitly used $x - z$ for the argument. For a given integral equation the functional form of $g(x - z)$

always is known explicitly. Of course, a given integral equation may contain several different types of integrals and kernels.

At this point two ancillary comments about terms (II-1) may be useful. First, by making these terms dimensionless wherever they appear in an integral equation, resulting solutions will be completely general. This adimensionalization is readily accomplished, and will be illustrated for a specific example later in this chapter. Second, the symbology we will employ is selected specifically to facilitate transformation of final results to FORTRAN. Thus, all lower case letters used in the mathematical development will appear later as the same upper case letters in FORTRAN. Moreover, these symbols will be consistent with the FORTRAN rules for integers and floating point numbers.

To obtain the solution of any equation in which terms (II-1) appear, it is necessary to develop approximate expressions for these terms. These approximate expressions will then replace terms (II-1) wherever they appear in a given integral equation. The result of this operation will be a system of algebraic equations, which is readily expressed in FORTRAN and solved on a digital computer.

The way in which these approximate expressions are developed is similar to that for conventional numerical evaluation of definite integrals, and is as follows. First, the range of integration is divided into an arbitrary number of equally spaced subintervals. Next, over each of these subintervals, an explicit (and arbitrary) functional form is assumed for the unknown function. Any constants involved in the definition of this function are then selected so that the function satisfies the integral equation at an arbitrary number of points along the subinterval. When this procedure is repeated for each successive subinterval, the result is a complete (approximate) description of the unknown function.

The approximation used for the unknown function is arbitrary, but most commonly polynomials are employed. Of course, the higher the degree of the polynomial, the more precise the approximation becomes. On the other hand, the complexity of the treatment increases greatly with the degree of the polynomial (see, for example, [17]), and therefore we have found the simple zero-order and first-order "polynomials" to be most useful. The first of these assumes that the unknown function can be approximated by a constant on each subinterval, and therefore we shall refer to that approximation as the "step function method." The second method approximates the unknown by a straight line on each subinterval. This approach is originally due to Huber [18], and therefore will be referred to hereafter by that name.

To apply the above ideas, we proceed as follows. First, the independent variable x in terms (II-1) is defined as

$$x = md. \qquad\qquad (II-2)$$

By treating m in Eq. (II-2) as an integer, m will represent a serial number which specifies a given subinterval of width d. Hence, substituting Eq. (II-2) into terms (II-1) divides the integration range into equally spaced subintervals

$$f(md); \quad \int_0^{md} f(z)g(md - z)dz. \tag{II-3}$$

The serial number m will assume the values $m = 1, 2, \ldots, m'$ where $m'$ specifies the complete domain of integration given by $x' = m'd$. For each value of m, the integer, i (with $i \leq m$), will be used to identify a specific integration interval $[(i - 1)d, id]$. On each of these subintervals the continuous function, $f(x)$, will be replaced by the approximation function labeled $f_i(x)$. We then define the numerical value of the approximate function at the end of an integration interval as

$$f_i = f_i(id). \tag{II-4}$$

With these definitions in mind, we next represent the integral in terms (II-3) as a sum of integrals over the individual subintervals:

$$f_m; \quad \sum_{i=1}^{m} \int_{(i-1)d}^{id} f_i(z)g(md - z)dz. \tag{II-5}$$

At this point it is possible to consider explicitly the approximations to be used for $f_i$. The following two sections are devoted to this problem.

### III. METHOD OF HUBER

#### A. Mathematical Development

It is logical to present Huber's method first, since the step function method is a limiting case of Huber's method. Huber's original treatment was applied to a specific Volterra integral equation. It is relatively simple to transpose his approach to the one we have adopted for this chapter.

In Huber's method the unknown function is approximated by connected line segments on the integration subintervals. The line segment on each interval may be viewed as one which is generated by the point-slope method of analytical geometry. In this case both the point and slope are selected so that the resulting line satisfies the integral equation on the subinterval. The point-slope formula for a straight line is

$$y(x) - y_0 = a(x - x_0) \tag{III-1}$$

where the known point is $(x_0, y_0)$, and the slope is a. Applying this approach to the present nomenclature, we obtain

$$f_i(x) - f_i = \left(\frac{f_i - f_{i-1}}{d}\right)(x - id). \qquad \text{(III-2)}$$

Figure 1 illustrates pictorially the approximation that is being employed.

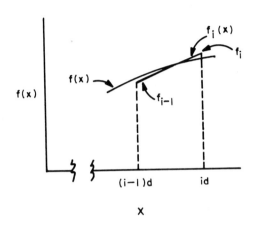

Fig. 1. Schematic illustration of Huber's method.

By analogy with Eq. (III-1) and reference to Fig. 1, the slope in Eq. (III-2) is

$$a_i = \frac{f_i - f_{i-1}}{d}. \qquad \text{(III-3)}$$

Equation (III-2) may now be written in the following form:

$$f_i(x) = a_i(x - id) + f_i. \qquad \text{(III-4)}$$

We next express $f_i$ in terms of the following identity:

$$f_i \equiv \sum_{j=1}^{i} (f_j - f_{j-1}) + f_0 \qquad \text{(III-5)}$$

where $f_0$ is the solution of the given integral equation evaluated at $x = 0$. Combining Eqs. (III-3) and (III-5), we obtain the following expression for $f_i$:

$$f_i = d \sum_{j-1}^{i} a_j + f_0. \qquad \text{(III-6)}$$

Elimination of $f_i$ between Eqs. (III-4) and (III-6) leads to

$$f_i(x) = a_i(x - id) + d \sum_{j=1}^{i} a_j + f_o \, . \tag{III-7}$$

Eq. (III-7) is the form of the linear approximation most easily incorporated in the integral of terms (II-5).

The approximate function represented by Eq. (III-7) depends only on the value of $a_i$ for a given value of d. Thus, the aim is to calculate values of these slopes, since with them the value of the unknown function can be calculated [Eq. (III-3) or (III-6)]. The expressions required for calculation of these slopes are obtained by substituting Eq. (III-7) into terms (II-5). The result is

$$\sum_{i=1}^{m} \int_{(i-1)d}^{id} \left[ a_i(z - id) + d \sum_{j=1}^{i} a_j + f_o \right] g(md - z) dz. \tag{III-8}$$

The following change of variable

$$z = md - y \tag{III-9}$$

converts expression (III-8) to

$$\sum_{i=1}^{m} \int_{(m-i)d}^{(m-i+1)d} \left\{ a_i[(m - i)d - y] + d \sum_{j=1}^{i} a_j \right\} g(y) dy + f_o \int_0^{md} g(y) dy. \tag{III-10}$$

For the immediate discussion it is convenient to carry only the first term of (III-8) and pick up the second term later. This first term can be written in the following equivalent form:

$$\sum_{i=1}^{m} \int_{(m-i)d}^{(m-i+1)d} a_i[(m - i)d - y]g(y)dy + d \sum_{j=1}^{m} \sum_{j=1}^{i} a_j \int_{(m-i)d}^{(m-i+1)d} g(y)dy. \tag{III-11}$$

Next, we reduce the double summation in terms (III-11) to a single summation. To accomplish this is simple enough, but not obvious, and therefore the appropriate steps are detailed in an appendix to this chapter. Applying the results of the Appendix simplifies terms (III-11) to

$$\sum_{i=1}^{m} \int_{(m-i)d}^{(m-i+1)d} a_i (m - i)d - y]g(y)dy + d \sum_{i=1}^{m} a_i \int_0^{(m-i+1)d} g(y)dy. \tag{III-12}$$

Expression (III-12) contains two types of definite integrals which, for convenience, we define with the following symbols:

$$t_k = \int_0^{kd} g(y)dy, \tag{III-13}$$

$$u_k = \int_0^{kd} y g(y) \, dy. \tag{III-14}$$

It may be useful at this point to remind the reader that since the kernel $g(y)$ is always known explicitly, Eqs. (III-13) and (III-14) are simply definite integrals.

Incorporating Eqs. (III-13) and (III-14) in terms (III-12) and including the term dropped from (III-10), we finally obtain

$$\sum_{i=1}^{m} a_i h_{m-i+1} + f_0 s_m \tag{III-15}$$

where the Huber function is

$$h_k = d[kt_k - (k-1)t_{k-1}] - u_k + u_{k-1}. \tag{III-16}$$

Since at any point in the solution of an integral equation the unknown quantity is the slope $a_m$, it is helpful to separate explicitly this term from (III-15),

$$a_m h_1 + \sum_{i=1}^{m-1} a_i h_{m-i+1} + f_0 s_m \tag{III-17}$$

Term (III-17) represents an approximation for the integral appearing in terms (II-1). Thus, given an integral equation, all integrals are simply replaced by term (III-17), and if $f(x)$ appears separately it is replaced by Eq. (III-6). Since the function $h_k$ can be calculated explicitly, the result of this operation is a system of algebraic equations in which the only unknowns are the slopes, $a_m$. These equations are then solved successively for $a_m$ (where $m = 1, \ldots, m'$), and the final solution of the equation is then obtained by substituting these values of $a_m$ in Eq. (III-3) or (III-6).

For most equations the function $f_0$ is zero. When this is not the case, however, the value of $f_0$ can be determined directly from the original integral equation, as described by Huber [18].

## B. Use of FORTRAN

It is convenient at this point to consider how operations such as those implied in terms (III-17) can be performed with a digital computer. As stated earlier, the mathematical development was intentionally structured so that results would transpose easily to FORTRAN coding.

To calculate the sum in terms (III-17) via FORTRAN involves the following simple steps. The integrations in Eqs. (III-13) and (III-14) are performed to provide explicit expressions for calculation of $t_k$ and $u_k$. Based on these expressions the appropriate FORTRAN statements are written to calculate and store $h_k$ from Eq. (III-16) as the subscripted variable H(K) for k = 1, ..., m′ (or K = 1, ..., MPRIME). The summation in (III-17) is then simply

SUM = 0.
DO 10 I = 1, M - 1
10 SUM = SUM + A(I) * H(M - I + 1).

Further examples of the ease with which equations presented in this chapter can be written in FORTRAN are contained in Section V.

## IV. STEP FUNCTION METHOD

The step function method is the simplest approach possible. The simplification is, of course, at the expense of accuracy. Nevertheless, with high-speed computers it often is practicable (or even desirable) to work with the simpler expressions and obtain the required accuracy by using small values of d (with either this or Huber's method the results become exact as d approaches zero).

With the step function method the unknown function is approximated by a constant on each subinterval,

$$f_i(x) = b_i. \tag{IV-1}$$

With this approximation the integral in terms (II-5) becomes

$$\sum_{i=1}^{m} b_i \int_{(i-1)d}^{id} g(md - z)dz. \tag{IV-2}$$

With the change of variable used before [Eq. (III-9)], term (IV-2) reduces to

$$\sum_{i=1}^{m} b_i \int_{(m-i)d}^{(m-i+1)d} g(y)dy. \tag{IV-3}$$

Term (IV-3) contains only one type of definite integral, namely the one called $t_k$ earlier. Incorporating the definition of $t_k$ [Eq. (III-13)] in (IV-3), we obtain the following approximation for the integral in terms (II-1),

$$\sum_{i-1}^{m} b_i s_{m-i+1} \tag{IV-4}$$

or

$$b_m s_1 + \sum_{i=1}^{m-1} b_i s_{m-i+1} \tag{IV-5}$$

where $s_k$ is analogous to the Huber function $h_k$,

$$s_k = t_k - t_{k-1}. \tag{IV-6}$$

As with Huber's method, a given integral equation is solved by replacing $f(x)$ with Eq. (IV-1), and integrals with term (IV-5). The result is a system of algebraic equations that can be solved for values of $b_m$ which in this case represent the solution of the integral equation directly [i.e., $b_m \approx f(md)$].

This method clearly provides the simplest possible expressions for numerical solution of integral equations. One manifestation of this fact is that only a single definite integral ($t_k$) has to be evaluated, whereas Huber's method requires evaluation of two integrals ($t_k$ and $u_k$). In some cases $t_k$ can be expressed analytically, but $u_k$ cannot. In these cases the step function method may be preferable, even at the sacrifice of accuracy.

## V. EXAMPLES

We now wish to illustrate the methods developed above by applying them to some specific integral equations. As examples we will use simple equations which have analytical solutions, but which possess features common to integral equations appearing in electrochemical theory.

### A. Simple Kernel, Step Function Method

The solution of the integral equation

$$x = \int_0^x \frac{f(z)dz}{\sqrt{x - z}} \tag{V-1}$$

is

$$f(x) = \frac{2\sqrt{x}}{\pi} \tag{V-2}$$

To solve Eq. (V-1) numerically, we first employ Eq. (II-2) to form the subintervals

$$md = \int_0^{md} \frac{f(z)dz}{\sqrt{md - z}}. \tag{V-3}$$

We then replace the integral in Eq. (V-3) by term (IV-5), so that

$$md = b_m s_1 + \sum_{i=1}^{m=1} b_i s_{m-i+1}. \tag{V-4}$$

Since $b_m$ is directly the approximate solution, we solve for $b_m$,

$$b_m = \frac{md}{s_1} - \frac{1}{s_1} \sum_{i=1}^{m-1} b_i s_{m-i+1}. \tag{V-5}$$

After a specific value for d has been selected, $b_m$ can be calculated iteratively for m = 1, 2, ..., m', where again m'd is the largest value of the independent variable x, for which values of f(x) are required. For example, if d is 0.1 (a typical choice) then $b_1$ is the approximate value of f(x) for x equal 0.1, $b_{205}$ is the approximate value of f(x) for x equal 20.5, and so on. To execute these calculations it is, of course, necessary to know $s_k$ explicitly, so we turn to that problem next.

From Eq. (IV-6) (renumbered here)

$$s_k = t_k - t_{k-1} \tag{V-6}$$

where from Eq. (III-13) (renumbered here)

$$t_k = \int_0^{kd} g(y)\,dy. \tag{V-7}$$

For Eq. (V-1) the kernel function is

$$g(x - z) = \frac{1}{\sqrt{x - z}}. \tag{V-8}$$

Hence, $t_k$ for the present case is

$$t_k = \int_0^{kd} \frac{dy}{\sqrt{y}}. \tag{V-9}$$

Performing the integration in Eq. (V-9), we obtain

$$t_k = 2\sqrt{y}\,\Big|_{y=0}^{y=kd} = 2\sqrt{d}\,\sqrt{k}. \tag{V-10}$$

Substituting this result in Eq. (V-6), we find

$$s_k = 2\sqrt{d}(\sqrt{k} - \sqrt{k - 1}). \tag{V-11}$$

All that remains now is to write a computer program that generates Eq. (V-5) for m = 1, 2, ..., m', and at the same time performs the operations indicated within Eq. (V-5). This trivial problem is described in the next section.

## B. FORTRAN Program

There seems little point in presenting entire programs, and therefore we will present only the program segment pertaining to the actual calculations required to solve Eq. (V-1). This program segment is sufficiently simple that we will not worry about trivial idiosyncrasies of the various FORTRAN dialects which the program segment violates.

TABLE 1

Program Segment for Solution of Eq. (V-1)

```
                           .
                           .
                           .
        CONST = 2.*SQRT(D)
        TEMP = 0.
        DO 10 K = 1, MPRIME
        SQ = SQRT(K)
        S(K) = CONST*(SQ-TEMP)
   10   TEMP = SQ
        DO 30 M = 1, MPRIME
        B(M) = M*D/S(1)
        DO 20 I = 1, M-1
   20   B(M) = B(M) - B(I)*S(M-I+1)/S(1)
        X = M*D
   30   PRINT X, B(M)
                           .
                           .
                           .
```

The FORTRAN segment for solving Eq. (V-1) by the step function method is presented in Table 1. Since this program has an essentially one-to-one correspondence with Eqs. (V-5) and (V-11), the program should require little explanation. Nevertheless, a few comments are in order.

First, the program segment assumes that values of d and $m'$ are stored as data called D and MPRIME, respectively. The DO LOOP ending with statement 10 calculates and stores all the required values of $s_k$ [called S(K)]. The main DO LOOP ending at statement 30 generates Eq. (V-5) for $m = 1, 2, \ldots, m'$. The nested DO LOOP ending at statement 20 performs the summation involved in Eq. (V-5). Statement 30 prints the independent variable, x, and the calculated value of f(x).

The reader may wonder why values of $s_k$ are calculated and stored as a subscripted variable prior to entering the main and nested DO LOOPS, since this approach unnecessarily increases the core requirements as compared to calculation of $s_k$ directly within the nested DO LOOP. The justification, which is not always valid, is as follows. The program as written calls for the library square-root routine MPRIME times. If $s_k$ were calculated as part of the nested DO LOOP the square-root routine would be called approximately (MPRIME)*(MPRIME-1) times. Even with high-speed computers, library routines are relatively slow, and therefore avoiding this unnecessary use of library routines can very significantly decrease execution time. This is particularly true when large values of MPRIME are required. For example, if MPRIME is 1000, then 1000 square roots are actually required,

TABLE 2

Solutions of Eq. (V-1) by the Huber (H) and Step (S) Methods for Several Values of d

| x | d = 1.0 | | 0.5 | | 0.1 | | 0.01 | | 0.001 | | $f(x)^a$ |
|---|---|---|---|---|---|---|---|---|---|---|---|
| | H | S | H | S | H | S | H | S | H | S | |
| 0.01 | — | — | — | — | — | — | 0.075 | 0.050 | 0.064 | 0.062 | 0.064 |
| 0.05 | — | — | — | — | — | — | 0.143 | 0.135 | 0.142 | 0.142 | 0.142 |
| 0.07 | — | — | — | — | — | — | 0.169 | 0.162 | 0.168 | 0.168 | 0.168 |
| 0.1 | — | — | — | — | 0.237 | 0.158 | 0.201 | 0.196 | 0.201 | 0.201 | 0.201 |
| 0.5 | — | — | 0.530 | 0.354 | 0.451 | 0.428 | 0.452 | 0.448 | 0.450 | 0.450 | 0.450 |
| 1.0 | 0.750 | 0.500 | 0.621 | 0.561 | 0.637 | 0.621 | 0.637 | 0.635 | — | — | 0.637 |
| 2.0 | 0.879 | 0.793 | 0.900 | 0.845 | 0.900 | 0.889 | 0.900 | 0.899 | — | — | 0.900 |
| 5.0 | 1.427 | 1.353 | 1.424 | 1.388 | 1.424 | 1.416 | 1.424 | 1.423 | — | — | 0.424 |

$^a$Calculated from Eq. (V-2).

but if $s_k$ were calculated in the nested DO LOOP the square-root subroutine would be called nearly one million times!

Values of $b_m$ calculated with a program incorporating the program segment of Table 1 are given in Table 2 for several values of d. Exact values of f(x) calculated from Eq. (V-2) are included for comparison. These data require little explanation, but it should be noted that relatively small values of d are required to provide reasonably high accuracy. It also should be noted that the relative accuracy is naturally dependent on the morphology of the unknown function, or equivalently on the value of x.

### C. Simple Kernel, Huber's Method

To use Huber's method to solve Eq. (V-1) is exactly analogous to the treatment just described for the step function method. Thus, the integral in Eq. (V-3) is replaced by Eq. (III-17), and the resulting equation is solved for $a_m$

$$a_m = \frac{md}{h_1} - \frac{1}{h_1} \sum_{i=1}^{m-1} a_i h_{m-i+1}. \qquad (V-12)$$

To solve this equation the procedure is formally identical to the one described in connection with Eq. (V-5). There are two practical differences. First, once values of $a_m$ are determined, values of f(x) have to be calculated with the aid of either Eq. (III-3) or Eq. (III-6). Second, of course, is the definition of the Huber function, $h_k$ [see Eq. (III-16)].

$$h_k = d[kt_{k-1} - (k-1)t_{k-1}] - u_k + u_{k-1}. \qquad (V-13)$$

The integral $t_k$ already has been evaluated [Eq. (V-10)]. The integral $u_k$ [Eq. (III-14)] for the present case is

$$u_k = \int_0^{kd} \sqrt{y} \; dy = \frac{2}{3}(kd)^{3/2}. \qquad (V-14)$$

Thus, $h_k$ in this case is

$$h_k = \frac{4}{3}d^{3/2}[k^{3/2} - (k-1)^{3/2}]. \qquad (V-15)$$

Except for the above definition of $h_k$ and the calculation of f(x) from $a_m$, the computer program would be identical with the one in Table 1 for the step function method. Values of f(x) calculated from Eqs. (V-13) and (III-3) are included in Table 2. The improved accuracy provided by Huber's method is obvious.

## D. More Complex Kernel, Method of Huber

As a final illustration we consider an integral containing a more complex kernel of a type common to electrochemical problems. The integral equation

$$te^{-ct} = \int_0^t \frac{e^{-c(t-\tau)}\psi(\tau)d\tau}{\sqrt{t-\tau}} \tag{V-16}$$

has the analytical solution

$$\psi(t) = \frac{2}{\pi}e^{-ct}\sqrt{t}. \tag{V-17}$$

Although Eq. (V-16) could be solved directly, for problems of this type it usually is advantageous to reduce the equation to dimensionless form first. For Eq. (V-16) this is accomplished by the following substitutions:

$$z = c\tau \tag{V-18}$$

$$x = ct \tag{V-19}$$

$$f(x) = \sqrt{c}\,\psi(t) \tag{V-20}$$

which transform Eq. (V-16) to

$$xe^{-x} = \int_0^x \frac{e^{-(x-z)}f(z)dz}{\sqrt{x-z}} \tag{V-21}$$

The solution of Eq. (V-21) clearly is

$$f(x) = \frac{2}{\pi}e^{-x}\sqrt{x}. \tag{V-22}$$

The advantage of adimensionalization should be apparent. Thus, to solve Eq. (V-16) numerically would require a new set of calculations for every value of c of interest. On the other hand, numerical solutions of Eq. (V-21) depend only on the independent variable x, and hence are completely general. Of course, calculated values of $f(x)$ are converted to $\psi(x)$ via Eq. (V-20).

To solve Eq. (V-21) we proceed as before, substituting Eq. (II-2) into Eq. (V-21),

$$md\,e^{-md} = \int_0^{md} \frac{e^{-(md-z)}f(z)dz}{\sqrt{md-z}}. \tag{V-23}$$

By applying Eq. (III-17) and rearranging to solve for $a_m$, we obtain

$$a_m = \frac{md\, e^{-md}}{h_1} - \frac{1}{h_1} \sum_{i=1}^{m-1} a_i h_{m-i+1}.$$

(V-24)

All that remains is to define the function $h_k$ explicitly, which again requires evaluation of $t_k$ and $u_k$. Although the integrations are not as straightforward in this case, both expressions are integrable with the following results:

$$t_k = \int_0^{kd} \frac{e^{-y} dy}{\sqrt{y}} = \sqrt{\pi}\ \mathrm{erf}(\sqrt{kd})$$

(V-25)

and

$$u_k = \int_0^{kd} \sqrt{y}\ e^{-y} dy$$

$$= \frac{\sqrt{\pi}}{2}\ \mathrm{erf}(\sqrt{kd}) - \sqrt{kd}\ e^{-kd}.$$

(V-26)

The operations from this point should be obvious from the preceding examples. Of course, an error function subroutine is required, but these are routinely available for digital computers [19].

## VI. ACCURACY AND OTHER PROBLEMS

Both methods described here result in convergence to the correct value as d is decreased. Thus, the pragmatic approach for evaluating accuracy is to perform a series of calculations for successively decreasing values of d, and note the relative convergence of calculated results. For example, even without prior knowledge of the correct answer (assuming, of course, that there are no programming errors), the data of Table 2 show that for x > 0.5, Huber's method with d = 0.1 gives answers accurate to ±0.002, or better. Finally, it should be mentioned that relative errors for both methods generally depend on the morphology of the unknown function being calculated, and that these errors are not cumulative. Indeed, the methods, by their nature, continually correct for errors in previously calculated values of the unknown. This important feature is nicely illustrated for Eq. (V-1) solved above. For example, with Huber's method for d = 1, values of f(x) for x = 1 and 2 possess significant errors. Nevertheless, values calculated with d = 1 for x > 100 (not shown in the table) are correct to within one part in 10, 000!

In some cases accuracy greater than that obtainable by the methods presented here may be required. The simplest approach is to use higher-order polynomials. The appropriate equations for a polynomial of arbitrary degree have been derived with a formalism and nomenclature essentially the same as that used here [17].

Several other miscellaneous comments about the methods described in this chapter also should be mentioned. First, to obtain numerical solutions of integral equations merely requires <u>numbers</u> for $t_k$ and $u_k$. Thus, even when $t_k$ and $u_k$ cannot be expressed analytically, the required numbers always can be obtained by straightforward numerical integrations (e. g., Simpson's rule). Indeed, this approach may be generally useful because it avoids working with complicated analytical expressions, and thereby avoids the necessity of having available uncommon library subroutines for the computer.

Another point worth mentioning is the fact that these numerical methods can be used to solve nonlinear integral equations. In this case expressions such as Eq. (V-4) will be nonlinear, so that terms such as $b_m$ cannot be solved for explicitly. Nevertheless, it (usually) is a simple matter to solve these transcendental equations by standard numerical methods (e. g., Newton-Raphson iteration). This approach has been used successfully to solve several electrochemical problems that take the form of nonlinear integral equations [20-24].

Finally, the methods described here can be used in conjunction with finite difference methods. This approach has been recently illustrated for a problem involving nonlinear partial differential equations [25-29].

## ACKNOWLEDGMENT

We are grateful to Virginia Anadahl of the National Institutes of Health, who generously helped us with the computer calculations.

## APPENDIX

The purpose of this appendix is to show that the double sum in terms (III-11) of the text,

$$\sum_{i=1}^{m} \sum_{j=1}^{i} a_j \int_{(m-i)d}^{(m-i+1)d} g(y)dy, \tag{A-1}$$

can be reduced to

$$\sum_{i=1}^{m} a_i \int_{0}^{(m-i+1)d} g(y)dy. \tag{A-2}$$

To make the writing easier we apply Eq. (III-13) of the text to term (A-1) to obtain

$$\sum_{i=1}^{m} \sum_{j=1}^{i} a_j (t_{m-i+1} - t_{m-i}). \tag{A-3}$$

Term (A-3) can then be rearranged to

$$\sum_{i=1}^{m} (t_{m-i+1} - t_{m-i}) \sum_{j=1}^{i} a_j \tag{A-4}$$

and then expanded to

$$\sum_{i=1}^{m} t_{m-i+1} \sum_{j=1}^{i} a_j - \sum_{i=1}^{m} t_{m-i} \sum_{j=1}^{i} a_j . \tag{A-5}$$

We next formally change the i index on the second term of (A-5):

$$\sum_{i=1}^{m} t_{m-i+1} \sum_{j=1}^{i} a_j - \sum_{i=2}^{m+1} t_{m-i+1} \sum_{j=1}^{i-1} a_j . \tag{A-6}$$

We then rearrange the first term of (A-6) and note from the definition of $t_k$ that $t_o$ in the second term is zero. These steps leave (A-6) in the following form

$$t_m a_1 + \sum_{i=2}^{m} t_{m-i+1} \sum_{j=1}^{i} a_j - \sum_{i=2}^{m} t_{m-i+1} \sum_{j=1}^{i-1} a_j . \tag{A-7}$$

We now expand the second term in (A-7) to obtain

$$t_m a_1 + \sum_{i=2}^{m} t_{m-i+1} \left[ a_i + \sum_{j=1}^{i-1} a_j \right] - \sum_{i=2}^{m} t_{m-i+1} \sum_{j=1}^{i-1} a_j . \tag{A-8}$$

Taking note of like terms which cancel, we find that (A-8) reduces to

$$t_m a_1 + \sum_{i=2}^{m} t_{m-i+1} a_i \tag{A-9}$$

or equivalently

$$\sum_{i=1}^{m} a_i t_{m-i+1} . \tag{A-10}$$

Finally, recalling the definition of $t_k$, we see that term (A-10) is

$$\sum_{i=1}^{m} a_i \int_{0}^{(m-i+1)} g(y) dy \tag{A-11}$$

which is the same as term (A-2) and the last member of terms (III-12) of the text.

## REFERENCES

[1].    R. V. Churchill, Operational Mathematics, 2nd ed., McGraw-Hill, New York, 1958.

[2].    W. H. Reinmuth, Anal. Chem., 34:1446 (1962).

[3].    R. S. Nicholson and I. Shain, Anal. Chem., 36:706 (1964).

[4].    P. Delahay, J. Amer. Chem. Soc., 75:1190 (1953).

[5].    F. G. Tricomi, Integral Equations, Interscience, New York, 1957.

[6].    A. Sevcik, Collection Czech. Chem. Commun., 13:340 (1948).

[7].    A. Y. Gokhshtein and Y. P. Gokhshtein, Dokl. Akad. Nauk SSSR, 131:601 (1960).

[8].    A. Y. Gokhshtein and Y. P. Gokhshtein, Dokl. Akad. Nauk SSSR, 128:985 (1959).

[9].    Y. P. Gokhshtein, Dokl. Akad. Nauk SSSR, 126:598 (1959).

[10].    Y. P. Gokhshtein and A. Y. Gokhshtein, Advances in Polarography, Vol. II (I. S. Longmuir, ed.), Pergamon Press, New York, 1960, p. 465.

[11].    Y. P. Gokhshtein and A. Y. Gokhshtein, Zh. Fiz. Khim., 34:1654 (1960).

[12].    R. S. Nicholson, J. M. Wilson, and M. L. Olmstead, Anal. Chem., 38:542 (1966).

[13].    R. S. Nicholson and I. Shain, Anal. Chem., 37:178 (1965).

[14].    W. T. deVries, J. Electroanal. Chem., 9:448 (1965).

[15].    M. L. Olmstead and R. S. Nicholson, J. Electroanal. Chem., 14:133 (1967).

[16].    T. G. McCord and D. E. Smith, Anal. Chem., 40:1959 (1968).

[17].    M. L. Olmstead and R. S. Nicholson, J. Electroanal. Chem., 16:145 (1968).

[18].    A. Huber, Monatsh. Math. Phys., 47:240 (1939).

[19].    C. Hastings, Jr., Approximations for Digital Computers, Princeton University Press, Princeton, New Jersey, 1955.

[20].    R. S. Nicholson, Anal. Chem., 37:667 (1965).

[21].    M. L. Olmstead and R. S. Nicholson, J. Phys. Chem., 72:1650 (1968).

[22].    W. T. deVries, J. Electroanal. Chem., 17:31 (1968).

[23]. W. T. deVries, J. Electroanal. Chem., 18:469 (1968).

[24]. J. R. Delmastro, Anal. Chem., 41:747 (1969).

[25]. M. L. Olmstead, R. G. Hamilton, and R. S. Nicholson, Anal. Chem., 41:260 (1969).

[26]. M. L. Olmstead and R. S. Nicholson, Anal. Chem., 41:851 (1969).

[27]. M. L. Olmstead and R. S. Nicholson, Anal. Chem., 41:862 (1969).

[28]. M. Mastragostino, L. Nadjo, and J. M. Saveant, Electrochim. Acta, 13:721 (1968).

[29]. D. T. Pence, J. R. Delmastro, and G. L. Booman, Anal. Chem., 41:737 (1969).

Chapter 6

LAPLACE PLANE ANALYSIS OF ELECTRODE KINETICS

Arthur A. Pilla

ESB Incorporated
Research Center
Yardley, Pennsylvania 19067

I. INTRODUCTION

The study of electrochemical systems using relaxation techniques is becoming widespread because of the inherent advantages of the use of time or frequency as a controlled experimental variable [1-3]. Thus, the behavior of any given electrode process may be examined in terms of what may essentially be considered its frequency response. This enables each reaction step to be detected in a much more direct manner than the corresponding steady state or dc study allows because of the relatively different manner in which each component (diffusion, charge transfer, etc.) of the total process

139

responds over specific frequency or time intervals. The above properties have long been recognized and it is in view of this that many very powerful techniques have been developed. Each of these involves perturbing the system with specific time functions of voltage [4-11], current [12-17], or charge [18-23], and observing the corresponding time-domain response function.

Examination of the working equations which have been developed to describe a given model for the electrode process reveals that even in the simplest cases, the time-domain solutions are relatively complex [1]. Thus certain approximations are often required for their use such as measurement at times short (or long) enough so that simple time functions (e. g., $\sqrt{t}$ or $1/\sqrt{t}$ ) may be employed. The necessity for approximation can be offset to some extent by the use of curve-fitting techniques readily accomplished through the use of a digital computer [24]. Recent studies [25-27] have also shown how to more fully utilize the closed-form working equations. However, it is often necessary in all of these cases to use four or more adjustable parameters, which can lead to some ambiguity in the interpretation of results. Note that when closed-form working equations are not available, digital simulation has been successfully employed [28, 29].

In addition to the above considerations, it has become increasingly apparent that heterogeneous reaction steps other than charge transfer play an important role in many electrode reactions [30-41]. Since the majority of attempts to detect processes in this category, such as adsorption, rely upon extrapolation (with or without computer-aided curve fitting) over time or frequency ranges in which measurements were not actually made [42, 43], it has become necessary to attempt to study these parameters at shorter and shorter times. To do this, step waveforms are normally employed (indeed, assumed [1] in most derivations of time-domain working equations). However, since most present instrumentation is such that the input pulse, of necessity, has a finite rise time, it is then clear that, unless this time is properly taken into account, valuable experimental information is being lost. Imperfections in the experimental apparatus (e. g., finite rise time) have not generally been considered, since the time-domain working equation assumes an even more complex form, rendering the unambiguous separation of appropriate kinetic parameters exceedingly difficult.

Because of the fact that any time-domain input perturbation results in a measurable time-domain response, a property of an electrochemical system, known as the transfer function [44], may be defined. This function, which exists in the frequency domain only, is the ratio of the output to the input voltage and is sufficient to describe the frequency behavior of the system. The transfer function may be obtained from a knowledge of the time-domain behavior of the electrode process if time-to-frequency domain conversion is carried out. Once this is performed, many of the above mentioned difficulties are markedly reduced.

It is thus the purpose of this study to present an approach to the interpretation of the time-domain response of an electrochemical system such that its frequency behavior may be obtained. It will be shown that this approach generates, for most systems of interest, working equations which are algebraic in nature, often containing kinetic parameters which are not as inseparably lumped together as in the classical time-domain expression. In addition, use may be made of experimental data as close to zero time as experimentally measurable response may be observed, thus allowing relatively higher frequencies to be studied than are normally compatible with the shortest times available in classical time-domain methods, and than are normally accessible with classical ac techniques. This approach involves the use of a digital computer for data reduction such that it becomes a powerful, flexible, and versatile research instrument. Thus the type of data reduction involved in this study necessitates the computer becoming under software (program) control, among other things, a spectrum analyzer, a signal averager, a lock-in amplifier, a phase meter and an oscilloscope. The digital computer available in so-called mini format (and, therefore, accessible to many research laboratories) changes, in a fundamental way, the manner in which a physical system, such as an electrochemical process, may be studied. Perhaps the most important reason for this is that new types of experiments are now possible because of the way in which the computer can be made to act as many instruments, some of which are not yet available in any other form, and which are employed and coupled in a manner not physically permissible for the actual instruments, were they available.

## II. TIME-TO-FREQUENCY DOMAIN CONVERSION

In order to examine the frequency response of an electrochemical system from a knowledge of its time-domain behavior, it is necessary to convert both the input perturbation and response from time functions into frequency functions. This may be accomplished through the use of an appropriate mathematical operation such as the Laplace, Fourier, or Z transforms [44]. The first two are, in fact, closely related if the single-sided Fourier transform is considered. The Z transform is designed for use with sampled-data systems which, in fact, is the case for most experimental measurements which result in a series of points later used to describe the complete experimental curve. A study of the various types of transforms reveals that the most useful is the Laplace transform since it results in frequency-domain equations for specific time functions which are algebraically simpler than can be obtained from other techniques. In this study, therefore, the Laplace transformation will be used exclusively.

The Laplace transformation operating upon the time variable t transforms a time function f(t) to a complex frequency function F(s) according to

$$F(s) = \int_0^\infty f(t) \exp(-st) \, dt \tag{1}$$

where s is the Laplace transform variable. The quantity s is a complex number having a real, $\sigma$, and an imaginary, $j\omega$, part (i.e., $s = \sigma + j\omega$) and the dimensions of reciprocal time (or frequency). The complex frequency function F(s) is thus described by a complex plane in which all absolute values of both $\sigma$ and $\omega$ may be employed provided the integral in (1) converges. This plane is known as the complex frequency or s plane. Under certain conditions, it is possible to perform the integration in (1) utilizing only the real or only the imaginary part of s. This will be termed, respectively, real-axis or imaginary-axis Laplace transformation.

In order to perform real-axis transformation, $s = \sigma$, and (1) now becomes

$$F(\sigma) = \int_0^\infty f(t) \exp(-\sigma t) \, dt. \tag{2}$$

Utilizing (2), the time function f(t) is transformed to the real-axis frequency function $F(\sigma)$, provided this definite integral exists. It can be seen that when $\sigma > 0$, the exponential function provides a relatively powerful means for forcing most time functions to zero with the result that the integral of the product in (2) converges. A time function which, when multiplied by $\exp(-\sigma t)$, does not converge is, for example, $\exp(bt^2)$. However, most electrochemical systems (and physical systems, in general) do not respond either in whole or in part according to this latter function.

As a specific illustration of the manner in which real-axis transformation is carried out, consider an exponential time function $\exp(-at)$. Substitution of this function for f(t) in (2) gives

$$F(\sigma) = \int_0^\infty \exp[-(\sigma + a)t] \, dt = \frac{1}{\sigma + a}. \tag{3}$$

The result is clearly a simpler function than the original transcendental function. It is to be noted that, since $\sigma$ is a real variable, $F(\sigma)$, which is now the corresponding frequency-domain function for the original exponential time function, is a real function and can be utilized in the same manner as a normal algebraic equation containing real variables. Thus a convenient method might be to plot $1/F(\sigma)$ vs. $\sigma$ which results in a straight line for all frequencies, the intercept of which is directly related to the time constant (1/a) involved in the exponential.

To further illustrate real-axis transformation, consider the voltage function V(t) actually applied to a working electrode via a typical potentiostat

having a finite rise time in the potential step method [4]. To a first approximation, the voltage can be considered to rise exponentially and is given
by

$$V(t) = V_0[1 - \exp(-ct)]$$ (4)

where $V_\theta$ is the value of the final voltage plateau and c is the reciprocal of
the time constant of the potentiostat. The real-axis Laplace transform of
(4) is, using (2),

$$V(\sigma) = \frac{V_0}{\sigma} - \frac{V_0}{\sigma + c}$$ (5)

in which the first term on the right-hand side (rhs) is the transform of the
constant portion of V(t), that normally used in the derivation of the working
equations for the potential or voltage step methods. The second term on the
rhs of (5) is the transform of the time-varying portion of V(t) which, as will
be shown below, can enable significant data to be obtained even before the
potentiostat (or any other perturbation instrument) has actually applied the
waveform required for classical time-domain analysis. This is one of the
most powerful aspects of treating transient data using frequency-domain
conversion and will be treated at length below.

For imaginary-axis Laplace transformation, $s = j\omega$, and (1) becomes

$$F(j\omega) = \int_0^\infty f(t) \exp(-j\omega t)dt$$ (6)

where $F(j\omega)$ is the imaginary-axis frequency function. Inspection of (6)
reveals that it is in reality the single-sided Fourier transform [44]. The
convergence requirements for this transform are much more severe than
those for real-axis transformation. Thus it is clear that (6) will not converge except for those time functions which actually go to zero as $t \to \infty$. An
apparent problem, therefore, exists for, e.g., the finite-rise voltage pulse
employed earlier [see Eq. (4)]. This function has an additive constant term
which, of course, will never attain a zero value and for which (6) will, therefore, never converge. The above problem may be avoided by considering the
manner in which $F(j\omega)$ behaves in terms of the complex frequency plane as
defined by s. In order to evaluate F(s) along the imaginary axis $j\omega$ of the s
plane, it is necessary to consider both the real and imaginary parts of
$F(j\omega)$. In the case of a constant A, it can be shown [45] that

$$F(j\omega) = \pi A\delta(\omega) + \frac{A}{j\omega}$$ (7)

in which the real part of $F(j\omega)$ is an impulse $\delta(\omega)$ of value $\pi A$. This result
is not obtainable from direct integration along the imaginary axis via Eq.
(6), but by consideration of the complete complex frequency function F(s) as
defined on the s plane.

Examination of Eq. (7) indicates that the impulse function represents transient frequency-domain behavior and need not, in fact, be retained if results are desired which describe the system as though the perturbation were a steady-state sinusoidal function. This is the most useful method by which to proceed, as will be discussed in detail below. In this case then, the imaginary-axis transformation for the constant A is $A/j\omega$. The complete imaginary-axis transform for the finite-rise-voltage input perturbation given in Eq. (4) is, therefore,

$$V(j\omega) = \frac{V_0}{j\omega} - \frac{V_0}{c + j\omega} . \tag{8}$$

This is the same result that would be obtained if the substitution $s = j\omega$ were performed on the complete function $V(s)$ as obtained by using (1). In actuality, the substitution $s = j\omega$ after (1) has been employed is permissible only for frequency behavior in the sinusoidal steady state [46]. Note that (6) can be expected to converge for the time-varying portion of $f(t)$ for most electrochemical systems since, as will be shown below, it usually can be represented by a series of exponential approximations for which $F(j\omega)$ can be obtained using (6).

The function $V(j\omega)$ is complex and, as such, has real and imaginary parts and the corresponding phase angle. Therefore, the use of (8) requires, as is well known, the use of the $F(j\omega)$ complex plane. In this respect, the separation of $V(j\omega)$ into its real and imaginary parts results in an algebraic equation for each part which is of higher order than that obtained for $V(\sigma)$ [see Eq. (5)]. This will normally be the case for any $F(\sigma)$ when compared to its corresponding $F(j\omega)$, indicating the advantage of using both functions to allow minimum ambiguity in the interpretation of the frequency response of an electrode process in terms of kinetic parameters.

Note at this point that when $f(t)$ is unknown analytically, as should be assumed in the application of time-to-frequency domain conversion in electrochemical systems, it is then virtually impossible to carry out the complete Laplace transformation according to (1). However, it is possible to carry out real- and/or imaginary-axis transformation, according to (2) and (6), in a relatively convenient manner, particularly with the aid of a digital computer. It is to be emphasized that obtaining both $F(\sigma)$ and $F(j\omega)$ involves operating upon the experimentally measurable time-domain function starting at $t = 0$ (e.g., along the rise of a potential "step"). The requirement that the analytic form of $f(t)$ be known is relaxed. It is simply necessary, as will be shown below, to describe the experimental curve using a sufficient number of data points.

### III. TRANSIENT IMPEDANCE USING

### FREQUENCY-DOMAIN CONVERSION

It has been long recognized that the study of an electrochemical system lends itself quite readily to the use of electrical equivalent circuits [18, 47-55]. A recent study [56] has shown that the use of frequency-independent (aperiodic) equivalent circuits may actually be considered as shorthand notation for the rigorous solution of the simultaneous differential equations used to describe the particular electrode model. This is a direct consequence of the possibility of considering the transfer function mentioned earlier as a basic property of such systems. The use of both real- and imaginary-axis transformation on both the input perturbation and its corresponding response allows a frequency function to be obtained which, when the system is studied under linear conditions, may be termed the impedance function. Thus, if both the input perturbation (e.g., a potential "step") and the corresponding response (current, in this case) are measurable in the time domain from as close to $t = 0$ as experimentally possible, both real- and imaginary-axis transformation may be carried on each function. The result is that for any arbitrary $V(t)$ and $I(t)$ known from experiment, $V(\sigma)$, $V(j\omega)$, $I(\sigma)$, and $I(j\omega)$ may be obtained. Knowledge of these functions allows the real-axis transient impedance to be obtained as

$$Z(\sigma) = V(\sigma)/I(\sigma) \tag{9}$$

while the imaginary-axis transient impedance is given by

$$Z(j\omega) = V(j\omega)/I(j\omega). \tag{10}$$

The quantity $Z(\sigma)$ is a real function and as such is the simplest of the impedance functions, $Z(j\omega)$ is a complex function having both real, $\text{Re}[Z(\omega)]$, and imaginary, $\text{Im}[Z(\omega)]$, parts and the corresponding phase angle $\theta(\omega)$. This function is mathematically identical in form to that employed in classical ac techniques [53, 54], where sinusoidal steady-state conditions prevail. It is to be remembered, however, that $Z(j\omega)$ in this study is always considered as obtained under nonsinusoidal conditions (i.e., from conversion of both the input and response time-domain functions).

Real-axis transformation was first suggested for electrochemical applications by Wijnen [57] who considered the current response to a potential step for a diffusion-coupled charge-transfer reaction. He did not, however, consider the aspect of transient impedance as outlined above. The scope of his approach was thus somewhat limited in nature and such problems as more complex reactions, double-layer charging, or the use of an imperfect potential step (e.g., with finite rise) were not taken into account. Real-axis transformation has since been considered by other workers [58, 59]. However, in many cases, their approach has been simply to consider some classical faradaic impedance expressions in terms of the complex frequency variable s. The most recent study [56] utilizing frequency-domain conversion has shown the usefulness of obtaining the complete transient impedance function, i.e. for $s = \sigma$ and $s = j\omega$, from knowledge of the experimental time-domain response. However, another study [60] considering only the imaginary axis does not take advantage of the full capabilities of this approach and presents an erroneous treatment of the coulostatic method in terms of the transient-impedance technique. The reasons for this will be considered in detail below.

In order to illustrate the advantage of obtaining the transient-impedance function rather than simply the real- or imaginary-axis transform of the appropriate time-domain response function, consider the equivalent circuit shown in Fig. 1. This circuit represents the simplest of electrochemical systems under uniform potential conditions wherein $R_e$ is the series ohmic

Fig. 1. Aperiodic equivalent circuit for an electrode exhibiting double-layer behavior only.

resistance (due to the electrolyte or the electrode configuration) and $C_d$ is the double-layer capacity. The impedance of this circuit is given by

$$Z(s) = R_e + \frac{1}{C_d s} \tag{11}$$

which may be utilized along the real axis ($s = \sigma$) or the imaginary axis ($s = j\omega$). For either axis, the functions involved are relatively simple and, in fact, allow straightforward diagnostic criteria for double-layer behavior to be developed. Thus, for $s = \sigma$, a plot of $Z(\sigma)$ vs. $1/\sigma$ is a straight line which may be treated as normally would be done for a real function of this type. For $s = j\omega$, the $\text{Re}\,[Z(j\omega)]$ and $\text{Im}\,[Z(j\omega)]$ are separately plotted vs. $1/\omega$. The former gives a horizontal line and the latter a straight line. The function $\theta(\omega)$ may be employed, if desired, resulting in, e.g., $\tan\theta = R_e C_d \omega$. There are thus four simple functional relationships available to verify whether or not double-layer behavior is being observed as well as to allow the evaluation of relevant electrochemical parameters when both $Z(\sigma)$ and $Z(j\omega)$ are considered. It is to be emphasized that $Z(\sigma)$ and $Z(j\omega)$ may be obtained independent of the type of perturbation to which the system is subjected. The only requirement is that it be of low enough signal level to enable linear conditions to be maintained.

Consider now the case in which only the transformed current response $I(s)$ is employed. For the above system, $I(s)$ is given by

$$I(s) = V(s) \frac{1}{R_e + 1/C_d s} \tag{12}$$

where $V(s)$ may be obtained from any arbitrary time function described by a series of experimental points. However, for the purpose of this treatment, the assumption will be made that it is a relatively simple exponential function. Then $I(s)$ becomes

$$I(s) = \left[\frac{V_o}{s} - \frac{V_o}{s + c}\right] \frac{1}{R_e + 1/C_d s} \tag{13}$$

using (5) and (8).

Inspection of (13) reveals that it is much more complex than the impedance function given in (11). Thus the simple functional relationships available when (11) is used for $s = \sigma$ or $\sigma = j\omega$ are no longer possible if (13) is employed. The situation becomes progressively more favorable for the use of the transient impedance function as both the input perturbation and the electrochemical system become more complex.

It can thus be seen that the use of frequency-domain conversion to generate the transient impedance of an electrochemical system allows a considerable amount of information to be obtained which is presented in the most easily manageable functional relationships. The use of both $Z(\sigma)$ and $Z(j\omega)$ permits four basic functions to be described, which allows considerable cross-checking, thereby providing less ambiguity in the elucidation of relatively complex systems.

## IV. TRANSIENT IMPEDANCE

### IN POTENTIOSTATIC TECHNIQUES

As a specific example of the manner in which frequency-domain conversion may be employed to markedly reduce instrumental errors and artifacts, particularly at high frequencies, the use of a potentiostat in potential control techniques will be considered. The specific features of this approach have been presented in detail elsewhere [56]; therefore, only a brief review will be given here.

Since a potentiostat is essentially a feedback device employed to control voltage appearing at a working electrode in a programmable manner, potentiostatic techniques readily lend themselves to frequency-domain analysis. For this, consider the circuit shown in Fig. 2. Here a single-amplifier potentiostat is used to control the voltage at the working electrode. It is important to note that the electrolytic cell forms an integral part of the feedback loop and, as such, must be considered when deriving the frequency-domain behavior of the system. For convenience, the circuit shown in Fig. 2 is redrawn in Fig. 3. in a more schematic manner wherein the representation

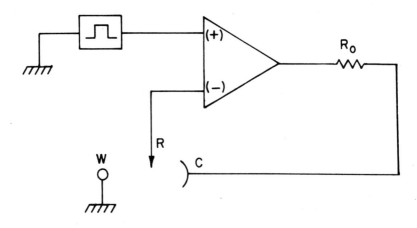

Fig. 2. Potentiostat circuit employing a single operational amplifier; shows how the electrolytic cell forms a part of the control system.

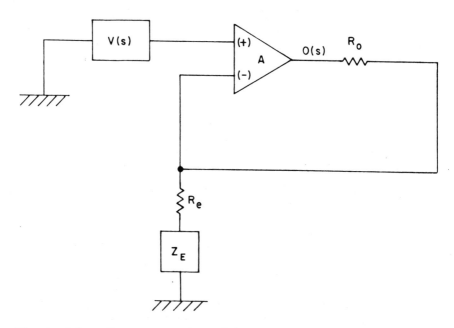

Fig. 3. Schematic representations of the circuit in Fig. 2.; a generalized arbitrary electrode impedance is assumed.

of all functions is made in terms of the Laplace transformation. Figure 3 shows that the output voltage O(s) of the potentiostat A is applied to the electrochemical cell where $R_e$ has its usual significance and $Z_E$ represents an arbitrary working electrode impedance about which no a priori assumptions are made. $R_o$ represents the output impedance of A, the counter electrode impedance (a good first approximation if it is of much larger surface area than the working electrode), and a current-measuring resistor, if present. The input perturbation V(s) appears across the electrochemical cell as E(s).

To derive the impedance of this potentiostat-electrolytic cell system, it is convenient to consider the transfer function of the complete system. This is given by   56]

$$\frac{O(s)}{V(s)} = \frac{[R_o + R_e + Z_E(s)]G(s)}{R_o + R_e + Z_E(s) + [R_e + Z_E(s)]G(s)} \tag{14}$$

where G(s) is the open-loop gain function of the potentiostat. Its value at high frequencies determines to a large extent whether measurable response can be obtained to study the system at these frequencies. Optimum gain functions for high-frequency applications have been considered elsewhere [61]. Briefly, however, the ideal potentiostat would be characterized by G(s) = K where K is the open-loop gain (usually greater than 1000). Unfortunately, all potentiostats presently available have a finite-rise time, and as such, cannot be described by this simple function. The best potentiostats have at least a first-order gain function given by

$$G(s) = \frac{K}{s\tau + 1} \tag{15}$$

where $\tau$ is the open-loop time constant of the potentiostat. Equation (15) gives, if $\tau$ is known, the highest frequency at which the potentiostat itself will have sufficient gain for observable response. For this study, no a priori assumptions need be made concerning the actual form of G(s). It is to be borne in mind, however, that to obtain the highest frequencies, G(s) must allow a signal of sufficient amplitude to be observed at the shortest times compatible with real- or imaginary-axis transformation.

The voltage function necessary to obtain the desired impedance value is E(s) which is related to O(s) by

$$E(s) = O(s) - I(s)R_o \tag{16}$$

from which, using (14),

$$E(s) = V(s) \frac{[R_e + Z_E(s)]G(s)}{[R_e + Z_E(s)][G(s) + 1] + R_o} \tag{17}$$

The current function I(s) is given by

$$I(s) + \frac{O(s)}{R_o + R_e + Z_E(s)} \tag{18}$$

from which, using (14),

$$I(s) = V(s) \frac{G(s)}{[R_e + Z_E(s)][G(s) + 1] + R_o} . \tag{19}$$

The impedance of the system represented schematically in Fig. 3 is then obtained by division of (17) by (19). The result is

$$Z(s) = E(s)/I(s) = R_e + Z_E(s). \tag{20}$$

Inspection of (20) indicates that both potentiostat parameters $G(s)$ and electrolytic cell parameters other than those related to the working electrode have been eliminated. Thus, $Z(s)$ is a function only of what appears across the potentiostat control points (i. e., between reference and working electrodes).

In a general manner $Z(s)$, evaluated for $s = \sigma$ or $s = j\omega$, may be obtained under conditions which allow the results to be relatively free of instrumental contributions. Thus, for high-speed potentiostatic studies in which an oscilloscope is normally employed, it is possible to obtain measurements of $V(t)$ and $I(t)$ with matched high-frequency probes and display the time-synchronous experimental curves on the same cathode-ray tube if the oscilloscope has dual beam capability. In this way, even if the measuring device has nonlinearities, these effects can be markedly reduced if they are as identical as possible for both $V(t)$ and $I(t)$. It is, therefore, possible to obtain $Z(s)$ relatively free from instrumental contributions both from the apparatus used to apply the input perturbation and from that employed to measure the experimental functions.

The above is a brief illustration of the manner in which knowledge of the transient impedance allows the characteristics of the electrochemical system to be isolated from the most troublesome instrumental artifacts when, e. g., a potentiostat is employed. The same conclusions are valid for any of the commonly employed relaxation techniques when used under linear conditions.

## V. TRANSIENT IMPEDANCE IN
## THE COULOSTATIC TECHNIQUE

One of the major advantages of the transient-impedance technique is the fact that meaningful results may be generated by use of data obtained as close to $t = 0$ as feasible. Thus it has been possible [56] to study electrochemical systems at frequencies as high as $10^8$ rad/sec. This is of interest when it is desired to evaluate the double layer in the presence of a faradaic

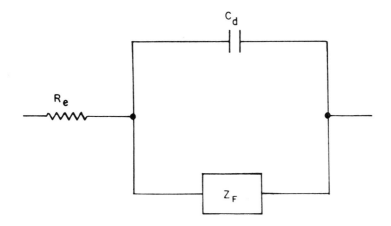

Fig. 4. General aperiodic equivalent circuit for an electrode at which faradaic and double-layer processes may be separated a priori.

reaction. In this respect, if the electrode reaction may be considered as a leak across the double-layer capacitor, then the system may be represented by the electrical equivalent circuit shown in Fig. 4, wherein $Z_F$ is a general arbitrary faradaic impedance. The impedance of this circuit may be written as

$$Z(s) = R_e + \frac{1}{1/Z_F(s) + C_d s}.$$ (21)

Inspection of (21) indicates that if a frequency range can be chosen such that $1/Z_F(s) \ll C_d s$, then $Z(s)$ becomes simply that for the $(R_e - C_d)$-series circuit considered earlier [see Eq. (11)]. For many electrochemical systems, frequencies above $10^6$ rad/sec are sufficient to render the above simplification possible. This is an important point since, as $Z_F(s)$ becomes relatively more complex, it becomes increasingly necessary to make an independent observation of the double-layer time constant so that it may be appropriately eliminated from the total electrode impedance. This then unmasks the faradaic impedance for further less ambiguous analysis.

When frequencies above $10^6$ rad/sec are employed, a serious experimental difficulty exists for the closed-circuit transient impedance method. In other words, for those techniques in which current flows in the external circuit, the relative value of the impedance of $R_e$ may cause considerable error in the evaluation of $C_d$. To illustrate this, consider the high-frequency simplification of (21) written for $s = \sigma$:

$$Z(\sigma) = R_e + \frac{1}{C_d \sigma}.$$ (22)

It can be seen from (22) that if $R_e \gg 1/C_d\sigma$, the double-layer impedance
will tend to be overwhelmed by the nonfaradaic resistance, and its unambiguous evaluation can become exceedingly difficult. These conditions may
easily exist in a typical electrolytic cell where even concentrated supporting
electrolyte can result in $R_e \cong 10\ \Omega$, since microelectrodes are usually employed. For these electrodes, $C_d \approx 1\ \mu F$. In this case, at $10^7$ rad/sec,
$1/C_d\sigma = 0.1\Omega$ which is relatively small compared to $R_e$ at this frequency.
It would thus be advantageous to further reduce the complexity of the general
electrical equivalent circuit given in Fig. 4 by the elimination of $R_e$.

The above conclusion has, of course, long been recognized and was
largely responsible for the development of the coulostatic technique [18-21]
in which a short burst of charge is applied to the system such that the potential across $C_d$ is changed by a few millivolts at the end of the charge burst.
The system is then allowed to relax under open-circuit conditions with the
observable experimental variable being $V(t)$. If the charge burst is of short
enough duration compared to the faradaic time constant(s), the input perturbation may be considered to be an impulse [20] given by

$$I(t) = q\delta(t) \tag{23}$$

where q is the value of the impulse which is equal to the quantity of charge
injected into $C_d$.

An input perturbation in the form of an impulse has the advantage that
its Laplace transformation is relatively simple. Thus, the transform of $I(t)$
given in (23) is

$$I(s) = q. \tag{24}$$

It is apparent that, if experimental conditions can be achieved such that (24)
is valid, i.e., the duration of the perturbation is, in reality, very short
compared to faradaic time constants, then the necessity of carrying out
either real- or imaginary-axis transformation according to (2) and (6) for
$I(t)$ is avoided. However, if (24) is employed, then the very advantage of the
coulostatic technique is eliminated. This is so, since the impulse is applied
to the system under closed-circuit conditions. In other words, the response
of the general circuit of Fig. 4 to an impulse in the frequency domain is, in
fact, given by

$$V(s) = q\left[R_e + \frac{1}{1/Z_F(s) + C_d s}\right]. \tag{25}$$

The resulting impedance is then

$$Z(s) = \frac{V(s)}{q}. \tag{26}$$

Inspection of (25) and (26) shows that $R_e$ is still one of the parameters in the
total impedance. A recent study [60] suggests the use of (26) for frequency-

.domain analysis (for $s = j\omega$) of coulostatic data. This approach is erroneous as can be seen from the above discussion. However, the experimentally observable contribution of $R_e$ is greatly diminished if $V(t)$ is observed not from $t = 0$ as it should be if $q$ is employed as the transform of $I(t)$, but from the time at which the impulse ends. In other words, for relatively large faradaic time constants, the impulse may be of longer "duration" and the contribution to $V(t)$ of its value during the impulse may be neglected in the evaluation of $V(s)$. In this case, $Z(s)$ evaluated for $s = \sigma$ or $s = j\omega$ can be expected to look as though $R_e$ were not present, but only for relatively low frequency values (certainly below $10^6$ rad/sec with present coulostatic instrumentation). Since this is so, independent observation of the double-layer impedance is not possible and the coulostatic method utilized in this way does not appear to be even as powerful as closed-circuit techniques.

The manner in which the coulostatic method may be employed to un-ambiguously eliminate $R_e$ from the total impedance will now be considered. If a true impulse is applied to an electrochemical system (i.e., input perturbation of duration $\leq$ 10 nsec) it is then valid to consider $t = 0$ to be at the end of the impulse such that the system is then truly at open circuit and relaxation has begun. At this $t = 0$, the charge $q$ applied to $C_d$ will begin discharging through $Z_F$. The internal current flowing out of $C_d$ is, therefore, given, in terms of $s$, by

$$I(s) = -C_d sV(s) + q \tag{27}$$

where $V(s)$ is the voltage function across $C_d$. Since $I(s)$ is not experimentally accessible, it is possible to eliminate it from (27) using the following:

$$I(s) = \frac{V(s)}{Z_F(s)} \tag{28}$$

which simply states that the total current also flows through $Z_F$. Substitution of (28) in (27) gives

$$V(s) = q\left[\frac{1}{C_d s + 1/Z_F(s)}\right]. \tag{29}$$

This relation may be employed to evaluate $C_d$, provided a frequency range can be attained such that $C_d s \gg 1/Z_F(s)$. This, of course, will be automatic if the perturbation applied to the system is truly an impulse function [i.e., its duration must be shorter and shorter as the time constants involved in $Z_F(s)$ are smaller]. The diagnostic plots for double-layer behavior are $V(\sigma)$ vs. $1/\sigma$ and $Im[V(j\omega)]$ vs. $1/\omega$, which allow the evaluation of $C_d$. Once this is performed, $I(s)$ may then be generated via the double-layer impedance. Thus

$$I(s) = V(s)C_d s \tag{30}$$

where $C_d$ is obtained by the appropriate use of (29), and $V(s)$ is obtained using $V(t)$, which is experimentally measurable [i.e., $V(s)$ is identical to that in (29)].

Once the above procedure is followed, $Z(s)$ may then be evaluated for $s = \sigma$ or $s = j\omega$. Note that under open-circuit conditions, $C_d$ is effectively in series with $Z_F$. The impedance is then

$$Z(s) = Z_F(s) + \frac{1}{C_d s}. \tag{31}$$

Use of this relation allows substantially different diagnostic functions to be obtained for specific electrode models than for $Z(s)$ obtained under closed-circuit conditions where $C_d$ and $Z_F$ are in parallel. Use of the coulostatic technique, therefore, provides an additional means whereby a less ambiguous choice may be made among the various models which may be proposed for a given electrochemical system.

## VI. COMPARISON OF FREQUENCY- AND TIME-DOMAIN
## ELECTROKINETIC STUDIES

In order to illustrate the advantage of frequency-domain over time-domain analysis, consider a linear-diffusion-coupled, charge-transfer electrode process. This is treated here as a classical system [14, 17], i.e., one in which a priori separation exists for faradaic and double-layer processes. The faradaic transient impedance for this process can be derived by utilizing the linearized rate equation [1] for the faradaic current $I_F(s)$, given by

$$I_F(s) = I_0 \left[ \frac{C_O(0,\ s)}{C_O} - \frac{C_R(0,\ s)}{C_R} - \frac{nF}{RT} E(s) \right] \tag{32}$$

in which $I_0$ is the exchange current density; $C_O$ and $C_R$ are the bulk concentrations of oxidized and reduced species, respectively; $C_O(0,\ s)$ and $C_R(0,\ s)$ represent the complex frequency dependence of these concentrations at the electrode surface; $E(s)$ is the voltage developed across the interphase region (i.e., across $C_d$); and the other quantities have their usual meaning. To utilize (32) specific expressions for the concentration ratios on its rhs must be obtained. The use of Fick's second law for linear diffusion for this is well known [1]. Briefly, however, using appropriate boundary conditions for semiinfinite character, these ratios may be expressed in terms of $I_F(s)$, as

$$\frac{C_i(0,\ s)}{C_i} = \frac{I_F(s)}{nFC_i \sqrt{D_i}\sqrt{s}} + \frac{1}{s} \tag{33}$$

in which $D_i$ is the diffusion coefficient of the diffusing species i. There are as many equations of this type as there are diffusing species. Using (33), Eq. (32) now becomes

$$I_F(s) = \frac{E(s)}{B/\sqrt{s} + RT/nFI_0} \tag{34}$$

where

$$B = \frac{RT}{n^2 F^2} \sum_i \frac{1}{C_i \sqrt{D_i}} \tag{35}$$

In order to employ (34) to develop the total time-domain voltage response for this system, double-layer charging current $I_{C_d}(t)$ must be taken into account. Since the faradaic impedance is assumed in this model to be in parallel with $C_d$, the current is given, in terms of s, by

$$I(s) = I_F(s) + I_{C_d}(s) \tag{36}$$

where

$$I_{C_d}(s) = E(s)C_d s. \tag{37}$$

Use of Eqs. (34) and (37) allows an expression for E(t) to be obtained provided that the form of the current input is known. For a true current step, $I(s) = I_0/s$, and E(t) is given by

$$E(t) = \frac{I_0}{C_d(a - b)} \left\{ \frac{a}{b^2} \left[ \exp(b^2 t)\, \mathrm{erfc}(bt^{1/2}) + \frac{2b\sqrt{t}}{\sqrt{\pi}} - 1 \right] \right.$$
$$\left. - \frac{b}{a^2} \left[ \exp(a^2 t)\, \mathrm{erfc}(at^{1/2}) + \frac{2a\sqrt{y}}{\sqrt{\pi}} - 1 \right] \right\} \tag{38}$$

where

$$a = \frac{nFI_0 B}{2RT} + \left[ \frac{n^2 F^2 I_0^2 B^2}{4R^2 T^2} - \frac{nF}{RT}\frac{I_0}{C_d} \right]^{1/2} \tag{39}$$

and b is identical to Eq. (39) except that it has a minus sign between the two terms on its rhs; B is given by Eq. (35). As can be seen by Eq. (38), analysis of this system in the time domain involves the manipulation of a relatively complex equation. The approach employed when a computer is not available is simplification such that a more useful $\sqrt{t}$ function is obtained. This requires that measurements be made at relatively short times, precisely the range in which instrumentation contribution may not be negligible. In addition, much short-time information is not usable since an imperfect (e.g., finite rise) current "step" was not considered. This factor can, of course, be taken into account by assuming some other (e.g., exponential) function for I(t) which, however, adds at least another time constant to a new and even more complex function for E(t) as compared to (38). The

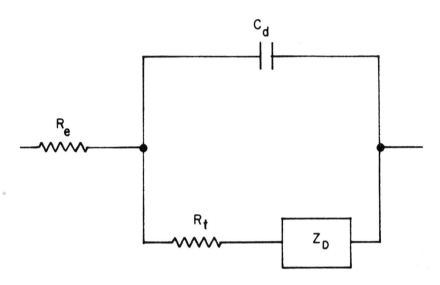

Fig. 5. Aperiodic equivalent circuit for an electrode at which the faradaic process is diffusion-coupled charge transfer.

separation of appropriate kinetic parameters then becomes even more difficult, if not virtually impossible.

Consideration of this system in terms of transient impedance allows the above mentioned problems to be avoided to a large extent. To illustrate this, use will be made of Fig. 5 which shows the frequency-independent (aperiodic) equivalent circuit for this electrochemical system [56]. The total impedance for this circuit is given by

$$Z(s) = R_e + \frac{1}{C_d s + [R_t + Z_D(s)]^{-1}} \tag{40}$$

where $R_t$ is the charge-transfer resistance [56, 62], and $Z_D$ the diffusion impedance. An explicit expression for the sum $[R_t + Z_D(s)]$ may be obtained through the use of (34). Thus

$$Z_F(s) = R_t + Z_D(s) = \frac{B}{\sqrt{s}} + \frac{RT}{nF} \frac{1}{I_0} \tag{41}$$

where B is given by (35). The last term on the rhs of (41) is $R_t$, and the first term is $Z_D(s)$.

Equation (40) may be employed for $s = \sigma$ and $s = j\omega$. As indicated earlier, if sufficiently high frequencies are attainable, the double-layer time constant $R_e C_d$ may be independently observed, leaving $Z_F(s)$ unmasked for analysis via (41). It is to be emphasized that once $R_e$ and $C_d$ are known,

$Z_F(s)$ may be examined throughout the complete frequency range. All of the necessary kinetic constants are, therefore, readily obtainable with no app-roximations or restrictions. In the case of real-axis transformation ($s = \sigma$), the diagnostic plot is $Z_F(\sigma)$ vs. $1/\sqrt{\sigma}$, the intercept of which gives $R_t$, while the slope contains $D_i$. Imaginary-axis transformation ($s = j\omega$) gives the well-known [50] results for $\text{Re}[F(j\omega)]$ and $\text{Im}[F(j\omega)]$, both of which are linear in $1/\sqrt{\omega}$, having identical slopes containing $D_i$. The intercept of $\text{Re}[F(j\omega)]$ contains $R_t$.

It can thus be seen that the manner in which kinetic parameters are made available for evaluation is substantially different for the time and frequency domains. In the former case, it is exceedingly difficult and am-biguous to employ short-time data since, even if the finite rise of the input perturbation were analytically describable, the lumping of parameters in all exponential and error function arguments renders their unambiguous separa-tion virtually impossible. Use of the frequency domain, however, allows much shorter times to be employed for the generation of significant infor-mation, permitting unhindered study of both double-layer and faradaic parameters over the whole accessible frequency range with no approximations necessary. Note that for the purposes of this study, $C_d$ is considered evalu-ated only when it is directly observed. Extrapolation over wide frequency ranges in which measurements were not actually made is considered too ambiguous to be acceptable. This point will be discussed at length below.

## VII. DIGITAL COMPUTER EVALUATION
### OF REAL- AND IMAGINARY-AXIS IMPEDANCE

As stressed throughout this study, the generation of both $Z(\sigma)$ and $Z(j\omega)$ can be carried out with a minimum of a priori assumptions concerning the actual physical model present. It is thus sufficient to experimentally observe $V(t)$ and $I(t)$ for any given technique (potential step, coulostatic, etc.) and produce their transforms from which the appropriate transient imped-ance may then be evaluated.

In order to discuss the digital computer programs which may be employed for this, it is appropriate to examine exactly how each transforma-tion is carried out. For the real-axis case, it will be recalled that the following integration is performed:

$$F(\sigma) = \int_0^\infty f(t) \exp(-\sigma t)\, dt \tag{42}$$

where $f(t)$ is an arbitrary, but experimentally known, time function. This operation is schematically illustrated in Fig. 6 for a single frequency ($\sigma$) value. Here, the time function is a typical current response to a finite-rise potential step. The time origin is the beginning of the experiment which, in

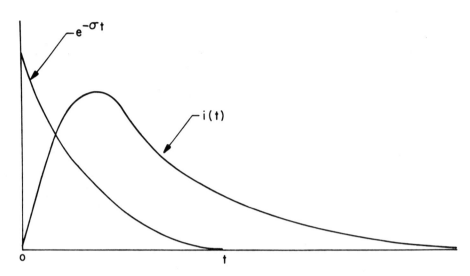

Fig. 6. Schematic representation of real-axis (s = σ) Laplace transformation.

practice, is that time at which the function has sufficient amplitude to be measurable with reasonable accuracy (this is typically 10 nsec for present state-of-the-art potentiostats). The experimental time function is multiplied by the exponential function exp(-σ t) and the area under the resulting curve is then equal to the integral defined in (42).

Since f(t) is known only via a succession of appropriately spaced experimental points, the integration in (42) can be carried out only if some assumption is made concerning the way in which f(t) behaves analytically between two successive time points. This is, of course, a well-known problem and is the reason for the existence of various numerical integration rules. Two of the most common [63] are the trapezoid rule, which assumes a straight line between two successive points, and Simpson's rules which assume parabolic behavior through an appropriate number of data points dependent upon whether their total number is even or odd. These approaches, however, are not the most physically meaningful due to the nature of the electrochemical system under linear conditions. Thus since impedance is the desired quantity, it can be expected that the over-all time behavior may be that due to the appropriate combination of the minimum number of time constants needed to describe the system.

In view of the above, it was considered that, provided the time interval between two successive data points is sufficiently small, an exponential function represents the most physically meaningful approximation. That this is so can be seen from the fact that, for an electrochemical system which can be described by an aperiodic equivalent circuit, only a single time constant will be predominant over a small enough time interval, and a single exponential will, therefore, prevail. The exponential approximation is also valid for diffusion-coupled systems exhibiting error function behavior. For two successive time points, therefore, the following may be written:

$$f(t_i) = A \exp(-a\, t_i) \tag{43}$$

and

$$f(t_{i+1}) = A \exp(-a\, t_{i+1}) \tag{44}$$

where $t_i$ and $t_{i+1}$ represent successive time points; A is the scale factor; and $1/a$ is the prevailing time constant. Equations (43) and (44) describe the limits of one segment of the time-domain function. This curve must then be multiplied by $\exp(-\sigma t)$ taken over the same time segment. The integral may then be performed over this time interval according to

$$F_i(\sigma) = \int_{t_i}^{t_{i+1}} A \exp[-(\sigma + a)t]\, dt. \tag{45}$$

This is performed over each time segment until f(t) is completely transformed. Note that the use of (43) and (44) allows the elimination of A and the evaluation of a. The total real-axis transform $F(\sigma)$ is then

$$F(\sigma) = \sum_i \frac{\left[f(t_{i+1}) \exp(-\sigma t_{i+1}) - f(t_i) \exp(-\sigma t_i)\right]\left[t_{i+1} - t_i\right]}{\left[\ln\, f(t_{i+1}) \exp(-\sigma t_{i+1})/f(t_i) \exp(-\sigma t_i)\right]} + \frac{f(t_\infty)}{\sigma} \tag{46}$$

where $t_\infty$ is the time at which the signal achieves steady state, if present. Inspection of (46) indicates that the segment-by-segment multiplication and summation need be performed only upon the time-varying portion of the signal since the transform of the constant portion, if it exists, is known a priori and may simply be added as $f(t_\infty)/\sigma$ to the result. The numerical summation in (46) is performed for every value of $\sigma$ desired for all voltage and current points giving $V(\sigma)$ and $I(\sigma)$, whereupon $Z(\sigma)$ may be obtained. It is to be noted that the exponential approximation may lead to ambiguous results for those systems in which two (or more) time constants are essentially identical. In this case, the system will behave as though only one time constant were present over the time interval considered, and separation will indeed be difficult. One possible solution to this problem is to vary the initial conditions of the experiment with respect to e.g., potential and/or concentration, which may alter the lumped time constants by sufficiently different degrees.

The digital computer program utilized to perform (46) and generate $Z(\sigma)$ when potentiostatic techniques are employed is reproduced in Table 1. It is written in Fortran IV compatible with an EAI 640 computer, which is one of the so-called midicomputers. Its characteristics are as follows:

TABLE 1

Digital Computer Program for Real Axis Laplace Transformation

```
C       REAL AXIS LAPLACE TRANSFORMATION FOR POTENTIOSTATIC
C       PULSE IMPEDANCE ELECTROCHEMICAL STUDIES
        DIMENSION T(100),C(100),V(100),S(100),B(36),VOFS(100),COFS(100)
        READ(2,10) SI,SF,NU
C       SI AND SF ARE THE DESIRED INITIAL AND FINAL REAL AXIS
C       FREQUENCY VALUES RESPECTIVELY; BOTH MUST BE THE START OF
C       A DECADE.
C       NU IS THE DEVICE NUMBER FOR THE DESIRED MODE OF DATA
C       OUTPUT(TELETYPE,HIGH SPEED PUNCH,LINE PRINTER,OR
C       MAGNETIC TAPE).
10      FORMAT(2E12.5,I3)
        READ(5,11) (B(M),M=1,36)
11      FORMAT(36A2)
        READ(5,12) NT,NCTM,NVTM
C       NT IS THE TOTAL NUMBER OF TIME DATA POINTS.
C       NCTM IS THE NUMBER OF CURRENT DATA POINTS UP TO AND
C       INCLUDING IT'S MAXIMUM VALUE.
C       NVTM IS THE NUMBER OF VOLTAGE DATA POINTS UP TO AND
C       INCLUDING IT'S FIRST CONSTANT VALUE.
12      FORMAT(3I5)
        IF(NVTM.LT.NT) GO TO 20
        READ(5,15)(T(K),C(K),V(K),K=1,NT)
15      FORMAT(3E12.5)
        GO TO 23
20      READ(5,21)(T(K),C(K),V(K),K=1,NVTM)
21      FORMAT(3E12.5)
        N=NVTM+1
        READ(5,22) (T(K),C(K),K=N,NT)
22      FORMAT(2E12.5)
C       THE FOLLOWING SERIES OF TESTS VERIFY PROPER SEQUENCING
C       FOR TIME, VOLTAGE AND CURRENT DATA INPUT. ONE MAXIMUM
C       IS ALLOWED FOR BOTH CURRENT AND VOLTAGE THUS LIMITING
C       THIS PROGRAM TO USE WITH LOW LEVEL POTENTIAL STEPS.
23      DO 25 K=2,NT
        IF(T(K).GT.T(K-1)) GO TO 25
        WRITE(1,24)
24      FORMAT(65H GI=GO:TIME INCREMENT  ERROR'/)
        GO TO 999
25      CONTINUE
        DO 35 K=2,NCTM
        IF(C(K).GT.C(K-1)) GO TO 35
        WRITE(1,31)
31      FORMAT(65H GI=GO:INCREASING CURRENT INCREMENT ERROR'/)
        GO TO 999
35      CONTINUE
        DO 40 K=2,NVTM
        IF(V(K).GT.V(K-1))GO TO 40
        WRITE(1,37)
37      FORMAT(65H GI=GO:INCREASING VOLTAGE INCREMENT ERROR'/)
        GO TO 999
40      CONTINUE
```

TABLE 1 (cont.)

```
        GO TO 76
71      VF=V(NVTM)
        DO 76 K=1,NVTM
        V(K)=V(K)-VF
76      CONTINUE
        DO 77 K=1,NS
        VOFS(K)=VF/S(K)
77      CONTINUE
        CF=C(NT)
        IF(CF.EQ.0.0) GO TO 79
        DO 78 K=1,NT
        C(K)=C(K)-CF
78      CONTINUE
        DO 79 K=1,NS
        COFS(K)=CF/S(K)
79      CONTINUE
C       THE FOLLOWING ROUTINES GENERATE THE REAL AXIS TRANSFORMATION
C       FOR BOTH CURRENT AND VOLTAGE, USING THE SUBROUTINE FOR TIME
C       VARYING DATA AND ADDING THIS RESULT TO THAT OBTAINED ABOVE
C       FOR THE CONSTANT PORTION OF BOTH FUNCTIONS, AT EACH DESIRED
C       REAL AXIS FREQUENCY.
        DO 125 K=1,NS
        CALL REAXTR(NT,T,C,S(K),CS)
        IF(CF.EQ.0.0) GO TO 105
        COFS(K)=CS+COFS(K)
        GO TO 106
105     COFS(K)=CS
106     IF(NVTM.LT.NT) GO TO 110
        CALL REAXTR(NT,T,V,S(K),VS)
        GO TO 120
110     CALL REAXTR(NVTM,T,V,S(K),VS)
120     VOFS(K)=VS+VOFS(K)
125     CONTINUE
        WRITE(NU,128)
128     FORMAT(//////////)
        WRITE(NU,130) (B(M),M=1,36)
130     FORMAT(36A2//)
        WRITE(NU,135)
135     FORMAT(7X,1HS,11X,3H1/S,6X,7HSR(1/S),6X,4HZ(S),8X,4HY(S)/)
```

TABLE 1 (cont.)

```
          J=NCTM+1
          DO 50 K=J,NT
          IF(C(K).LT.C(K-1)) GO TO 50
          WRITE(1,46)
46        FORMAT(65H GI=GO:DECREASING CURRENT INCREMENT ERROR'/)
          GO TO 999
50        CONTINUE
          IF(NVTM.LT.NT) GO TO 57
          J=NVTM+1
          DO 57 K=J,NT
          IF(V(K).LT.V(K-1)) GO TO 57
          WRITE(1,56)
56        FORMAT(65H GI=GO:DECREASING VOLTAGE INCREMENT ERROR'/)
          GO TO 999
57        CONTINUE
          WRITE(1,61)
61        FORMAT(65H DATA INPUT CORRECT,CALCULATION PROCEEDING'/)
          TS=1.0/T(2)
          DO 300 K=1,NT
          T(K)=T(K)*TS
300       CONTINUE
C         THE FOLLOWING ROUTINE GENERATES THE DESIRED REAL AXIS
C         FREQUENCY VALUES.
          SI=SI*1.0/TS
          SF=SF*1.0/TS
          S(1)=SI
          A=ALOG10(SF/S(1))
          I=A+0.5
          NS=I*14
          N=0
          SINC=0.2*S(1)
          DO 65 K=2,NS
          S(K)=S(K-1)+SINC
          N=N+1
          IF(N.EQ.5)GO TO 62
          IF(N.EQ.7)GO TO 63
          IF(N.EQ.14)GO TO 64
          GO TO 65
62        SINC=2.5*SINC
          GO TO 65
63        SINC=2.0*SINC
          GO TO 65
64        SINC=2.0*SINC
          N=0
65        CONTINUE
C         THE FOLLOWING ROUTINES SUBTRACT THE CONSTANT VALUE OF
C         VOLTAGE AND CURRENT (IF PRESENT) FROM ALL PRECEEDING
C         RESPECTIVE DATA POINTS, AND GENERATE THE REAL AXIS
C         TRANSFORMATION FOR THESE CONSTANTS AT EACH DESIRED FREQUENCY.
          IF(NVTM.LT.NT) GO TO 71
          VF=V(NT)
          DO 70 K=1,NT
          V(K)=V(K)-VF
70        CONTINUE
```

TABLE 1 (cont.)

```
C       THE FOLLOWING ROUTINE CALCULATES THE REAL AXIS PULSE
C       IMPEDANCE AND ADMITTANCE
        N=0
        DO 140 K=1,NS
        IF(COFS(K).EQ.0.0) GO TO 140
        ZOFS=VOFS(K)/COFS(K)
        SK=S(K)*TS
        SIV=1.0/SK
        SRSIV=SQRT(SIV)
        YS=1.0/ZOFS
        N=N+1
        WRITE(NU,150) SK,SIV,SRSIV,ZOFS,YS
        IF(N.EQ.14) GO TO 139
        GO TO 140
139     WRITE(NU,137)
137     FORMAT(2X)
        N=0
140     CONTINUE
150     FORMAT(1P3E12.3,1P2E12.5)
999     END
C       SUBROUTINE FOR REAL AXIS LAPLACE TRANSFORMATION
C       DIVIDE VERSION
        SUBROUTINE REAXTR(NT,T,X,S,ANS)
        DIMENSION T(100),X(100),FTS(100)
        DO 20 N=1,NT
        ST=S*T(N)
        IF(ST.GE.20.0) GO TO 10
        FTS(N)=X(N)*EXP(-ST)
        GO TO 20
10      FTS(N)=0.0
20      CONTINUE
        ANS=0.0
        DO 30 J=2,NT
        DT=T(J)-T(J-1)
        FTSF=FTS(J)
        FTSI=FTS(J-1)
        IF(FTSF.EQ.0.0.OR.FTSI.EQ.0.0) GO TO 25
        DFTS=FTSF-FTSI
        RFTS=FTSF/FTSI
        ALRFTS=ALOG(RFTS)
        FS=DT*DFTS/ALRFTS
        GO TO 27
25      FS=0.5*DT*(FTSF+FTSI)
27      ANS=ANS+FS
30      CONTINUE
        RETURN
        END
```

16-bit word length; memory cycle time of 1.65 $\mu$ sec; a remarkably good
set of basic machine instructions; and relatively powerful addressing
modes.

The general flow of real-axis impedance computation is as follows:
(1) data input which, for this case, is prepared on paper tape, is checked
for proper sequencing; (2) time points must, of course, be continuously increasing; (3) voltage and current are both allowed one maximum, which
generally covers the case for low-level potential steps including allowance
for nonfaradaic resistance compensation [61]; (4) in addition, both functions
may have a constant value at the end of their time-varying portions; (5) both
voltage and current are checked to assure that they are monotonic before and
after their respective maximum values.

It is to be emphasized that real-axis transformation operates upon the
appropriate time-domain functions from $t = 0$. This means that, since all
available pulse instrumentation does not instantaneously deliver the required
power to, e.g., charge the double layer, but starts at $t = 0$ with no signal
level, it is permissible and necessary to include in the data a zero-time
data point for which $V(t) = 0$, and $I(t) = 0$. In addition, examination of Fig. 6
indicates that the use of large $\sigma$ values heavily weights the short-time
behavior of $f(t)$ since $\exp(-\sigma t)$ is a very rapidly decaying function in this
case. As $\sigma$ becomes smaller, short-time data contribute relatively less
to the total transform. It is thus necessary to adequately describe the
experimental curve with correctly spaced data points. This is particularly
so for very-short-time data (e.g., along the rise of a potential step) for
which the signal is rapidly varying with time and, because of instrumental
nonlinearities, may require approximation by a large number of successive
exponential functions. It has been found that a 10-20% variation in signal
level between any two data points is usually sufficient to ensure that computation according to (46) results in minimum error.

Once the data have been verified, the quantities necessary to perform
calculations according to the algorithm in (46) for the desired frequency
range are generated. Since the same calculations are performed for both
$V(t)$ and $I(t)$, the summation in (46) was programmed in the form of a subroutine which returns the partial values of either $V(\sigma)$ or $I(\sigma)$ to the main
program, to each of which the transform of their constant values $[f(t_\infty)/\sigma]$
is added. Both the real-axis impedance $Z(S)$ and admittance $Y(S)$ are then
evaluated and presented as the output of this program, an example of which
is shown in Table 2. In addition to these values, the real-axis frequency S,
its reciprocal $1/S$, and the square root of its reciprocal $SR(1/S)$ are given.
These quantities were chosen to enable both double layer and diffusion behavior to be readily diagnosed [see Eqs. (11) and (41)]. The results given
in Table 2, which is an actual computer readout, were obtained for a series
$R_e C_d$ test circuit for which $R_e = 0.5\ \Omega$ and $C_d = 0.1\ \mu F$, and are presented
for the highest frequency range attainable with current instrumentation (to

## TABLE 2

### Laplace Transformation Real Axis Test Data

| S | 1/S | SR(1/S) | Z(S) | Y(S) |
|---|-----|---------|------|------|
| 1.00E+05 | 1.00E-05 | 3.16E-03 | 1.0067E+02 | 9.9328E-03 |
| 1.20E+05 | 8.33E-06 | 2.88E-03 | 8.3980E+01 | 1.1907E-02 |
| 1.40E+05 | 7.14E-06 | 2.67E-03 | 7.2055E+01 | 1.3878E-02 |
| 1.60E+05 | 6.25E-06 | 2.50E-03 | 6.3111E+01 | 1.5845E-02 |
| 1.80E+05 | 5.55E-06 | 2.35E-03 | 5.6154E+01 | 1.7807E-02 |
| 2.00E+05 | 5.00E-06 | 2.23E-03 | 5.0589E+01 | 1.9767E-02 |
| 2.50E+05 | 4.00E-06 | 2.00E-03 | 4.0571E+01 | 2.4647E-02 |
| 3.00E+05 | 3.33E-06 | 1.82E-03 | 3.3893E+01 | 2.9504E-02 |
| 4.00E+05 | 2.50E-06 | 1.58E-03 | 2.5545E+01 | 3.9145E-02 |
| 5.00E+05 | 2.00E-06 | 1.41E-03 | 2.0536E+01 | 4.8692E-02 |
| 6.00E+05 | 1.66E-06 | 1.29E-03 | 1.7197E+01 | 5.8146E-02 |
| 7.00E+05 | 1.42E-06 | 1.19E-03 | 1.4812E+01 | 6.7509E-02 |
| 8.00E+05 | 1.25E-06 | 1.11E-03 | 1.3023E+01 | 7.6782E-02 |
| 9.00E+05 | 1.11E-06 | 1.05E-03 | 1.1632E+01 | 8.5965E-02 |
| 1.00E+06 | 1.00E-06 | 1.00E-03 | 1.0519E+01 | 9.5061E-02 |
| 1.20E+06 | 8.33E-07 | 9.12E-04 | 8.8499E+00 | 1.1299E-01 |
| 1.40E+06 | 7.14E-07 | 8.45E-04 | 7.6573E+00 | 1.3059E-01 |
| 1.60E+06 | 6.25E-07 | 7.90E-04 | 6.7629E+00 | 1.4786E-01 |
| 1.80E+06 | 5.55E-07 | 7.45E-04 | 6.0673E+00 | 1.6481E-01 |
| 2.00E+06 | 5.00E-07 | 7.07E-04 | 5.5108E+00 | 1.8146E-01 |
| 2.50E+06 | 4.00E-07 | 6.32E-04 | 4.5090E+00 | 2.2177E-01 |
| 3.00E+06 | 3.33E-07 | 5.77E-04 | 3.8412E+00 | 2.6033E-01 |
| 4.00E+06 | 2.50E-07 | 5.00E-04 | 3.0065E+00 | 3.3261E-01 |
| 5.00E+06 | 2.00E-07 | 4.47E-04 | 2.5056E+00 | 3.9909E-01 |
| 6.00E+06 | 1.66E-07 | 4.08E-04 | 2.1717E+00 | 4.6045E-01 |
| 7.00E+06 | 1.42E-07 | 3.77E-04 | 1.9332E+00 | 5.1726E-01 |
| 8.00E+06 | 1.25E-07 | 3.53E-04 | 1.7544E+00 | 5.6997E-01 |
| 9.00E+06 | 1.11E-07 | 3.33E-04 | 1.6153E+00 | 6.1906E-01 |
| 1.00E+07 | 1.00E-07 | 3.16E-04 | 1.5040E+00 | 6.6486E-01 |
| 1.20E+07 | 8.33E-08 | 2.88E-04 | 1.3371E+00 | 7.4786E-01 |
| 1.40E+07 | 7.14E-08 | 2.67E-04 | 1.2179E+00 | 8.2105E-01 |
| 1.60E+07 | 6.25E-08 | 2.50E-04 | 1.1285E+00 | 8.8609E-01 |
| 1.80E+07 | 5.55E-08 | 2.35E-04 | 1.0590E+00 | 9.4427E-01 |
| 2.00E+07 | 5.00E-08 | 2.23E-04 | 1.0034E+00 | 9.9660E-01 |
| 2.50E+07 | 4.00E-08 | 2.00E-04 | 9.0331E-01 | 1.1070E+00 |
| 3.00E+07 | 3.33E-08 | 1.82E-04 | 8.3663E-01 | 1.1952E+00 |
| 4.00E+07 | 2.50E-08 | 1.58E-04 | 7.5329E-01 | 1.3274E+00 |
| 5.00E+07 | 2.00E-08 | 1.41E-04 | 7.0332E-01 | 1.4213E+00 |
| 6.00E+07 | 1.66E-08 | 1.29E-04 | 6.7003E-01 | 1.4924E+00 |
| 7.00E+07 | 1.42E-08 | 1.19E-04 | 6.4625E-01 | 1.5473E+00 |
| 8.00E+07 | 1.25E-08 | 1.11E-04 | 6.2834E-01 | 1.5914E+00 |
| 9.00E+07 | 1.11E-08 | 1.05E-04 | 6.1454E-01 | 1.6272E+00 |

$10^8$ rad/sec). Inspection of the results shows that the maximum error due to the computational procedure occurs in the range $10^7 \leq \sigma \leq 10^8$, attaining $\pm 1\%$ at $10^8$ rad/sec.

In the case of imaginary-axis transformation, it will be recalled that the following integration is performed:

$$F(j\omega) = \int_0^\infty f(t) \exp(-j\omega t)\, dt. \tag{47}$$

This operation can be carried out in a numerical fashion by rewriting (47) as

$$F(j\omega) = \int_0^\infty f(t) \cos(\omega t)\, dt - j \int_0^\infty f(t) \sin(\omega t)\, dt \tag{48}$$

in which $F(j\omega)$ is expressed in terms of its real and imaginary parts. It can be seen that for both cases, multiplication of the original time function by a trigonometric function is involved. This is illustrated schematically in Fig. 7 where f(t) is, as for the previous case, the typical current response to a potential step input. This is multiplied, from t = 0 to the time at which i → 0, by a sine or cosine function of given frequency $\omega$. The area under the resulting curve is, as in real-axis transformation, equal to the appropriate integral in (48). As noted earlier, (48) will not converge unless f(t) actually becomes zero as t → ∞. These integrations, therefore, may not be performed numerically unless the constant portion of f(t), if it exists, is subtracted from the remainder of the experimental curve.

Examination of Fig. 7 shows that, as for the case of real-axis transformation, the use of large $\omega$ values emphasizes short-time system response. This is so since the major contribution to the product $f(t) \cdot \sin(\omega t)$ for large

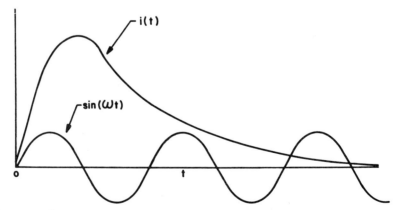

Fig. 7.  Schematic representation of imaginary-axis (s = $j\omega$) Laplace transformation.

$\omega$ occurs only in those time intervals for which f(t) changes significantly over one frequency cycle. This means essentially that only small time constants will be observable for large $\omega$, as could be expected. Correspondingly, as $\omega$ decreases, larger time constants (i. e., longer-time data) contribute relatively more to (48). It is, therefore, equally necessary for imaginary-axis transformation to obtain correctly spaced data points.

The numerical integrations corresponding to (48) are performed by again assuming exponential behavior between two successive data points for V(t) or I(t). Thus, (43) and (44) are employed and give, for each segment of the real part of F(j$\omega$),

$$\text{Re}_i \left[ F(j\omega) \right] = \int_{t_i}^{t_{i+1}} A \exp(-at) \cos(\omega t) \, dt \tag{49}$$

which is performed for each segment of f(t) and for all desired $\omega$ values, until the total real part is generated by the addition of the transform of each segment. The total algorithm for this summation is

$$\text{Re}\left[F(j\omega)\right] = \sum_i \frac{1}{a^2 + \omega^2} \{ f(t_{i+1}) [a \cos(\omega t_{i+1}) + \omega \sin(\omega t_{i+1})]$$

$$- f(t_i) \, a \cos(\omega t_i) + \omega \sin(\omega t_i)] \} \tag{50}$$

where

$$a = \frac{\ln[ f(t_{i+1})/f(t_i)]}{[t_{i+1} - t_i]}. \tag{51}$$

The imaginary part of F(j$\omega$) is generated in a similar manner. Thus for each segment,

$$\text{Im}_i(F(j\omega)] = - \int_{t_i}^{t_{i+1}} A \exp(-at) \sin(\omega t) \, dt \tag{52}$$

which is utilized to obtain the total imaginary part according to the following algorithm:

$$\text{Im}[F(j\omega)] = -\sum_i \frac{1}{a^2 + \omega^2} \{ f(t_{i+1})[a \sin(\omega t_{i+1}) - \omega \cos(\omega t_{i+1})]$$

$$- f(t_i)[a \sin(\omega t_i) - \omega \cos(\omega t_i)] \} - \frac{f(t_\infty)}{\omega} \tag{53}$$

where a is given by (51). Note that the imaginary-axis transform of the constant portion of f(t) is, as stated earlier, given by $f(t_\infty)/j\omega$ which is then appropriately added to the result of the summation in (53) to generate the complete Im $[F(j\omega)]$. Equations (50) and (53) are employed to numerically

TABLE 3

Digital Computer Program for Imaginary Axis Laplace Transformation

```
C       IMAGINARY AXIS LAPLACE TRANSFORMATION FOR POTENTIOSTATIC
C       PULSE IMPEDANCE ELECTROCHEMICAL STUDIES
        DIMENSION T(100),C(100),V(100),W(100),B(36),SINWT(100)
        DIMENSION COSWT(100),ZR(100),ZI(100)
        READ(2,10) WI,WF,NU
C       WI AND WF ARE THE DESIRED INITIAL AND FINAL IMAGINARY AXIS
C       FREQUENCY VALUES RESPECTIVELY; BOTH MUST BE THE START OF
C       A DECADE.
C       NU IS THE DEVICE NUMBER FOR THE DESIRED MODE OF DATA
C       OUTPUT(TELETYPE,HIGH SPEED PUNCH,LINE PRINTER,OR
C       MAGNETIC TAPE).
10      FORMAT(2E12.5,I3)
        READ(5,11) (B(M),M=1,36)
11      FORMAT(36A2)
        READ(5,12) NT,NCTM,NVTM
C       NT IS THE TOTAL NUMBER OF TIME DATA POINTS.
C       NCTM IS THE NUMBER OF CURRENT DATA POINTS UP TO AND
C       INCLUDING IT'S MAXIMUM VALUE.
C       NVTM IS THE NUMBER OF VOLTAGE DATA POINTS UP TO AND
C       INCLUDING IT'S FIRST CONSTANT VALUE.
12      FORMAT(3I5)
        IF(NVTM.LT.NT) GO TO 20
        READ(5,15)(T(K),C(K),V(K),K=1,NT)
15      FORMAT(3E12.5)
        GO TO 23
20      READ(5,21)(T(K),C(K),V(K),K=1,NVTM)
21      FORMAT(3E12.5)
        N=NVTM+1
        READ(5,22) (T(K),C(K),K=N,NT)
22      FORMAT(2E12.5)
C       THE FOLLOWING SERIES OF TESTS VERIFY PROPER SEQUENCING
C       FOR TIME, VOLTAGE AND CURRENT DATA INPUT. ONE MAXIMUM
C       IS ALLOWED FOR BOTH CURRENT AND VOLTAGE THUS LIMITING
C       THIS PROGRAM TO USE WITH LOW LEVEL POTENTIAL STEPS.
23      DO 25 K=2,NT
        IF(T(K).GT.T(K-1)) GO TO 25
        WRITE(1,24)
24      FORMAT(65H GI=GO:TIME INCREMENT  ERROR'/)
        GO TO 999
25      CONTINUE
        DO 35 K=2,NCTM
        IF(C(K).GT.C(K-1)) GO TO 35
        WRITE(1,31)
31      FORMAT(65H GI=GO:INCREASING CURRENT INCREMENT ERROR'/)
        GO TO 999
35      CONTINUE
        DO 40 K=2,NVTM
        IF(V(K).GT.V(K-1))GO TO 40
        WRITE(1,37)
37      FORMAT(65H GI=GO:INCREASING VOLTAGE INCREMENT ERROR'/)
```

TABLE 3 (cont.)

```
        GO TO 999
40      CONTINUE
        J=NCTM+1
        DO 50 K=J,NT
        IF(C(K).LT.C(K-1)) GO TO 50
        WRITE(1,46)
46      FORMAT(65H GI=GO:DECREASING CURRENT INCREMENT ERROR'/)
        GO TO 999
50      CONTINUE
        IF(NVTM.LT.NT) GO TO 57
        J=NVTM+1
        DO 57 K=J,NT
        IF(V(K).LT.V(K-1)) GO TO 57
        WRITE(1,56)
56      FORMAT(65H GI=GO:DECREASING VOLTAGE INCREMENT ERROR'/)
        GO TO 999
57      CONTINUE
        WRITE(1,61)
61      FORMAT(65H DATA INPUT CORRECT,CALCULATION PROCEEDING'/)
        TS=1.0/T(2)
        DO 300 K=1,NT
        T(K)=T(K)*TS
300     CONTINUE
C       THE FOLLOWING ROUTINE GENERATES THE DESIRED IMAGINARY AXIS
C       FREQUENCY VALUES.
        WI=WI*1.0/TS
        WF=WF*1.0/TS
        W(1)=WI
        A=ALOG10(WF/W(1))
        I=A+0.5
        NW=I*14
        N=0
        WINC=0.2*W(1)
        DO 65 K=2,NW
        W(K)=W(K-1)+WINC
        N=N+1
        IF(N.EQ.5)GO TO 62
        IF(N.EQ.7)GO TO 63
        IF(N.EQ.14)GO TO 64
        GO TO 65
62      WINC=2.5*WINC
        GO TO 65
63      WINC=2.0*WINC
        GO TO 65
64      WINC=2.0*WINC
        N=0
65      CONTINUE
        IF(NVTM.LT.NT) GO TO 71
        VF=V(NT)
        DO 70 K=1,NT
        V(K)=V(K)-VF
```

TABLE 3 (Cont.)

```
70      CONTINUE
        GO TO 76
71      VF=V(NVTM)
        DO 76 K=1,NVTM
        V(K)=V(K)-VF
76      CONTINUE
        CF=C(NT)
        IF(CF.EQ.0.0) GO TO 78
        DO 78 K=1,NT
        C(K)=C(K)-CF
78      CONTINUE
        NTA=NT-1
        DO 125 K=1,NW
        DO 104 M=2,NTA
        WT=W(K)*T(M)
        SINWT(M)=SIN(WT)
        COSWT(M)=COS(WT)
104     CONTINUE
        CALL WAXTR(NT,T,C,W(K),SINWT,COSWT,CR,CI)
        IF(CF.EQ.0.0) GO TO 105
        COFWI=-CF/W(K)
        CI=CI+COFWI
105     IF(NVTM.LT.NT) GO TO 110
        CALL WAXTR(NT,T,V,W(K),SINWT,COSWT,VR,VI)
        GO TO 120
110     CALL WAXTR(NVTM,T,V,W(K),SINWT,COSWT,VR,VI)
120     VOFWI=-VF/W(K)
        VI=VI+VOFWI
        CCW=(CR*CR)+(CI*CI)
        ZR(K)=((VR*CR)+(VI*CI))/CCW
        ZI(K)=((CR*VI)-(VR*CI))/CCW
125     CONTINUE
        WRITE(NU,128)
128     FORMAT(////////////)
        WRITE(NU,130) (B(M),M=1,36)
130     FORMAT(36A2//)
        WRITE(NU,135)
135     FORMAT(5X,1HW,8X,3H1/W,4X,7HSR(1/W),5X,5HRE(Z),7X,5HIM(Z),
       1 7X,4HTAN0/)
        N=0
        DO 140 K=1,NW
        ZRK=ZR(K)
        ZIK=ZI(K)
        TAN0=ZIK/ZRK
        WK=W(K)*TS
        WIV=1.0/WK
        SRWIV=SQRT(WIV)
        N=N+1
        WRITE(NU,150) WK,WIV,SRWIV,ZRK,ZIK,TAN0
        IF(N.EQ.14) GO TO 139
        GO TO 140
```

TABLE 3 (Cont.)

```
139       WRITE(NU,137)
137       FORMAT(2X)
          N=0
140       CONTINUE
150       FORMAT(1PE9.2,1P2E10.3,1P4E12.5)
999       END
C         SUBROUTINE FOR IMAGINARY AXIS LAPLACE TRANSFORMATION
          SUBROUTINE WAXTR(NT,T,X,W,SINWT,COSWT,ANSR,ANSI)
          DIMENSION T(100),X(100),SINWT(100),COSWT(100)
          NTA=NT-1
          WT2=W*T(2)
          WTNT=W*T(NT)
          WTNTA=W*T(NTA)
          BNT=(X(NT)-X(NTA))/(W*(T(NT)-T(NTA)))
          WAIMNT=(BNT*SIN(WTNT))-(BNT*WTNT*COS(WTNT))-(BNT*SIN(WTNTA))
     1    +(BNT*WTNTA*COS(WTNTA))
          WARENT=(BNT*COS(WTNT))+(BNT*WTNT*SIN(WTNT))-(BNT*COS(WTNTA))
     1    -(BNT*WTNTA*SIN(WTNTA))
          B1=(X(2)-X(1))/(W*T(2))
          WAIMTI=(B1*SIN(WT2))-(B1*WT2*COS(WT2))
          WARETI=(B1*COS(WT2))+(B1*WT2*SIN(WT2))-B1
          ANSR=0.0
          ANSI=0.0
          WAIMNT=-WAIMNT
          WAIMTI=-WAIMTI
          ANSR=WARENT+WARETI
          ANSI=WAIMNT+WAIMTI
          W2=W*W
          DO 10 K=3,NTA
          XF=X(K)
          XI=X(K-1)
          A=ALOG(XF/XI)/(T(K)-T(K-1))
          A2=A*A
          SINF=SINWT(K)
          SINI=SINWT(K-1)
          COSF=COSWT(K)
          COSI=COSWT(K-1)
          AWI=1.0/(W2+A2)
          WAIM=(AWI*XF*A*SINF)-(AWI*XF*W*COSF)-(AWI*A*XI*SINI)+
     1    (AWI*XI*W*COSI)
          WARE=(AWI*XF*A*COSF)+(AWI*XF*W*SINF)-(AWI*XI*A*COSI)-
     1    (AWI*XI*W*SINI)
          WAIM=-WAIM
          ANSR=ANSR+WARE
          ANSI=ANSI+WAIM
10        CONTINUE
          RETURN
          END
```

TABLE 4

Laplace Transformation Imaginary Axis Test Data

| W | 1/W | SR(1/W) | RE(Z) | IM(Z) | TANØ |
|---|---|---|---|---|---|
| 1.0E+05 | 1.00E-05 | 3.16E-03 | 5.0297E-01 | -1.0017E+02 | -1.9916E+02 |
| 1.2E+05 | 8.33E-06 | 2.88E-03 | 5.0297E-01 | -8.3480E+01 | -1.6597E+02 |
| 1.4E+05 | 7.14E-06 | 2.67E-03 | 5.0297E-01 | -7.1555E+01 | -1.4226E+02 |
| 1.6E+05 | 6.25E-06 | 2.50E-03 | 5.0297E-01 | -6.2610E+01 | -1.2448E+02 |
| 1.8E+05 | 5.55E-06 | 2.35E-03 | 5.0297E-01 | -5.5653E+01 | -1.1064E+02 |
| 2.0E+05 | 5.00E-06 | 2.23E-03 | 5.0297E-01 | -5.0088E+01 | -9.9584E+01 |
| 2.5E+05 | 4.00E-06 | 2.00E-03 | 5.0297E-01 | -4.0070E+01 | -7.9667E+01 |
| 3.0E+05 | 3.33E-06 | 1.82E-03 | 5.0297E-01 | -3.3392E+01 | -6.6389E+01 |
| 4.0E+05 | 2.50E-06 | 1.58E-03 | 5.0297E-01 | -2.5044E+01 | -4.9792E+01 |
| 5.0E+05 | 2.00E-06 | 1.41E-03 | 5.0297E-01 | -2.0035E+01 | -3.9833E+01 |
| 6.0E+05 | 1.66E-06 | 1.29E-03 | 5.0297E-01 | -1.6696E+01 | -3.3194E+01 |
| 7.0E+05 | 1.42E-06 | 1.19E-03 | 5.0297E-01 | -1.4310E+01 | -2.8452E+01 |
| 8.0E+05 | 1.25E-06 | 1.11E-03 | 5.0297E-01 | -1.2522E+01 | -2.4895E+01 |
| 9.0E+05 | 1.11E-06 | 1.05E-03 | 5.0297E-01 | -1.1130E+01 | -2.2129E+01 |
|  |  |  |  |  |  |
| 1.0E+06 | 1.00E-06 | 1.00E-03 | 5.0297E-01 | -1.0017E+01 | -1.9916E+01 |
| 1.2E+06 | 8.33E-07 | 9.12E-04 | 5.0297E-01 | -8.3479E+00 | -1.6597E+01 |
| 1.4E+06 | 7.14E-07 | 8.45E-04 | 5.0298E-01 | -7.1553E+00 | -1.4225E+01 |
| 1.6E+06 | 6.25E-07 | 7.90E-04 | 5.0297E-01 | -6.2608E+00 | -1.2447E+01 |
| 1.8E+06 | 5.55E-07 | 7.45E-04 | 5.0298E-01 | -5.5652E+00 | -1.1064E+01 |
| 2.0E+06 | 5.00E-07 | 7.07E-04 | 5.0298E-01 | -5.0086E+00 | -9.9578E+00 |
| 2.5E+06 | 4.00E-07 | 6.32E-04 | 5.0298E-01 | -4.0068E+00 | -7.9660E+00 |
| 3.0E+06 | 3.33E-07 | 5.77E-04 | 5.0299E-01 | -3.3389E+00 | -6.6380E+00 |
| 4.0E+06 | 2.50E-07 | 5.00E-04 | 5.0300E-01 | -2.5040E+00 | -4.9781E+00 |
| 5.0E+06 | 2.00E-07 | 4.47E-04 | 5.0301E-01 | -2.0030E+00 | -3.9820E+00 |
| 6.0E+06 | 1.66E-07 | 4.08E-04 | 5.0302E-01 | -1.6690E+00 | -3.3179E+00 |
| 7.0E+06 | 1.42E-07 | 3.77E-04 | 5.0303E-01 | -1.4304E+00 | -2.8435E+00 |
| 8.0E+06 | 1.25E-07 | 3.53E-04 | 5.0304E-01 | -1.2514E+00 | -2.4877E+00 |
| 9.0E+06 | 1.11E-07 | 3.33E-04 | 5.0304E-01 | -1.1122E+00 | -2.2109E+00 |
|  |  |  |  |  |  |
| 1.0E+07 | 1.00E-07 | 3.16E-04 | 5.0304E-01 | -1.0008E+00 | -1.9895E+00 |
| 1.2E+07 | 8.33E-08 | 2.88E-04 | 5.0304E-01 | -8.3368E-01 | -1.6572E+00 |
| 1.4E+07 | 7.14E-08 | 2.67E-04 | 5.0305E-01 | -7.1424E-01 | -1.4198E+00 |
| 1.6E+07 | 6.25E-08 | 2.50E-04 | 5.0306E-01 | -6.2462E-01 | -1.2416E+00 |
| 1.8E+07 | 5.55E-08 | 2.35E-04 | 5.0307E-01 | -5.5489E-01 | -1.1030E+00 |
| 2.0E+07 | 5.00E-08 | 2.23E-04 | 5.0307E-01 | -4.9909E-01 | -9.9207E-01 |
| 2.5E+07 | 4.00E-08 | 2.00E-04 | 5.0303E-01 | -3.9855E-01 | -7.9230E-01 |
| 3.0E+07 | 3.33E-08 | 1.82E-04 | 5.0294E-01 | -3.3143E-01 | -6.5898E-01 |
| 4.0E+07 | 2.50E-08 | 1.58E-04 | 5.0252E-01 | -2.4720E-01 | -4.9191E-01 |
| 5.0E+07 | 2.00E-08 | 1.41E-04 | 5.0203E-01 | -1.9605E-01 | -3.9052E-01 |
| 6.0E+07 | 1.66E-08 | 1.29E-04 | 5.0183E-01 | -1.6150E-01 | -3.2182E-01 |
| 7.0E+07 | 1.42E-08 | 1.19E-04 | 5.0182E-01 | -1.3681E-01 | -2.7262E-01 |
| 8.0E+07 | 1.25E-08 | 1.11E-04 | 5.0157E-01 | -1.1841E-01 | -2.3607E-01 |
| 9.0E+07 | 1.11E-08 | 1.05E-04 | 5.0099E-01 | -1.0395E-01 | -2.0749E-01 |

evaluate $V(j\omega)$ and $I(j\omega)$ from which $Z(j\omega)$ can then be obtained by their appropriate division.

The digital computer program utilized to carry out the above calculations when potentiostatic techniques are employed is reproduced in Table 3. The general flow of the program is essentially identical to that employed for real-axis transformation with the major difference being in the subroutine which is employed to perform the summations in (50) and (53). These subroutines return $Re[V(j\omega)]$, $Re[I(j\omega)]$, $Im[V(j\omega)]$, and $Im[I(j\omega)]$ to the main program whereupon RE(Z) and IM(Z) are evaluated and form part of the output of this program. In addition to these quantities, imaginary-axis frequency W, its reciprocal 1/W, the square root of its reciprocal SR(1/W), and the tangent of the phase angle TAN$\varphi$ also are given. The program was tested on the same series $R_eC_d$ test circuit ($R_e = 0.5\ \Omega$, $C_d = 0.1\ \mu F$) as was used with real-axis transformation. The actual computer printout is given in Table 4. The results show that maximum computational error occurs in the range $10^7 \leq \omega \leq 10^8$ attaining $\pm 4\%$ for IM(Z) and $<1\%$ for RE(Z) at $10^8$ rad/sec.

It can be seen that both programs contribute very little error in terms of the computational procedure involved in real- and imaginary-axis Laplace transformation. In particular, the exponential approximation appears to be a reasonable choice for typical electrochemical systems provided that linear conditions are maintained. The programs are speedy, taking approximately 2 min for 7 decades of frequency for $s = \sigma$ and 5 min for $\sigma = j\omega$ on the EAI 640 computer employed in this study. As mentioned earlier, this is of the midicomputer variety normally employed in real-time experimental work. Data for studies utilizing the transform technique are presently obtained by observation with an oscilloscope with subsequent transcription to cards. Work is presently underway in this laboratory to directly record data in digital fashion. This will be performed using very fast sample and hold modules to record appropriately spaced initial time points, e.g., along the rise of a potential step which normally occurs within $10^{-7}$ to $5 \times 10^{-7}$ sec for the fastest potentiostats available. Longer-time data will be recorded at computer cycle time (1.65 $\mu$ sec) after which the sample and hold modules will be appropriately read into memory.

## VIII. EXPERIMENTAL VERIFICATION OF THE TRANSIENT IMPEDANCE

### TECHNIQUE

In order to more fully test the transformation technique, the equivalent circuit shown in Fig. 8 was examined. This circuit represents an electrode exhibiting double-layer behavior along with charge transfer and adsorption [56], i.e., a system with two time constants. The circuit components chosen were: $R_e = 5\ \Omega$, $C_d = 0.15\ \mu F$, $R_t = 50\ \Omega$, $C_a = 1.5\ \mu F$, and $R_a = 10\ k\Omega$. The value of $R_a$ was chosen to be relatively large in order to

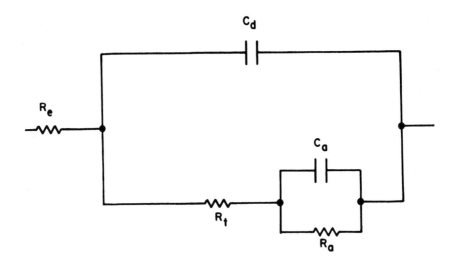

Fig. 8. Aperiodic equivalent circuit for an electrode at which the faradaic process is adsorption-coupled charge transfer.

represent irreversible adsorption for which $R_a$ can then be essentially neglected. An example of this is the hydrogen-platinum system at potentials anodic to the RHE which has recently been studied in this laboratory with the aid of the transient impedance method [64].

The total impedance of this system in terms of its equivalent circuit (Fig. 8) is given, neglecting $R_a$, by

$$Z(s) = R_e + \frac{R_t C_a s + 1}{R_t C_a C_d s^2 + (C_d + C_a)s} \tag{54}$$

which becomes, at sufficiently high frequencies,

$$Z(s) = R_e + \frac{1}{C_d s} \tag{55}$$

thus illustrating that, for this case, there is a frequency range over which only double-layer behavior will be observed. At sufficiently low frequencies, Eq. (54) becomes

$$Z(s) = R_e + \frac{R_t C_a}{C_d + C_a} + \frac{1}{(C_d + C_a)s}. \tag{56}$$

Examination of Eqs. (55) and (56) indicates that there are two frequency ranges over which both $Z(\sigma)$ and $\text{Im}[Z(j\omega)]$ exhibit inverse frequency

behavior. It is thus necessary to examine a wide frequency range for unambiguous study of this system.

For the actual measurements, the general potentiostat circuit shown in
Fig. 3, where the components of $Z_E$ are shown in Fig. 8, was employed.
The potentiostat A (Fig. 3) is the ultrafast-rise Tacussel Model PIT-20-2X
provided with external phase compensation to correctly match the potentiostat with the electrochemical cell for fast, smooth, and easily measurable
response. The step voltage source V(s) (Fig. 3) is the Monsanto 300A pulse
generator which can be correctly mated to the input of the potentiostat
through the use of its variable rise and fall times. All voltage and current
time functions were measured on a Tektronix 556 dual beam oscilloscope
equipped with two 1A5 preamplifiers. Active probes, Tektronix P6045 or
P6046 probes, were used exclusively since the signal is detected at the
probe tip at relatively high impedance allowing faithful signal reproduction
at rise times as short as 3 nsec, effectively eliminating the influence of
cable length. Identical probes were utilized so that correct signal matching
could be maintained. These probes could be matched to better than ±5% by

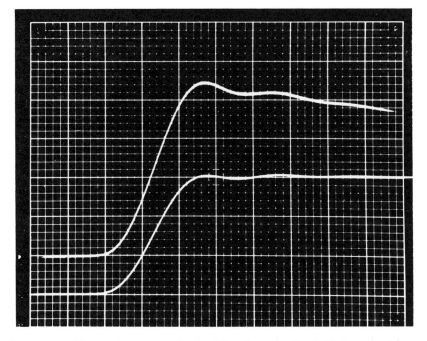

Fig. 9. Oscilloscopic curves obtained for the potentiostatic transient impedance technique. The time scale is 50 nsec per major horizontal division;
lower curve is voltage at 2 mV per major division; and upper curve is
current at 7 mA per major division.

appropriate peaking of the high-frequency circuits within the respective probe amplifiers to obtain as closely identical pulse shapes from the same signal source as possible. In this way, nonlinearities which may be present in signal measurements are nearly identical for both current and voltage and cancel out of the resulting impedance expression. The photograph shown in Fig. 9 illustrates the oscilloscopic recording of current and voltage time traces at 50 nsec per major horizontal division for the circuit given above. It can be seen that an initial voltage or current point at 10 nsec is readily accessible. This is sufficient, along with adequate point spacing, to enable frequencies as high as $10^8$ rad/sec to be examined.

Using data obtained from the above experiments, $Z(\sigma)$ was evaluated numerically and diagnostic plots of $Z(\sigma)$ vs. $1/\sigma$ are shown in Figs. 10-12. Overall behavior is shown in Fig. 10 in the frequency range $10^6$ to $10^4$ rad/sec which covers the crossover region between only double-layer behavior and combined double-layer and faradaic behavior.

It can already be seen that there is some indication that the system may be diagnosed according to either Eq. (55) or (56); however, linearity in $1/\sigma$,

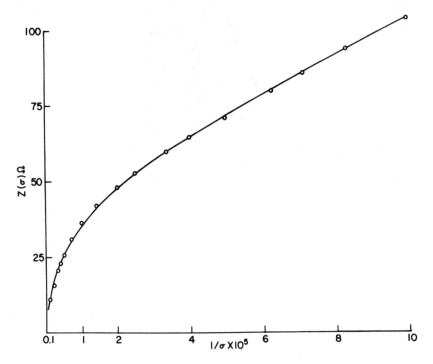

Fig. 10. Over-all real-axis impedance behavior for electrode represented by circuit in Fig. 8.

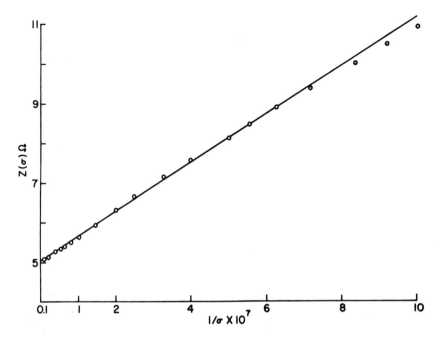

Fig. 11. Real-axis impedance behavior at high frequencies for electrode represented by circuit in Fig. 8.

as presented, is too short in either region to be unambiguous. Figure 11, therefore, shows the high-frequency behavior of this system over the range $10^8$ to $10^6$ rad/sec. The plot is linear over nearly this complete frequency range and allows the evaluation of $R_e$ and $C_d$ which were within ±5% of their known values. Figure 12 shows the low-frequency behavior of this circuit from $10^4$ to $10^2$ rad/sec. As can be seen, this plot is again linear in $1/\sigma$ over this frequency range. The values of $R_t$ and $C_a$ calculated from this behavior were within ±5% of their known values. Essentially similar results were obtained from $Z(j\omega)$.

It is to be noted that the time constants of this example were chosen so that double-layer behavior can be observed only if relatively high frequencies can be obtained. An electrochemical system exhibiting this behavior is the platinum-hydrogen case [16, 64, 65]. Linearity of $Z(\sigma)$ vs. $1/\sigma$ in the relatively low frequency range begins at approximately $10^5$ rad/sec, normally a high frequency for most impedance studies. Since this plot is a good diagnostic for double-layer behavior, it might be assumed that this system did not exhibit adsorption properties, if relatively higher frequencies were not examined. This then illustrates the importance of making frequency studies over as wide a frequency range as possible, especially if adsorption is suspected.

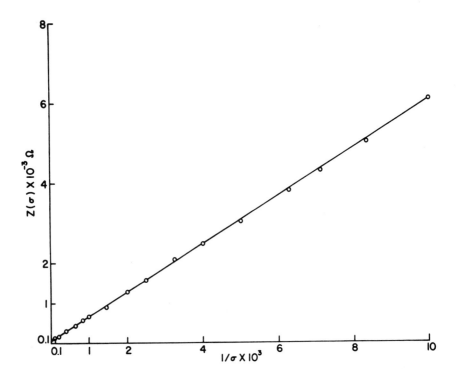

Fig. 12.  Real-axis impedance behavior at low frequencies for electrode represented by circuit in Fig. 8.

## REFERENCES

[1].    P. Delahay, in Advances in Electrochemistry and Electrochemical Engineering, vol. 1 (P. Delahay, ed.), Wiley-Interscience, New York, 1961.

[2].    W. H. Reinmuth, Anal. Chem., 36:211 (1964); 38:270R (1966).

[3].    A. A. Pilla, Bull. Soc. Franc. Elec., 4:24 (1963).

[4].    H. Gerischer and W. Vielstich, Z. Physik. Chem., 3:16 (1955).

[5].    W. Vielstich and H. Gerischer, Z. Physik. Chem., 4:10 (1955).

[6].    W. Vielstich and P. Delahay, J. Am. Chem. Soc., 79:1874 (1957).

[7].    F. C. Anson, Anal. Chem., 38:54 (1966).

[8].    H. A. Laitinen, R. P. Tischer, and D. K. Roe, J. Electrochem. Soc., 107:546 (1960).

[9]. R. S. Nicholson and I. Shain, Anal. Chem., 36:705 (1964); 37:178, 190 (1965).

[10]. R. S. Nicholson, Anal. Chem., 37:667, 1351 (1965).

[11]. J. M. Saveant and E. Vianello, Electrochim. Acta, 8:905 (1963); 12:629 (1967).

[12]. R. W. Murray and C. N. Reilley, J. Electroanal. Chem., 3:182 (1962).

[13]. H. B. Herman and A. J. Bard, J. Phys. Chem., 70:396 (1966).

[14]. T. Berzins and P. Delahay, J. Am. Chem. Soc., 77:6448 (1955).

[15]. M. Bonnemay, G. Bronoel, E. Levart, and A. A. Pilla, J. Electroanal. Chem., 13:44 (1967).

[16]. M. Rosen, D. R. Flinn, and S. Schuldiner, J. Electrochem. Soc., 116:1112 (1969).

[17]. R. L. Birke and D. K. Roe, Anal. Chem., 37:450, 455 (1965).

[18]. G. C. Barker, in Transactions of the Symposium on Electrode Processes (E. Yeager, ed.), J. Wiley, New York 1961; Pure Appl. Chem., 15:239 (1967).

[19]. P. Delahay, J. Phys. Chem., 16:2204 (1962).

[20]. W. H. Reinmuth and C. E. Wilson, Anal. Chem., 34:1159 (1962).

[21]. F. C. Anson, Anal. Chem., 38:1924 (1966).

[22]. S. Schuldiner and C. H. Presbrey, J. Electrochem. Soc., 111:457 (1964).

[23]. W. D. Weir and C. G. Enke, J. Phys. Chem., 71:275, 280 (1967).

[24]. G. Lauer, R. Abel and F. C. Anson, Anal. Chem., 39:765 (1967).

[25]. D. J. Kooijman and J. H. Sluyters, J. Electroanal. Chem., 13:152 (1967).

[26]. D. J. Kooijman, J. Electroanal. Chem., 18:81 (1968).

[27]. K. J. Oldham and R. A. Osteryoung, J. Electroanal. Chem., 11:397 (1966).

[28]. S. W. Feldberg, in Electroanalytical Chemistry, Vol. 3 (A. J. Bard, ed.), Marcel Dekker, New York, 1969.

[29]. K. B. Prater and A. J. Bard, J. Electrochem. Soc., 117:207, 355 (1970).

[30]. B. Timmer, M. Sluyters-Rehbach, and J. H. Sluyters, J. Electroanal. Chem., 15:343 (1967).

[31]. G. C. Barker and F. A. Bolzan, Z. Anal. Chem., 216:215 (1966).

[32]. F. C. Anson, J. H. Christie, and R. A. Osteryoung, J. Electroanal. Chem., 13:343 (1967).

[33]. F. C. Anson and D. J. Barclay, Anal. Chem., 40:1791 (1968).

[34]. D. J. Barclay and F. C. Anson, J. Electrochem. Soc., 116:438 (1969).

[35]. Z. Kowalski and F. C. Anson, J. Electrochem. Soc., 116:1208 (1969).

[36]. D. M. Mohilner, in Electroanalytical Chemistry, Vol. 1 (A. J. Bard, ed.), Marcel Dekker, New York, 1966.

[37]. P. Delahay, Double Layer and Electrode Kinetics, Wiley-Interscience, New York, 1965, p. 197.

[38]. P. Delahay, J. Electrochem. Soc., 113:967 (1966).

[39]. K. Holub, G. Tessari, and P. Delahay, J. Phys. Chem., 71:2612 (1967).

[40]. H. A. Laitinen and J. E. B. Randles, Trans. Faraday Soc., 51:54 (1955).

[41]. A. M. Baticle and F. Perdu, J. Electroanal. Chem., 12:15 (1966).

[42]. R. de Levie, Electrochim. Acta, 10:395 (1965).

[43]. J. H. Sluyters, Rec. Trav. Chim., 79:1092 (1960).

[44]. D. K. Cheng, Analysis of Linear Systems, Addison-Wesley, Reading, Mass., 1959, pp. 155, 278, 303.

[45]. E. A. Guillemin, Theory of Linear Physical System, Wiley, New York, 1963, p. 369.

[46]. M. E. Van Valkenberg, Network Analysis, Prentice-Hall, Englewood Cliffs, New Jersey, 1964, p. 331.

[47]. B. Breyer and H. H. Bauer, Alternating Current Polarography and Tensammetry, Wiley-Interscience, New York, 1963.

[48]. D. C. Grahame, J. Electrochem. Soc., 99:370C (1952).

[49]. P. Dolin and B. V. Ershler, Acta Physiocochim, U.R.S.S., 13:747 (1940).

[50]. J. E. B. Randles, Discussions Faraday Soc., 1:11 (1947).

[51]. H. Gerischer, Z. Physik. Chem., 198:286 (1951).

[52]. T. Berzins and P. Delahay, Z. Elektrochem., 59:792 (1955).

[53]. M. Sluyters-Rehbach and J. H. Sluyters, Rec. Trav. Chim., 82:525 (1963).

[54]. D. E. Smith, in Electroanalytical Chemistry, Vol. 1 (A. J. Bard, ed.), Marcel Dekker, New York, 1966.

[55]. A. A. Pilla, Communication presented at the A.I.Ch.E. National Meeting, St. Louis (1968).

[56]. A. A. Pilla, J. Electrochem. Soc., 117:467 (1970).

[57]. M. D. Wijnen, Rec. Trav. Chim., 79:1203 (1960).

[58]. E. Levart and E. Poirier d'Ange d'Orsay, J. Electroanal. Chem., 19:335 (1968).

[59]. A. A. Pilla, D.Sc. Thesis, Univ. of Paris, 1965.

[60]. H. P. Van Leeuwen, D. J. Kooijman, M. Sluyters-Rehbach, and J. H. Sluyters, J. Electroanal. Chem., 23:475 (1969).

[61]. A. A. Pilla, R. B. Roe, and C. C. Herrmann, J. Electrochem. Soc., 116:1105 (1969).

[62]. K. J. Vetter, Electrochemical Kinetics, Academic Press, New York, 1967.

[63]. M. L. James, G. M. Smith, and J. C. Wolford, Analog and Digital Computer Methods in Engineering Analysis, International Textbook Co., Scranton, Pa., 1965, pp. 303-320.

[64]. A. A. Pilla, J. A. Christopulos, and G. J. DiMasi, Communication presented at the Electrochemical Society Meeting, New York (1969).

[65]. R. Payne, J. Electroanal. Chem., 19:1 (1968).

Part II

SIMULATION METHODS

A. Introduction

Chapter 7

# DIGITAL SIMULATION OF ELECTROCHEMICAL SURFACE BOUNDARY PHENOMENA: MULTIPLE ELECTRON TRANSFER AND ADSORPTION[1]

Stephen W. Feldberg

Brookhaven National Laboratory
Upton, New York 11973

## I. INTRODUCTION

The basic concepts of digital simulation of electrochemical phenomena have been discussed in detail by several authors [2-7]. The validity and usefulness of the technique have been indicated by application to a variety of problems [2-19].

The outstanding virtue of this technique is the ease with which one can translate phenomena into operational mathematical (arithmetic would be a better word!) terms. Perhaps the most difficult aspect of the translation involves calculation of the surface boundary conditions — particularly when there are multiple-electron transfers and/or adsorption phenomena. In this chapter I shall discuss in detail the treatment of surface boundary phenomena involving multiple electron transfer and adsorption of reactant or product. The electrode potential will be assumed to be the independent variable and for purposes of demonstration simulated cyclic voltammograms will be presented. It is impossible, of course, to present a comprehensive discussion of these problems; I do think, however, that the ideas presented here can be modified easily and applied to problems of specific interest.

Except for a brief synopsis of the digital simulation technique, I shall discuss only the calculations dealing with surface boundary phenomena. I shall assume that the reader is familiar with the general concepts of digital simulation, particularly as discussed in Ref. [5].

The mathematical notation is presented with the understanding that the equations will be programmed for a digital computer. Actual computer programs are not included for several reasons: differences in computers, systems, and programming languages; and primarily to avoid a cookbook approach to solving these problems.

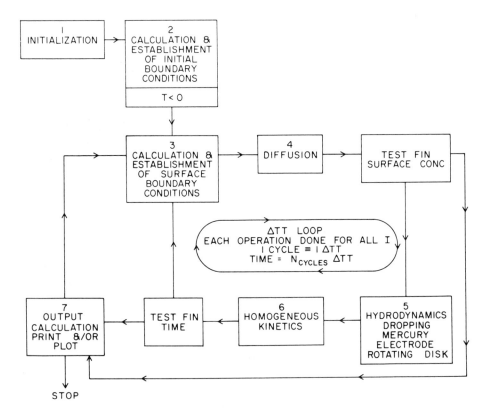

Fig. 1.  Flow chart for digital simulation. Execution of operations 3-6 constitutes a single cycle or time unit.

## II.  SYNOPSIS OF DIGITAL SIMULATION

A flow chart for digital simulation is shown in Fig. 1.  Because there are a large number of cycles, the change per cycle effected by any given operation is small.  Thus, each operation may be considered independently of all others.  Since most electrochemical phenomena are initiated by a perturbation at the electrode surface, * proper treatment of the surface boundary conditions is critical.  In certain cases initial boundary conditions (conditions extant prior to any perturbation) will have to be modified to allow for starting conditions other than the usual single reactant homogeneously distributed in solution, e. g., both components of a redox couple initially present in solution, or a species in solution in equilibrium with its adsorbed form.  The other operations (diffusion, hydrodynamics, chemical kinetics) are carried out as described in Ref. [5].

## III.  MODEL OF THE SURFACE REGION FOR

## SURFACE BOUNDARY CALCULATIONS

Figure 2 is a schematic drawing of the electrode surface and the volume elements adjacent to it.  The concentrations $c_1$, $c_2$, ..., $c_I$ represent the average concentration of a species in the 1st, 2nd, ..., I-th volume elements, and $c_0$ is the concentration at the electrode surface (really the outer Helmholtz plane).  The surface concentration $\Gamma$ of an adsorbed species is a function of: the activity or concentration of the soluble form of that species at $x = 0$; $c_0$; the nature of the isotherm; and the equilibrium constant (which is probably potential-dependent).  The thickness of the adsorbed film will be considered infinitely small compared to the thickness of the volume element.

The concentration gradient between any two volume elements is

$$\frac{\Delta c}{\Delta x} = \frac{c_{I+1} - c_I}{\Delta x} \qquad (1)$$

where $\Delta x$ is the thickness of a volume element and therefore the distance between centers of adjacent volume elements.  The concentration gradient at the surface may be approximated as

$$\left(\frac{\Delta c}{\Delta x}\right)_{x=0} = \frac{2(c_1 - c_0)}{\Delta x}. \qquad (2)$$

The 2 appears in Eq. (2) because the points at which $c_1$ and $c_0$ are measured are separated by a distance $\Delta x/2$.  It is clear that the gradient thus calculated is closer to being at a distance $\Delta x/4$ from the surface than at zero; the results of test calculations on systems having known mathematical solutions

---

*An exception is the case where the electric field in the solution is large enough to necessitate consideration of migration effects [19].

## VOLUME ELEMENTS

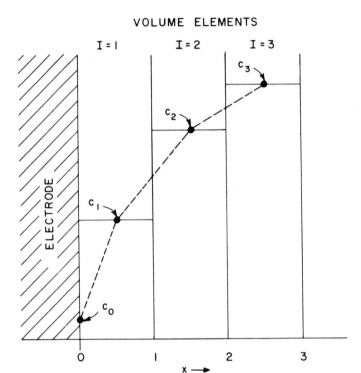

Fig. 2. Schematic drawing of electrode surface and adjacent volume elements.

indicate that the approximation is adequate. The term $\Delta x$ can be assumed to be unity, and will not appear explicitly in the ensuing discussions.

### A. Review of Simple One-Electron Transfer

Simulation of the simple one-electron transfer

$$e^- + A \underset{k_{hb_1}}{\overset{k_{hf_1}}{\rightleftharpoons}} B \tag{R1}$$

has been thoroughly discussed [5]. I shall present here, however, a generalized approach for developing the appropriate surface boundary relationships. A set of equations may be written describing the surface boundary conditions:

$$f_A = 2D_A (A_1 - A_0) \tag{3}$$

$$f_A = k_{hf_1} A_0 - k_{hb_1} B_0 \tag{4}$$

$$f_B = 2D_B(B_1 - B_0) \tag{5}$$

$$f_B = -f_A \tag{6}$$

$$f_{FAR} = f_A. \tag{7}$$

Equations (3) and (5) are representations of Fick's law:

$$(f)_{x=0} = D\left(\frac{dc}{dx}\right)_{x=0}. \tag{8}$$

The unknowns in these equations are $f_A$, $f_B$, $A_0$, $B_0$. The rate constants are calculated for each cycle according to the usual relationships:

$$k_{hf_1} = k_{hs_1} \exp\left[-\alpha_1 \frac{F(E - E_1^\circ)}{RT}\right] \tag{9}$$

$$k_{hb_1} = k_{hs_1} \exp\left[(1 - \alpha_1) \frac{F(E - E_1^\circ)}{RT}\right]. \tag{10}$$

Equations (3)-(7) comprise the complete set of equations for the surface boundary conditions. The flux terms are evaluated and then incorporated in the equations calculating diffusion (during a given time cycle) between volume elements 1 and 2, and for calculating diffusion "through" the electrode surface. Since species may be converted (by oxidation or reduction) at the surface, there is a flux at $x = 0$. The diffusion equation for the volume element 1 is (written for species A)

$$\Delta A_1 = D(A_2 - A_1) - f_A. \tag{11}$$

The generalized diffusion equation for volume elements 2 to N is (for planar diffusion)

$$\Delta A_I = D_Z(A_{I+1} - 2A_I + A_{I-1}). \tag{12}$$

Evaluation of the flux terms can be done in several ways. By relatively simple algebraic manipulation of Eqs. (3)-(7), one can obtain simple, explicit solutions for $f_A$, $f_B$, and $f_{FAR}$. In multielectron transfers these manipulations become cumbersome; by using the method of determinants, however, it is easy to handle these equations in a routine manner. I suggest the following procedure:

1. Eliminate the surface concentration terms; e.g., from Eq. (3) obtain

$$A_0 = A_1 - \frac{f_A}{2D_A} \tag{13}$$

and substitute in Eq. (4). Similarly rearrange Eq. (5) and substitute in Eq. (4).

2. Rearrange the resulting equations so that the flux terms and their coefficients form the left-hand side (LHS) of an equation and constants and/or knowns form the right-hand side (RHS).

3. Solve for each flux term using the method of determinants [only if necessary! In the case of simple one-electron transfer, once $f_A$ is calculated it is then trivial to obtain $f_B$ and $f_{FAR}$ from Eqs. (6) and (7)].

Pursuing this procedure through step 2 gives the following two equations (obtained from Eqs. (3)-(6) for the simple one-electron transfer):

$$f_A\left(1 + \frac{k_{hf_1}}{2D_A}\right) - f_B \cdot \frac{k_{hb_1}}{2D_B} = k_{hf_1} A_1 - k_{hb_1} B_1 \tag{14}$$

$$f_A + f_B = 0. \tag{15}$$

Solving for $f_A$ by the method of determinants gives

$$f_A = \frac{\begin{vmatrix} k_{hf_1} A_1 - k_{hb_1} B_1 & -\dfrac{k_{hb_1}}{2D_B} \\ 0 & 1 \end{vmatrix}}{\begin{vmatrix} 1 + \dfrac{k_{hf_1}}{2D_D} & -\dfrac{k_{hb_1}}{2D_B} \\ 1 & 1 \end{vmatrix}} \tag{16}$$

or

$$f_A = \frac{k_{hf_1} A_1 - k_{hb_1} B_1}{1 + \dfrac{k_{hf_1}}{2D_A} + \dfrac{k_{hb_1}}{2D_B}}. \tag{17}$$

The remaining equations for $f_B$ and $f_{FAR}$ have in fact already been established: Eqs. (6) and (7). These three equations (17), (6), and (7) are all that are needed to calculate the flux terms for the one-electron transfer. The equations are completely general and allow one to calculate cyclic voltammograms for any degree of reversibility (i.e., any value of $k_{hs_1}$) and for any value of the transfer coefficient $\alpha_1$. Examples are shown in Figs. 3-5.

## B. Multiple-Electron Transfer

When there is a sequence of electron transfers (or when several one-electron transfers are coupled by infinitely rapid chemical equilibria) all heterogeneous reactions must be considered in the surface boundary

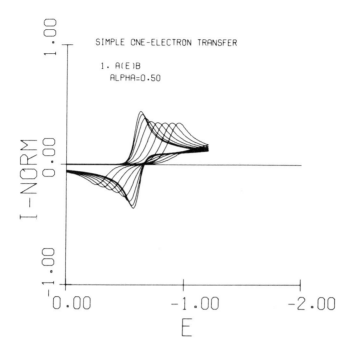

Fig. 3. Simulated cyclic voltammogram demonstrating the effect of varying the electrochemical transfer coefficient $\alpha$ (ALPHA) and the normalized
standard heterogeneous rate constant $*k_{hs_1}$ [RATEHS(1)]. The value of $*k_{hs_1}$
corresponding to each of the 8 curves is (in order of decreasing reversibility): $10^7$, 1.0, 0.3, 0.1, 0.03, 0.01, 0.003, 0.001. Ordinate, I-NORM =
$f_{FAR}/\{A_{bulk}[D(F/RT)(dE/dt)]^{1/2}\}$ and abscissa, E = volts. $|\Delta E|$ = 0.0024 V
(for $*k_{hs_1} \geq 1$) and 0.0048 V (for $*k_{hs_1} \leq 0.3$).

calculations. In certain cases simplification is obvious, e.g., when the
$E^o$'s for two reversible one-electron reductions are separated by ±100 mV
they may be considered as a single reversible 2e process ($E_2^o > E_1^o + 0.10$ V)
or as two separate 1e processes ($E_2^o < E_1^o - 0.10$ V). But simplifications
are not necessary.

Polcyn and Shain [20] have treated several examples of two-electron
transfers. It is important to note that in their work and in the treatment
presented here the question of homogeneous-electron transfers is ignored.
When all reactions being considered are reversible that question becomes
irrelevant. If, however, there exists some degree of electrochemical
irreversibility then the homogeneous-electron transfer reactions may provide a viable alternate path toward establishment of the redox equilibria and

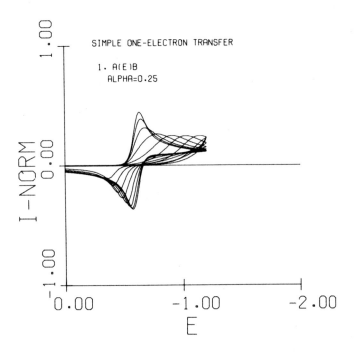

Fig. 4. See legend, Fig. 3.

must be considered. Also, if the electrochemical reactions are reversible but there are irreversible chemical reactions of any of the redox species, then too, the homogeneous-electron transfers must be considered [10, 11].

Consider the following sequence of two-electron transfers:

$$e^- + A \underset{k_{hb_1}}{\overset{k_{hf_1}}{\rightleftharpoons}} B \tag{R1}$$

$$e^- + B \underset{k_{hb_2}}{\overset{k_{hf_2}}{\rightleftharpoons}} C. \tag{R2}$$

The pertinent surface boundary equations are:

$$f_A = 2D_A (A_1 - A_0) \tag{18}$$

$$f_A = k_{hf_1} A_0 - k_{hb_1} B_0 \tag{19}$$

$$f_C = 2D_C (C_1 - C_0) \tag{20}$$

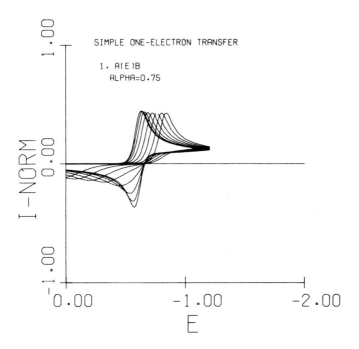

Fig. 5. See legend, Fig. 3.

$$f_C = k_{hb_2} C_0 - k_{hf_2} B_0 \tag{21}$$

$$f_B = 2D_B (B_1 - B_0) \tag{22}$$

$$f_B = -f_A - f_C \tag{23}$$

$$f_{FAR} = k_{hf_1} A_0 - k_{hb_1} B_0 + k_{hf_2} B_0 - k_{hb_2} C_0 = f_A - f_C. \tag{24}$$

Following the outlined procedure a set of three equations is obtained:

$$f_A \left(1 + \frac{k_{hf_1}}{2D_A}\right) - f_B \frac{k_{hb_1}}{2D_B} = k_{hf_1} A_1 - k_{hb_1} B_1 \tag{25}$$

$$f_A = f_B + f_C = 0 \tag{26}$$

$$-f_B \frac{k_{hf_2}}{2D_B} + f_C \left(1 + \frac{k_{hb_2}}{2D_C}\right) = k_{hb_2} C_1 - k_{hf_2} B_1. \tag{27}$$

Solving for $f_A$ gives

$$f_A = \frac{\begin{vmatrix} k_{hf_1}A_1 - k_{hb_1}B_1 & -\dfrac{k_{hb_1}}{2D_B} & 0 \\[1.5em] 0 & 1 & 1 \\[1.5em] k_{hb_2}C_1 - k_{hf_2}B_1 & -\dfrac{k_{hf_2}}{2D_B} & 1+\dfrac{k_{hb_2}}{2D_C} \end{vmatrix}}{\begin{vmatrix} 1+\dfrac{k_{hf_1}}{2D_A} & -\dfrac{k_{hb_1}}{2D_B} & 0 \\[1.5em] 1 & 1 & 1 \\[1.5em] 0 & -\dfrac{k_{hf_2}}{2D_B} & 1+\dfrac{k_{hb_2}}{2D_C} \end{vmatrix}} \tag{28}$$

$$f_A = \frac{(k_{hf_1}A_1 - k_{hb_1}B_1)\left(1+\dfrac{k_{hb_2}}{2D_C}+\dfrac{k_{hf_2}}{2D_B}\right) - \dfrac{k_{hb_1}}{2D_B}(k_{hb_2}C_1 - k_{hf_2}B_1)}{\left(1+\dfrac{k_{hf_1}}{2D_A}\right)\left(1+\dfrac{k_{hb_2}}{2D_C}+\dfrac{k_{hf_2}}{2D_B}\right)+\dfrac{k_{hb_1}}{2D_B}\left(1+\dfrac{k_{hb_2}}{2D_C}\right)} \tag{29}$$

$$f_A = \frac{k_{hf_1}A_1 - k_{hb_1}B_1 - \dfrac{k_{hb_1}}{2D_B}\left(\dfrac{k_{hb_2}C_1 - k_{hf_2}B_1}{1+\dfrac{k_{hb_2}}{2D_C}+\dfrac{k_{hf_2}}{2D_B}}\right)}{1+\dfrac{k_{hf_1}}{2D_A}+\dfrac{k_{hb_1}}{2D_B}\left(\dfrac{1+\dfrac{k_{hb_2}}{2D_C}}{1+\dfrac{k_{hb_2}}{2D_C}+\dfrac{k_{hf_2}}{2D_B}}\right)}. \tag{30}$$

The reader will note that Eq. (30) reduces to (17) when $k_{hf_2} = k_{hb_2} = 0$. Similarly one can calculate

$$f_C = \frac{\begin{vmatrix} 1 + \dfrac{k_{hf_1}}{2D_A} & -\dfrac{k_{hb_1}}{2D_B} & k_{hf_1}A_1 - k_{hb_1}B_1 \\ 1 & 1 & 0 \\ 0 & -\dfrac{k_{hf_2}}{2D_B} & k_{hb_2}C_1 - k_{hf_2}B_1 \end{vmatrix}}{\begin{vmatrix} 1 + \dfrac{k_{hf_1}}{2D_A} & -\dfrac{k_{hb_1}}{2D_B} & 0 \\ 1 & 1 & 1 \\ 0 & -\dfrac{k_{hf_2}}{2D_B} & 1 + \dfrac{k_{hb_2}}{2D_C} \end{vmatrix}} \tag{31}$$

$$f_C = \frac{(k_{hb_2}C_1 - k_{hf_2}B_1)\left(1 + \dfrac{k_{hf_1}}{2D_A} + \dfrac{k_{hb_1}}{2D_B}\right) - \dfrac{k_{hf_2}}{2D_B}(k_{hf_1}A_1 - k_{hb_1}B_1)}{\left(1 + \dfrac{k_{hf_1}}{2D_A} + \dfrac{k_{hb_1}}{2D_B}\right)\left(1 + \dfrac{k_{hb_2}}{2D_C}\right) + \dfrac{k_{hf_2}}{2D_B}\left(1 + \dfrac{k_{hf_1}}{2D_A}\right)} \cdot \tag{32}$$

Note that the denominators for the RHS of Eqs. (29) and (32) are mathematically equivalent [note Eqs. (28) and (31)].

$$f_C = \frac{k_{hb_2}C_1 - k_{hf_2}B_1 - \dfrac{k_{hf_2}}{2D_B}\left(\dfrac{k_{hf_1}A_1 - k_{hb_1}B_1}{1 + \dfrac{k_{hf_1}}{2D_A} + \dfrac{k_{hb_1}}{2D_B}}\right)}{1 + \dfrac{k_{hb_2}}{2D_C} + \dfrac{k_{hf_2}}{2D_B}\left(\dfrac{1 + \dfrac{k_{hf_1}}{2D_A}}{1 + \dfrac{k_{hf_1}}{2D_A} + \dfrac{k_{hb_1}}{2D_B}}\right)} \cdot \tag{33}$$

The terms $f_B$ and $f_{FAR}$ can be evaluated from Eqs. (23) and (24).

The most stringent test of these equations is to introduce a very large value for $k_{hs_1}$ and $k_{hs_2}$ (effectively invoking reversibility) and then note the effect of shifting $E_1^\circ - E_2^\circ$ from positive values (where a cyclic would exhibit two distinct one-electron transfers) to negative values (where a cyclic would exhibit a single wave for a reversible two-electron transfer [21]. The results of such a test are shown in Figs. 6 and 7.

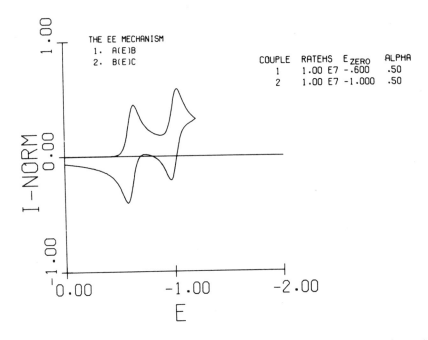

Fig. 6. Simulated cyclic voltammogram demonstrating the effect of varying $E_1^\circ$ [$E_{ZERO}$ (1)], $E_2^\circ$ [$E_{ZERO}$ (2)], $^*k_{hs_1}$ [RATEHS (1)], and $^*k_{hs_2}$ [RATEHS (2)] in a two-electron transfer process. Ordinate I-NORM = $f_{FAR}/\{A_{bulk}$ [D(F/RT)(dE/dt)]$^{1/2}\}$ and abscissa, E = volts. $|\Delta E| = 0.0048\,V.$

It is also interesting to observe the effect of diminishing the standard rate constants and varying the $E^\circ$'s as before. When the normalized $k_{hs}$'s are reduced to a value of 1, and $E_2^\circ \ll E_1^\circ$, the cyclic still looks like two sequential reversible one-electron transfers (Fig. 8). But when $E_2^\circ \gg E_1^\circ$ (Fig. 9) the cyclic has the characteristics of a highly irreversible one-electron transfer where (1) two electrons are transferred in the reduction and oxidation processes and (2) the heterogeneous rate parameters controlling the reduction wave are $k_{hs_1}$, $\alpha_1$, and $E_1^\circ$, while parameters controlling the oxidative wave are $k_{hs_2}$, $\alpha_2$, and $E_2^\circ$. This sort of phenomenon has been discussed previously [22, 23].

The surface boundary calculations for systems with n-coupled-electron transfers involve the same approach as that described for the one- and two-electron transfers. The order of the resulting determinant will be n + 1, and thus the algebraic expressions may be cumbersome; but a large number of the terms in the determinant will be zero.

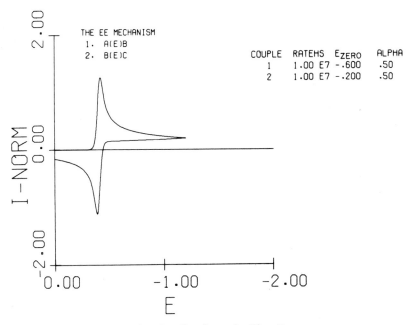

Fig. 7.  See legend, Fig. 6.

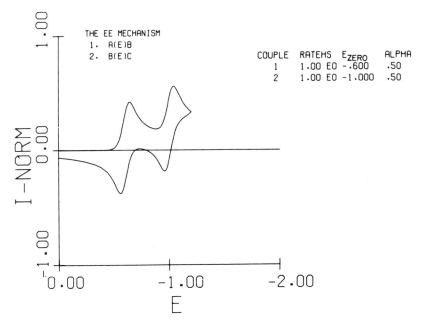

Fig. 8.  See legend, Fig. 6.

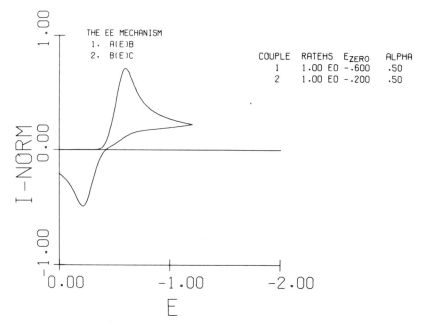

Fig. 9.  See legend, Fig. 6.  $\Delta E = 0.0024$.

## C.  Electron Transfer Phenomena Involving Adsorption

Discussions of the implications of adsorption phenomena are too numerous to reference in detail.  Wopschall and Shain [24, 25] have discussed the implications of adsorption in cyclic voltammetry.  Their treatment is limited to the Langmuirian adsorption and considers that only one species (product or reactant) is significantly adsorbed.  The treatment I shall develop here for digital simulation is not limited to Langmuirian adsorption (I shall discuss Frumkin adsorption by way of example), and I shall consider that there are two redox couples:  the adsorbed and the nonadsorbed couples.  I shall assume, however, as did Wopschall and Shain, that adsorption equilibrium obtains at all times and that one of the species is only slightly adsorbed.  The capacitive current will be considered negligible.

The system to be treated is

$$e^- + A \underset{k_{hb_1}}{\overset{k_{hf_1}}{\rightleftharpoons}} B \tag{R1}$$

$$e^- + \Gamma_A \underset{k_{hb_3}}{\overset{k_{hf_3}}{\rightleftharpoons}} \Gamma_B \tag{R3}$$

$$A \xrightleftharpoons{K_A} \Gamma_A \tag{R4}$$

$$B \xrightleftharpoons{K_B} \Gamma_B. \tag{R5}$$

The constants $K_A$ and $K_B$ define the adsorption equilibria in terms of the Frumkin isotherm [26]. For species A,

$$K_A A_0 = \frac{\Gamma_A}{\Gamma_{max} - \Gamma_A} \exp\left[a \frac{\Gamma_A}{\Gamma_{max}}\right]; \tag{34}$$

and for species B (assuming that very little B adsorbs and that an adsorbed molecule (ion) of species B acts as if it were A),

$$K_B B_0 = \frac{\Gamma_B}{\Gamma_{max} - \Gamma_A} \exp\left[a \frac{\Gamma_A}{\Gamma_{max}}\right]. \tag{35}$$

Combining Eqs. (34) and (35) gives

$$\frac{K_A A_0}{K_B B_0} = \frac{\Gamma_A}{\Gamma_B}. \tag{36}$$

With the same assumptions Eq. (36) could be derived from most of the common isotherms [27]: Henry, Volmer, Van der Waals, Virial, Langmuir, modified Hefland, Frisch, and Lebowitz. If all electron transfer reactions are in equilibrium then

$$k_{hf_1} A_0 = k_{hb_1} B_0 \tag{37}$$

and

$$k_{hf_3} \Gamma_A = k_{hb_3} \Gamma_B \tag{38}$$

and one deduces

$$\frac{\Gamma_A}{\Gamma_B} = \frac{k_{hb_3} k_{hf_1} A_0}{k_{hf_3} k_{hb_1} B_0}. \tag{39}$$

Since Eq. (36) is valid even if the equilibria of Eqs. (37) and (38) do not obtain, Eq. (39) is similarly free of these constraints and thus

$$\frac{K_A}{K_B} = \frac{k_{hf_1} k_{hb_3}}{k_{hb_1} k_{hf_3}} = \exp\left[\frac{F}{RT}(E_1^0 - E_3^0)\right]. \tag{40}$$

The surface boundary equations are similar in form to those derived for nonadsorbing systems:

$$f_A = 2D_A(A_1 - A_0) \tag{41}$$

$$f_A = a_s(k_{hf_1}A_0 - k_{hb_1}B_0) + \Delta\Gamma_A + k_{hf_3}\Gamma_A - k_{hb_3}\Gamma_B \tag{42}$$

$$f_B = 2D_B(B_1 - B_0) \tag{43}$$

$$f_B = -f_A + \Delta\Gamma_A + \Delta\Gamma_B \tag{44}$$

$$f_{FAR} = f_A - \Delta\Gamma_A. \tag{45}$$

The term $a_s$ represents the activity of the surface available for electron transfer by nonadsorbed species. In the absence of any adsorption $a_s = 1$. In the presence of adsorption, however, the evaluation of $a_s$ is a bit more complex. Mohilner [28] has discussed this problem and gives the following relationship for the Frumkin isotherm:

$$a_s = \frac{\Gamma_{max} - \Gamma_A}{\Gamma_{max}} \exp\left[-\lambda a \frac{\Gamma_A}{\Gamma_{max}}\right] \tag{46}$$

where $0 < \lambda < 1$. When $\Gamma_A = 0$, $a_s = 1$; and when $a = 0$ (or when $\lambda = 0$), Eq. (46) reduces to the expression for the Langmuir isotherm

$$a_s = 1 - \theta. \tag{47}$$

The term $a_s$ is easily calculated for any given simulation cycle; values for $\Gamma_{max}$ and $\lambda$ will have to be assumed.

The reason for using the terms $k_{hf_3}\Gamma_A$ and $k_{hb_3}\Gamma_B$ rather than $k_{hf_3}a_{\Gamma_A}$ and $k_{hb_3}a_{\Gamma_B}$ is that the first pair gives the rate of change of surface coverage in moles (unit area)$^{-1}$ (unit time)$^{-1}$ while the second pair gives the rate of change of surface activity in activity (unit time)$^{-1}$. If activity coefficients are defined as

$$\gamma_{\Gamma_A} = \frac{a_{\Gamma_A}}{\Gamma_A} \tag{48}$$

and

$$\gamma_{\Gamma_B} = \frac{a_{\Gamma_B}}{\Gamma_B} \tag{49}$$

the implication of Eqs. (34) and (35) is that

$$\gamma_{\Gamma_A} = \gamma_{\Gamma_B}. \tag{50}$$

Thus the rate of change of surface coverage in moles (unit area)$^{-1}$ (unit time)$^{-1}$

that is due to reaction (R3) only is

$$\Delta\Gamma_A = \frac{\Delta a_{\Gamma_A}}{\gamma_{\Gamma_A}} = -k_{hf_3}\frac{a_{\Gamma_A}}{\gamma_{\Gamma_A}} + k_{hb_3}\frac{a_{\Gamma_B}}{\gamma_{\Gamma_B}} \tag{51}$$

or after combining Eqs. (48)-(51)

$$\Delta\Gamma_A = -k_{hf_3}\Gamma_A + k_{hb_3}\Gamma_B. \tag{52}$$

Although $\gamma_{\Gamma_A}$ is a function of time, I assume that during any given time unit it is effectively constant.

Determination of the flux terms will generally require an iterative process since most isotherms are nonlinear. The form of Eq. (42) can be simplified by replacing $k_{hf_1}$, $k_{hb_1}$, $k_{hf_3}$, and $k_{hb_3}$ with $k_{hs_1}$, $k_{hs_3}$, and the appropriate exponential [see Eqs. (9) and (10)].

$$f_A = a_s k_{hs_1}\left(A_0 \exp\left[-\alpha_1\frac{F}{RT}(E - E_1^o)\right] - B_0 \exp\left[(1 - \alpha_1)\frac{F}{RT}(E - E_1^o)\right]\right) + \Delta\Gamma_A$$

$$+ k_{hs_3}\left(\Gamma_A \exp\left[-\alpha_3\frac{F}{RT}(E - E_3^o)\right] - \Gamma_B \exp\left[(1 - \alpha_3)\frac{F}{RT}(E - E_3^o)\right]\right). \tag{53}$$

Substituting from Eqs. (36) and (40) and assuming $\alpha_1 = \alpha_3 = \alpha$ gives

$$f_A = \left[a_s k_{hs_1} + \frac{\Gamma_A}{A_0}\left(\frac{K_A}{K_B}\right)^{-\alpha} k_{hs_3}\right]$$

$$\cdot \left(A_0 \exp\left[-\alpha\frac{F}{RT}(E - E_1^o)\right] - B_0 \exp\left[(1 - \alpha)\frac{F}{RT}(E - E_1^o)\right]\right) + \Delta\Gamma_A. \tag{54}$$

Assuming the Frumkin isotherm and Mohilner's definition for $a_s$ [Eq. (46)] gives

$$f_A = a_s\left(k_{hs_1} + \Gamma_{max}K_A^{(1-\alpha)}K_B^\alpha k_{hs_3}\exp\left[(\lambda - 1)a\frac{\Gamma_A}{\Gamma_{max}}\right]\right)$$

$$\cdot \left(A_0 \exp\left[-\alpha\frac{F}{RT}(E - E_1^o)\right] - B_0 \exp\left[(1 - \alpha)\frac{F}{RT}(E - E_1^o)\right]\right) + \Delta\Gamma_A. \tag{55}$$

Equation (55) may be rewritten as

$$f_A = k'_{hf}A_0 - k'_{hb}B_0 + \Delta\Gamma_A \tag{56}$$

where

$$k_{hf}' = a_s\left(k_{hs_1} + \Gamma_{max}K_A^{(1-\alpha)}K_B^{\alpha}k_{hs_3}\ \exp\left[(\lambda-1)a\frac{\Gamma_A}{\Gamma_{max}}\right]\right)$$
$$\cdot\exp\left[-\alpha\frac{R}{RT}(E-E_1^{\circ})\right] \tag{57}$$

and

$$k_{hb}' = k_{hf}'\cdot\exp\left[\frac{F}{RT}(E-E_1^{\circ})\right]. \tag{58}$$

Combining Eqs. (41), (43), (44), and (56) and solving for $\Delta\Gamma_A(\Delta\Gamma_B \to 0)$ gives

$$\Delta\Gamma_A = \frac{f_A\left(1 + \dfrac{k_{hf}'}{2D_A} + \dfrac{k_{hf}'}{2D_B}\right) + k_{hb}'B_1 - k_{hf}'A_1}{1 + \dfrac{k_{hb}'}{2D_B}}. \tag{59}$$

The isotherm is brought into the calculation by combining Eqs. (34), (41), and (59) to give

$$\frac{\Gamma_A + \Delta\Gamma_A}{K_A\left(A_1 - \dfrac{A}{2D_A}\right)(\Gamma_{max} - \Gamma_A - \Delta\Gamma_A)}\ \exp\left[a\frac{(\Gamma_A + \Delta\Gamma_A)}{\Gamma_{max}}\right] = 1. \tag{60}$$

A simple iterative procedure known as the "bisection method" has been described by Hamming [29]. Prerequisites for application of the method are that a function of a single variable be monotonic in the range of interest and that the sign of the derivative be known; and that the selection of the correct value of the variable give a known value for the function (e.g., 1.0000). Equation (60), combined with Eq. (59), fulfills these requirements. The iterative procedure which follows is computed once per cycle (i.e., once per time unit) of the simulation:

a. Evaluate $a_s$ [Eq. (46)].

b. Evaluate $k_{hf}'$ and $k_{hb}'$ [Eqs. (57) and (58)]. The term $\Gamma_{max}K_A^{(1-\alpha)}K_B^{\alpha}k_{hs_3}$ is treated as a single constant. Note that the units of this constant are $LT^{-1}$.

c. Determine a maximum and minimum value for $f_A$:

$$(f_A)_{max} = 2D_A A_1; \qquad (f_A)_{min} = -2D_B B_1 - \Gamma_A.$$

d. $f_A = \frac{1}{2}[(f_A)_{max} + (f_A)_{min}]$.

e. Evaluate $\Delta\Gamma_A$ [Eq. (59)].

f. If $(\Delta\Gamma_A + \Gamma_A) < 0$ then $(f_A)_{min} = f_A$ and go to operation d above; or if $(\Delta\Gamma_A + \Gamma_A) > \Gamma_{max}$ then $(f_A)_{max} = f_A$ and go to operation d above.

g. Evaluate the LHS of Eq. (60) and if LHS $< 0.9999$ then $(f_A)_{min} = f_A$ and go to operation d above; or if LHS $> 1.0001$ then $(f_A)_{max} = f_A$ and go to operation d above.

h. If this step is reached then $0.9999 <$ LHS $< 1.0001$ and the correct values for $f_A$ and $\Delta\Gamma_A$ have been found.

i. Calculate the new value for $\Gamma_A$,

$$(\Gamma_A)_{new} = (\Gamma_A)_{old} + \Delta\Gamma_A,$$

and calculate the remaining flux terms $f_B$ and $f_{FAR}$ according to Eqs. (44) and (45).

The initializing operation for the entire simulation must, of course, include the establishment of the equilibrium between the bulk concentrations of species A and $\Gamma_A$. The method used for this calculation may also have to be iterative (depending on the nature of the isotherm). If so, then the approach just described can be applied. The equations are simplified since all fluxes are zero and $A_0 = A_{bulk}$. I shall not present the details of this calculation.

Although these equations were derived for a system with reactant adsorption, the equations for a system with product adsorption are identical. Simply by changing the initial conditions — starting with species B initially present in solution and scanning potential from negative to positive — simulation of a system with product adsorption is effected.

Equation (55) has several interesting implications: if either $k_{hs_1}$ or $k_{hs_3}$ is very large the cyclic voltammogram should be reversible; if the electron transfer process is controlled by only one of the parameters $k_{hs_1}$ or $\Gamma_{max} K_A^{(1-\alpha)} K_B^{\alpha} k_{hs_3} \exp[(\lambda - 1)a(\Gamma_A/\Gamma_{max})]$ the character of the irreversibility will differ, depending upon which of the two parameters is controlling, but only if the Frumkin exponential term has the value $a(\lambda - 1) \neq 0$.

Simulated cyclic voltammograms of a system with reactant adsorption are shown in Figs. 10-19. The voltammogram for Langmuir adsorption (Frumkin exponential term $a = 0$) and reversible electron transfer ($k_{hs_1}$ or $k_{hs_3} \rightarrow \infty$) is shown in Fig. 10. The effect of introducing positive and negative values for the Frumkin exponential term a (with $\lambda = 0$) is shown in Figs. 11-18. The range $-3.0 \leq a \leq 3.0$ is probably extreme, although Damaskin [30] has reported a value of -1.97 for pyridine adsorption on mercury. Since the electron transfer rates are virtually infinite the value of $\lambda$ is irrelevant. For the same reason there is no distinction between control by reaction (R1) or control by reaction (R2).

One can estimate that a cyclic voltammogram will begin to show signs of irreversibility when

$$\frac{{}^{*}k_{hs_1} + {}^{*}k_{hs_3}}{K_A A_{bulk} \exp\left[(\lambda - 1)a\dfrac{\Gamma_A}{\Gamma_{max}}\right]} < 1. \tag{61}$$

Simulated voltammograms obtained for Frumkin adsorption (a = 3.0, -3.0, 0.6, -0.6; and $\lambda = 0.5$) with nonreversible electron transfer indicate that there is no simple qualitative way of distinguishing electron transfer control by reaction (R1) from control by reaction (R3). A typical nonreversible simulated voltammogram is shown in Fig. 19. A more detailed theoretical

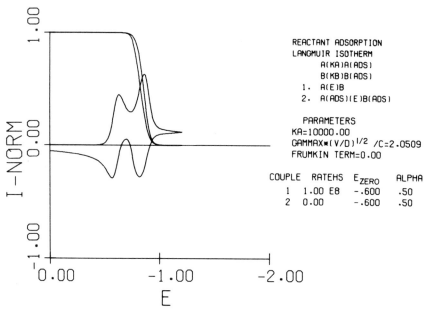

REACTANT ADSORPTION
LANGMUIR ISOTHERM
  A(KA)A(ADS)
  B(KB)B(ADS)
1.  A(E)B
2.  A(ADS)(E)B(ADS)

PARAMETERS
KA=10000.00
GAMMAX*(V/D)$^{1/2}$ /C=2.0509
FRUMKIN TERM=0.00

| COUPLE | RATEHS | $E_{ZERO}$ | ALPHA |
|--------|--------|------------|-------|
| 1 | 1.00 E8 | -.600 | .50 |
| 2 | 0.00 | -.600 | .50 |

Fig. 10.  Simulated cyclic voltammogram for a process with Langmuir adsorption of reactant obtained for ${}^{*}k_{hs_1} + {}^{*}k_{hs_3} \to \infty$. RATEHS(1) = ${}^{*}k_{hs_1}$, and RATEHS(2) = ${}^{*}k_{hs_3}$. Ordinate I-NORM = $f_A / A_{bulk}[D(F/RT)(dE/dt)]^{1/2}$ and abscissa E = volts. $|\Delta E| = 0.0048$ V. Additional curve shows variation of $\Gamma_A/\Gamma_{max}$ (= $\theta$) as a function of potential.

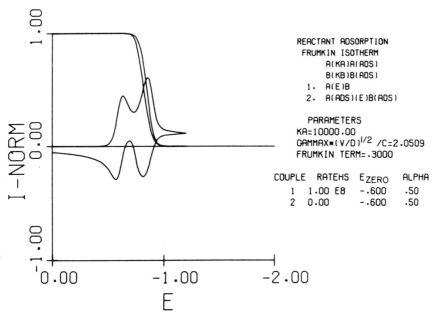

Fig. 11. Simulated cyclic voltammogram demonstrating the effect of varying the Frumkin term a for a system with Frumkin adsorption and reversible electron transfer. Ordinate, abscissa, and $|\Delta E|$ are same as for Fig. 10.

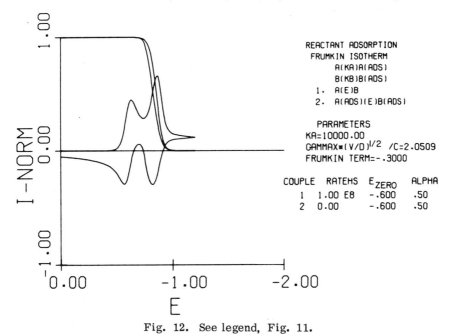

Fig. 12. See legend, Fig. 11.

Stephen W. Feldberg

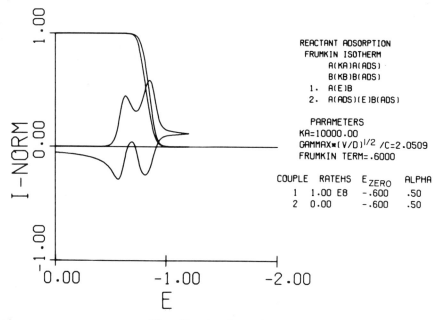

Fig. 13. See legend, Fig. 11.

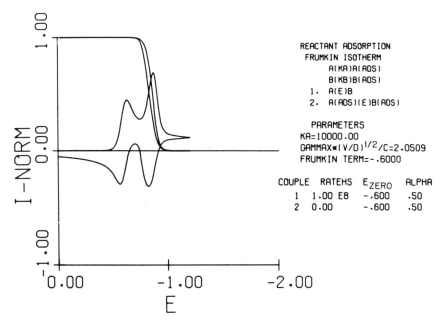

Fig. 14. See legend, Fig. 11.

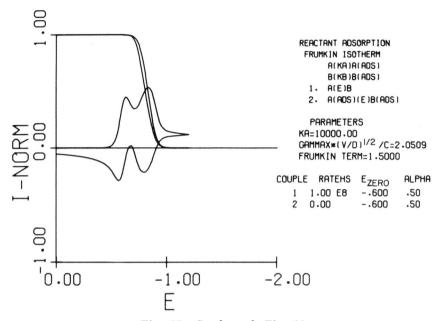

Fig. 15. See legend, Fig. 11.

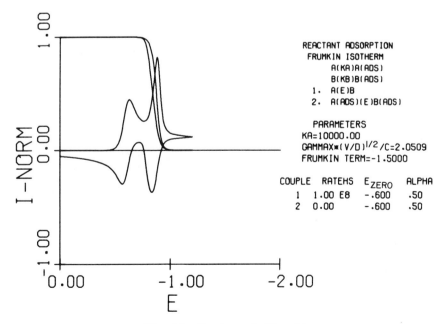

Fig. 16. See legend, Fig. 11.

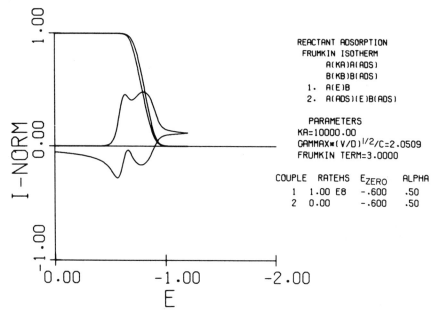

Fig. 17.  See legend, Fig. 11.

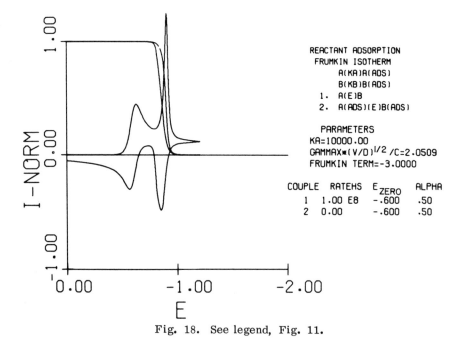

Fig. 18.  See legend, Fig. 11.

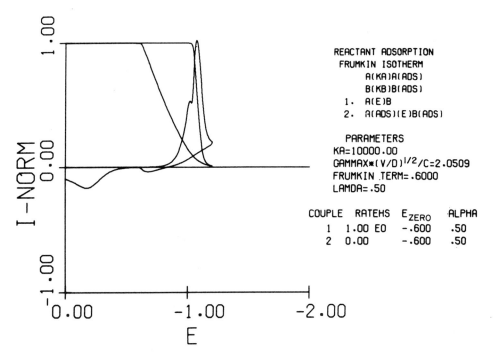

REACTANT ADSORPTION
FRUMKIN ISOTHERM
     A(KA)A(ADS)
     B(KB)B(ADS)
  1. A(E)B
  2. A(ADS)(E)B(ADS)

  PARAMETERS
KA=10000.00
GAMMAX*(V/D)$^{1/2}$/C=2.0509
FRUMKIN TERM=.6000
LAMDA=.50

| COUPLE | RATEHS | $E_{ZERO}$ | ALPHA |
|--------|--------|------------|-------|
| 1 | 1.00 E0 | -.600 | .50 |
| 2 | 0.00 | -.600 | .50 |

Fig. 19. Simulated cyclic voltammogram demonstrating nonreversible electron transfer for a system with Frumkin adsorption. RATEHS(2) = $^*k_{hs_3}$. Ordinate, abscissa, and $|\Delta E|$ are same as for Fig. 10.

study is warranted.

Throughout this discussion several simplifying assumptions have been made. Double-layer effects on the concentration of species A and B in the outer Helmholtz plane have been ignored. If the supporting electrolyte concentration is large enough, then $A_0 = A_{bulk}$ and $B_0 = B_{bulk}$ [27]. More serious is the assumption that the isotherm parameters are constants. The equilibrium constants $K_A$ and $K_B$ and the exponential term a are functions of potential [30]. Furthermore, there is no good reason to assume that $\Gamma_{max}$ and $\lambda$ are not also functions of potential. I think it is clear from the treatment presented that were all of these terms to be expressed as functions of potential it would in no way complicate the calculation of the surface boundary conditions. The complications will arise in quantitating the potential dependences of the functions. Damaskin [30] has discussed an approach to evaluating these functions from the differential capacitance data. In deducing the values of these parameters from capacitance data, usually measured under conditions where $c_{x=0} \propto c_{bulk}$, the term $\theta = \Gamma/\Gamma_{max}$ appears in the expression for the isotherm. Consequently variation in $\Gamma_{max}$ with potential would simply be observed as part of the potential dependence of a. Under dynamic conditions, such as those generated by cyclic voltammetry, it is necessary to calculate the actual surface coverage — thus one has an additional dynamic relationship between $c_{bulk}$ and $\Gamma_{max}$:

$$P_{\Gamma_{max}} = \frac{\Gamma_{max}\left(\frac{F}{DRT}\frac{dE}{dt}\right)^{1/2}}{c_{bulk}}. \tag{62}$$

This relationship in addition to all the classical relationships is a basis for quantitating surface coverage. It depends, of course, on a faradaic reaction of the adsorbed and/or adsorbing species. Anson and colleagues [31-34] have effectively applied this concept using chronocoulometry rather than cyclic voltammetry. A similar relationship arises even in the absence of a faradaic process when one considers diffusion-controlled adsorption and adsorption kinetics [35-37]. The effects, of course, will be much more subtle, manifesting themselves as potential and time-dependent changes in the double-layer capacitance. Simulation of diffusion-controlled adsorption processes in the absence of faradaic processes can be obtained directly from the treatment I have just presented by setting $^*k_{hs_1}$ and $^*k_{hs_3}$ equal to zero. The observed current would then be calculated by deducing the double-layer capacitance from the isotherms and capacitance data.

### D. Adsorption Kinetics (in the Absence of Faradaic Reaction)

The previous discussion dealt with diffusion-controlled adsorption coupled with faradaic processes. Generally the adsorption processes are fast and the assumption that adsorption equilibrium conditions obtain is reasonable. A few electrochemical techniques (e.g., the coulostatic technique) are perhaps fast enough to measure adsorption kinetics. Delahay and Mohilner [36] have treated the problem of adsorption kinetics assuming the Temkin isotherm, and Delahay [37] has treated the associated mass transfer problem for coulostatic charging.

I shall present the basis for digital simulation of any adsorption system, and then show briefly the development of the equations for Frumkin adsorption.

The system is

$$A \underset{k_{ba}}{\overset{k_{fa}}{\rightleftharpoons}} \Gamma_A \tag{R4}$$

where

$$\frac{k_{fa}}{k_{ba}} = K_A.$$ (63)

The surface boundary equations are

$$f_A = 2D_A(A_1 - A_0)$$ (64)

$$f_A = \frac{k_{fa} a_S A_0}{\gamma_{\Gamma_A}} - k_{ba}(\Gamma_A + \Delta\Gamma_A)$$ (65)

$$f_A = \Delta\Gamma_A.$$ (66)

The activity coefficient $\gamma_{\Gamma_A}$ appears in Eq. (65) in order that the rate process be expressed as a rate of change of surface coverage in units $SL^{-2}T^{-1}$ rather than in units of $T^{-1}$. At this point it is necessary to assume an isotherm. For a Frumkin isotherm, according to Mohilner [28],

$$a_{\Gamma_A} = \frac{\Gamma_A}{\Gamma_{max}} \exp\left[(1 - \lambda) a \frac{\Gamma_A}{\Gamma_{max}}\right]$$ (67)

since

$$\gamma_{\Gamma_A} = \frac{a_{\Gamma_A}}{\Gamma_A}$$ (68)

by substituting from Eqs. (46), (67), and (68), Eq. (65) can be rewritten

$$f_A = k_{fa} A_0(\Gamma_{max} - \Gamma_A - \Delta\Gamma_A) \exp\left[-a \frac{\Gamma_A + \Delta\Gamma_A}{\Gamma_{max}}\right] - k_{ba}(\Gamma_A + \Delta\Gamma_A).$$ (69)

Combining Eqs. (63), (64), (66), and (69) yields

$$f_A = \Delta\Gamma_A = k_{fa}\left(A_1 - \frac{f_A}{2D_A}\right)(\Gamma_{max} - \Gamma_A - f_A) \exp\left[-a \frac{\Gamma_A + f_A}{\Gamma_{max}}\right]$$

$$- \frac{k_{fa}}{K_A}(\Gamma_A + f_A).$$ (70)

Equation (70) may be solved using the iterative procedure described before. Once an isotherm is established on the basis of thermodynamic double-layer data (electrocapillary or capacitance measurements), $K_A$ (and the variable a in the Frumkin isotherm) can be expressed as a function of charge density only [36]. Hopefully $k_{fa}$ and $k_{ba}$ are also functions of charge density only, although it is not obvious just how $k_{fa}$ and $k_{ba}$ will vary with charge.

Delahay and Mohilner [36] have suggested that the charge density dependence
of the rates of adsorption and desorption might be analogous to the potential
dependence of heterogeneous-electron transfer rates.

It is clearly possible, however, to calculate current-time (potentiostatic
technique) or potential-time (coulostatic or galvanostatic technique) relationships for a variety of theoretical models.

## IV. CONCLUSIONS

In the preceding sections of this chapter, I have discussed the digital
simulation of the surface boundary conditions for multiple-electron transfer
and adsorption phenomena. Simultaneously, and independently, one can
introduce additional mechanistic complexities such as homogeneous chemical
kinetics [5], or "technique" complexities such as hydrodynamic phenomena
associated with a rotating disk or dropping mercury electrode [5].

### LIST OF SYMBOLS

| Symbol | Definition | Units[‡] |
|--------|------------|-------|
| $c_0$ | Concentration of species in solution at $x = 0$ | $SL^{-3}$ |
| $c_1$, $c_2$, ..., $c_I$ | Concentration of species in the I-th volume element | $SL^{-3}$ |
| $N$ | Maximum number of volume elements required during a given cycle in the simulation | 0 |
| $A_0$, $B_0$, $C_0$ | Concentration of species A, B, or C, in solution at $x = 0$ | $SL^{-3}$ |
| $A_1$, $B_1$, $C_1$ | Concentration of species A, B, or C in 1st volume element | $SL^{-3}$ |
| $f_A$, $f_B$, $f_C$ | Flux of species A, B, or C at $x = 0$. | $SL^{-2}T^{-1}$ |
| $f_{FAR}$ | Flux of electrons through the electrode surface (positive for reduction, negative for oxidation) | $SL^{-2}T^{-1}$ |
| $D_A$, $D_B$, $D_C$ | Diffusion coefficient for species A, B, C | $L^2T^{-1}$ |
| $E_1^0$, $E_2^0$, $E_3^0$ | Standard potential for heterogeneous reactions (R1), (R2), and (R3) | volts |
| $|\Delta E|$ | Change in potential per unit time ($\Delta t$) | volts $T^{-1}$ |

[‡] Definitions of units: S, species units; L, distance units; and T, time
units. A "0" denotes a dimensionless parameter. If the reader wishes to
check the equations for dimensional consistency, please note that $\Delta x$ and $\Delta t$
do not appear explicitly since they have been assigned numerical values of 1.

| Symbol | Definition | Units |
|---|---|---|
| $k_{hs_1}$, $k_{hs_2}$ | Standard heterogeneous rate constant for reactions (R1) and (R2) | $LT^{-1}$ |
| $k_{hs_3}$ | Standard heterogeneous rate constant for reaction (R3) | $T^{-1}$ |
| $^*k_{hs_1}$ | Normalized standard heterogeneous rate constants: $$^*k_{hs} = k_{hs} \Big/ \Big(\pi D \frac{F}{RT}\frac{dE}{dt}\Big)^{1/2}.$$ | 0 |
| $^*k_{hs_3}$ | Normalized standard heterogeneous rate constant for reaction (R3): $$^*k_{hs_3} = \Gamma_{max} K_A^{(1-\alpha)} K_B^{\alpha} k_{hs_3} \Big/ \Big(\pi D \frac{F}{RT}\frac{dE}{dt}\Big)^{1/2}$$ | 0 |
| $\alpha_1$, $\alpha_2$, $\alpha_3$ | Transfer coefficient | 0 |
| $k_{hf_1}$, $k_{hf_2}$ $k_{hb_1}$, $k_{hb_2}$ | Heterogeneous rate constants for reactions (R1) and (R2) defined according to Eqs. (9) and (10) | $LT^{-1}$ |
| $k_{hf_3}$, $k_{hb_3}$ | Heterogeneous rate constants for reaction (R1), defined according to Eqs. (9) and (10) | $T^{-1}$ |
| $\Gamma_A$, $\Gamma_B$ | Surface concentration of adsorbed species | $SL^{-2}$ |
| $\Delta\Gamma_A$, $\Delta\Gamma_B$ | Change in surface concentration during a single computation cycle (computer time unit) | $SL^{-2}$ |
| $\Gamma_{max}$ | Maximum surface concentration of adsorbed species | $SL^{-2}$ |
| $a_{\Gamma_A}$, $a_{\Gamma_B}$ | Activity of adsorbed species | 0 |
| $a_S$ | Activity of electrode sites | 0 |
| $\gamma_{\Gamma_A}$, $\gamma_{\Gamma_B}$ | Activity coefficients of adsorbed species | $L^2 S^{-1}$ |
| $K_A$, $K_B$ | Equilibrium constants of adsorption isotherm | $L^3 S^{-1}$ |
| $a$ | Exponential term of Frumkin isotherm [see Eqs. (34) and (35)] | 0 |
| $\lambda$ | See Eq. (46) | 0 |
| $\theta$ | Fraction of surface covered: $$\theta = \frac{\Gamma}{\Gamma_{max}}.$$ | 0 |

| Symbol | Definition | Units |
|--------|------------|-------|
| $k_{fa}$ | Rate constant for adsorption process | $L^3 S^{-1} T^{-1}$ |
| $k_{ba}$ | Rate constant for desorption process | $T^{-1}$ |
| $\lvert \Delta E \rvert$ | Change in potential per unit time (per simulation cycle) | volts |

## ACKNOWLEDGMENTS

I wish to thank Dr. D. M. Mohilner for several enlightening discussions on the topic of surface adsorption. Mrs. Grace Searles is also thanked for her invaluable assistance in the preparation of this manuscript.

## REFERENCES

[1].    Work done under the auspices of the United States Atomic Energy Commission.

[2].    S. W. Feldberg and C. Auerbach, Anal. Chem., 36:505 (1964).

[3].    G. L. Booman and D. T. Pence, Anal. Chem., 37:1366 (1965).

[4].    F. A. Covitz, "Computer Simulation of Cyclic Voltammetry," in The Synthetic Mechanistic Aspects of Electroorganic Chemistry, U.S. Army Research Office, Durham, North Carolina, 1968.

[5].    S. Feldberg, "Digital Simulation: A General Method for Solving Electrochemical Diffusion-Kinetic Problems," in Electroanalytical Chemistry, Vol. 3 (A. J. Bard, ed., Marcel Dekker, New York, 1969, pp. 199-295.

[6].    D. T. Pence, J. R. Delmastro, and G. L. Booman, Anal. Chem., 41:737 (1969).

[7].    J. R. Delmastro, Anal. Chem., 41:747 (1969).

[8].    K. B. Prater and A. J. Bard, J. Electrochem. Soc., 117:207 (1970).

[9].    K. B. Prater and A. J. Bard, J. Electrochem. Soc., 117:335 (1970).

[10].    M. D. Hawley and S. W. Feldberg, J. Phys. Chem., 70:3459 (1966).

[11].    R. N. Adams, M. D. Hawley, and S. W. Feldberg, J. Phys. Chem., 71:851 (1967).

[12].    S. W. Feldberg, J. Am. Chem. Soc., 88:390 (1966).

[13].    S. W. Feldberg, J. Phys. Chem., 70:3928 (1966).

[14].    S. W. Feldberg, J. Phys. Chem., 73:1238 (1969).

[15]. G. Kissel and S. W. Feldberg, J. Phys. Chem., 73:3082 (1969).

[16]. L. S. Marcoux, R. N. Adams, and S. W. Feldberg, J. Phys. Chem., 73:2611 (1969).

[17]. R. F. Nelson and S. W. Feldberg, J. Phys. Chem., 73:2623 (1969).

[18]. L. Papauchado, R. N. Adams, and S. W. Feldberg, J. Electroanal. Chem., 21:408 (1969).

[19]. S. W. Feldberg, J. Phys. Chem., 74:87 (1970).

[20]. D. S. Polcyn and I. Shain, Anal. Chem., 38:370 (1966).

[21]. R. S. Nicholson and I. Shain, Anal. Chem., 36:706 (1964).

[22]. R. M. Hurd, J. Electrochem. Soc., 109:327 (1962).

[23]. D. M. Mohilner, J. Phys. Chem., 68:623 (1964).

[24]. R. H. Wopschall and I. Shain, Anal. Chem., 39:1514 (1967).

[25]. R. H. Wopschall and I. Shain, Anal. Chem., 39:1527 (1967).

[26]. A. N. Frumkin, Z. Physik, 35:792 (1926).

[27]. D. M. Mohilner, "The Electrical Double Layer," in Electroanalytical Chemistry, Vol. 1 (A. J. Bard, ed.), Marcel Dekker, New York, 1966, p. 368.

[28]. D. M. Mohilner, J. Phys. Chem., 73:2652 (1969).

[29]. R. W. Hamming, Numerical Methods for Scientists and Engineers, McGraw-Hill, 1962, p. 352.

[30]. B. B. Damaskin, Electrochim. Acta, 9:231 (1964).

[31]. F. C. Anson, Anal. Chem., 36:932 (1964).

[32]. J. H. Christie, G. Lauer, and R. A. Osteryoung, J. Electroanal. Chem., 7:60 (1964).

[33]. R. A. Osteryoung and F. C. Anson, Anal. Chem., 36:975 (1964).

[34]. F. C. Anson, J. H. Christie, and R. A. Osteryoung, J. Electroanal. Chem., 13:343 (1967).

[35]. R. Parsons, "The Structure of the Electrical Double Layer and Its Influence on the Rates of Electrode Reactions", Advances in Electrochemistry and Electrochemical Engineering, Vol. 1 (P. Delahay, ed.), Interscience-Wiley, New York, 1961, pp. 1-64.

[36]. P. Delahay and D. M. Mohilner, J. Am. Chem. Soc., 84:4247 (1962).

[37]. P. Delahay, J. Phys. Chem., 67:135 (1962).

B. Example Simulations

Chapter 8

DIGITAL SIMULATION OF THE ROTATING RING-DISK ELECTRODE

Keith B. Prater

University of Texas at El Paso
El Paso, Texas

I. INTRODUCTION

The rotating ring-disk electrode* (RRDE) [1] presents an excellent
example of an electrochemical technique which can be treated only with
difficulty by the usual mathematical approaches [2-11], but which can easily
be treated by digital simulation [12-14]. Only three modifications of the
general digital simulation technique introduced by Feldberg (see [15, 16])
are necessary for the simulation of the RRDE. The first of the modifications
is the redefinition of the model.

*See Appendix I for a description.

## .II. THE MODEL

For the RRDE, the symmetry of the system suggests the following model. The solution is first divided into thin parallel layers which are $\Delta x$ cm thick. The electrode is placed in the center of the first layer, parallel to the planes dividing the layers. Then each layer is divided into a cylindrical box of radius $\Delta r/2$ cm, which is centered about the axis of rotation, and a series of concentric annular boxes $\Delta r$ cm wide (Fig. 1). If $\Delta x$ and $\Delta r$ are sufficiently small, then the solution in each box may be considered homogeneous and a single concentration for each chemical species may be associated with each box. In particular, this concentration is associated with the center of each box so that the distance between two points of known concentration will be a multiple of $\Delta x$ or $\Delta r$.

The parameter $\Delta r$ is chosen in such a way that the pertinent radii of the electrode $r_1$ (the radius of the disk electrode), $r_2$ (the radius to the inner edge of the ring electrode), and $r_3$ (the radius to the outer edge of the ring electrode) can be satisfactorily approximated by

$$r_1 = (IR1 - 0.5)\Delta r \tag{1a}$$

$$r_2 = (IR2 - 0.5)\Delta r \tag{1b}$$

$$r_3 = (IR3 - 0.5)\Delta r \tag{1c}$$

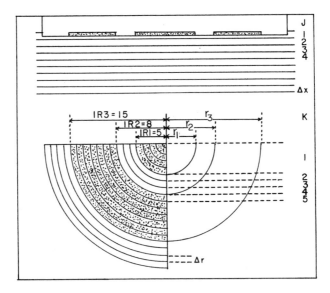

Fig. 1. The digital simulation model of the rotating ring-disk electrode.

where IR1, IR2, and IR3 are integers and correspond to the number of boxes necessary to represent the particular radial distances. Clearly, the smaller $\Delta r$ is (i.e., the greater the number of annular boxes), the better the approximation can be. Unfortunately, the number of boxes must be weighed against computation time and computer storage requirements and some compromise reached. It has been found that most electrodes can be adequately modeled using fewer than 200 annular boxes.

In the interest of minimizing computation time and storage space, one can usually assume that the disk electrode is a uniformly accessible surface [17]. This means that within each layer the solution in the cylindrical volume of radius $(IR1 - 0.5)\Delta r$ centered about the axis of rotation is homogeneous and can be reasonably represented by one large cylindrical box.

For the sake of convenience the following notation is used. Each box is referred to by its layer number J and its radial box number K. The layer containing the electrode is the $(J = 1)$ layer and all other layers are numbered outward from the electrode. Within any layer the large cylindrical box is the $(K = 1)$ box and all annular boxes are numbered outward from it. The large cylindrical box in the first layer $(J = 1, K = 1)$ corresponds to the surface of the disk electrode. Those $(J = 1)$ boxes for which $(IR2 - IR1) + 1 < K < (IR3 - IR1) + 2$ correspond to the surface of the ring electrode. The other $(J = 1)$ boxes represent either the surface of the insulating gap or the outer insulation.

## III. INITIAL AND BOUNDARY CONDITIONS

Once the model is defined, the initial and boundary conditions must be specified. If, for example, the solution initially contains only species A, then the fractional concentration of A, FA, in each box is unity. That is,

$$FA(J,K) = C_A(J,K)/C_A^o = 1.0 \tag{2}$$

where $C_A(J,K)$ is the concentration of A in the J, K box and $C_A^o$ is the bulk concentration of A. The fractional concentrations of all other species which may be produced by subsequent chemical or electrochemical reactions are initially zero in all boxes.

If, in the mechanism to be simulated, B is produced from A at the disk at a diffusion-limited rate, and B is consumed at the ring to produce A at a diffusion-limited rate, then the boundary conditions for A and B at the disk surface are

$$FA(1,1) = 0.0 \tag{3a}$$

$$FB(1,1) = 1.0. \tag{3b}$$

For those $(J = 1)$ boxes which correspond to the ring electrode, the boundary conditions are

$$FA(1, K_r) = 1.0 \tag{4a}$$

$$FB(1, K_r) = 0.0 \tag{4b}$$

where $K_r$ represents all values of K corresponding to the ring electrode. The initial conditions in those boxes corresponding to the gap and outer insulating region are the same as those in the bulk of the solution.

If the mechanism to be simulated involves a first-order decomposition of B to yield electroinactive species C, then the above boundary conditions at the ring and disk electrodes are valid only at the first short time interval of electrolysis, $\Delta t$. After that, the condition at the disk electrode surface after each $\Delta t$ interval can be expressed in FORTRAN notation by

$$FA(1, 1) = 0.0. \tag{5a}$$

$$FB(1, 1) = FB(1, 1) + DMA * (FA(2, 1) - FA(1, 1))$$
$$+ DMB * (FB(2, 1) - FB(1, 1)). \tag{5b}$$

The boundary condition for B states that the fractional concentration of B at the disk electrode after $\Delta t$ seconds have passed is given by the previous fractional concentration plus the quantity of A which has diffused to the electrode and been converted to B in the $\Delta t$ time interval, less the quantity of B which has diffused into the second-layer box as a result of the prevailing concentration gradient. Note that FA(1, 1) is defined to be zero and so is usually omitted. The parameters DMA and DMB are dimensionless diffusion coefficients which are defined by the equations

$$DMA = D_A \Delta t/(\Delta x)^2 \tag{6a}$$

$$DMB = D_B \Delta t/(\Delta x)^2, \tag{6b}$$

where $D_A$ and $D_B$ are the diffusion coefficients of A and B, respectively.

The fractional concentration of species C could be calculated using the boundary condition

$$FC(1, 1) = FC(1, 1) - DMC * (FC(1, 1) - FC(2, 1)). \tag{7}$$

However, since in this case C is considered to be electroinactive, its concentration would not usually be calculated.

At the ring electrode, the boundary conditions for this mechanism after the first $\Delta t$ interval would be

$$FA(1, K_r) = FA(1, K_r) + DMB * (FB(2, K_r) - FB(1, K_r))$$
$$+ DMA * (FA(2, K_r) - FA(1, K_r)) \tag{8a}$$

$$FB(1, K_r) = 0.0 \tag{8b}$$

$$FC(1, K_r) = FC(1, K_r) + DMC * (FC(2, K_r) - FC(1, K_r)). \tag{8c}$$

Finally, the boundary conditions for species A at the nonelectroactive gap and outer insulation surfaces will be given by

$$FA(1, K_I) = FA(1, K_I) + DMA * (FA(2, K_I) - FA(1, K_I)) \qquad (9)$$

where $K_I$ represents all appropriate values of K associated with these regions. Analogous equations would be written for all other chemical species.

## IV. DIFFUSION

After the initial conditions have been set up in all boxes and the boundary conditions applied to the (J = 1) boxes, the effect of diffusion normal to the electrode for a period of $\Delta t$ seconds upon the fractional concentrations of A in each box is calculated using the following general equation,

$$FA(J, K) = FA(J.K) + DMA * (FA(J + 1, K) - FA(J, K))$$

$$- DMA * (FA(J, K) - FA(J - 1, K)) \qquad (10)$$

or

$$FA(J, K) = FA(J, K) + DMA * (FA(J + 1, K) - 2.0 * FA(J, K)$$

$$+ FA(J - 1, K)). \qquad (11)$$

Care must be taken that all fractional concentrations on the right-hand side (rhs) of the above equations represent the concentration at the same point in time. Similar equations are used for all other species. Note that if $FA(J, K) = FA(J + 1, K) = FA(J - 1, K)$, then $FA(J, K)$ is unchanged and thus the calculation need not be made. In principle, the effects of diffusion in the radial direction should be calculated. However, it is found [3] that these effects are negligible when compared with the effects of convection in the radial direction and so are neglected.

## V. TIME AND CONVECTION

It is necessary to relate the passage of time within the simulation to the passage of time in the system being simulated. This can be accomplished by defining a parameter $t_k$, which is a time-dimensioned variable known in the real system, by

$$t_k = L\Delta t \qquad (12)$$

where L is the total number of $\Delta t$ intervals needed to simulate the time $t_k$. The form of $t_k$ will depend upon the system which is being simulated. In this case it is derived from a consideration of the effects of convection upon the concentration array previously generated.

Consider the equation for the velocity of convective fluid flow in the direction normal to the electrode. This is given to a good approximation by [18]

$$v_x = -0.51\omega^{3/2}\nu^{-1/2}x^2 \tag{13}$$

where $\omega$ is the rotation rate in rad sec$^{-1}$, $\nu$ is the kinematic viscosity in cm$^2$ sec$^{-1}$, and x is the distance from the electrode surface in cm. The velocity in the x direction may be represented by the derivative dx/dt. Thus

$$\frac{dx}{dt} = -0.51\omega^{3/2}\nu^{-1/2}x^2. \tag{14}$$

Rearranging, integrating, and evaluating between times $t_2$ and $t_1$, one obtains

$$-\frac{1}{x_2} + \frac{1}{x_1} = -0.51\,\omega^{3/2}\nu^{-1/2}\,(t_2 - t_1). \tag{15}$$

If we define

$$t_2 - t_1 = \Delta t, \tag{16}$$

then $x_2$ and $x_1$ represent the distance of a particular solution volume from the electrode surface at the end and beginning of the time interval $\Delta t$. In terms of the layer number J, the distance from the electrode surface to the center of any box is given by $(J - 1)\Delta x$. If we define

$$XJ = J - 1 \tag{17}$$

and let

$$x_2 = XJ \cdot \Delta x, \tag{18}$$

then $x_1$ is the position in the existing concentration array of the solution volume which will be in the center of the J-th box at the end of the time interval $\Delta t$. It is convenient to let

$$x_1 = XJJ \cdot \Delta x \tag{19}$$

where XJJ is a nonintegral distance from the electrode expressed in terms of box numbers. Equation (15) can now be rewritten as

$$\frac{1}{XJJ} - \frac{1}{XJ} = -0.51\omega^{3/2}\nu^{-1/2}\Delta x\Delta t \tag{20}$$

or

$$XJJ = \frac{XJ}{1 - XJ \cdot (0.51\omega^{3/2}\nu^{-1/2}\Delta x\Delta t)} \tag{21}$$

The product $0.51\omega^{3/2}\nu^{-1/2}\Delta x\Delta t$ is dimensionless and is designated VNAUT; that is,

$$VNAUT = 0.51\omega^{3/2}\nu^{-1/2}\Delta x\Delta t. \tag{22}$$

From the definition of DMA [Eq. (6a)] it follows that

$$\Delta x = \left(\frac{D_A\Delta t}{DMA}\right)^{1/2} \tag{23}$$

From the definition of $t_k$ [Eq. (12)],

$$\Delta t = t_k/L. \tag{24}$$

Thus,

$$VNAUT = 0.51\omega^{3/2}\nu^{-1/2}D_A^{1/2}DMA^{-1/2}t_k^{3/2}L^{-3/2}. \tag{25}$$

If $t_k$ is now defined by

$$t_k = (0.51)^{-2/3}\omega^{-1}\nu^{1/3}D_A^{-1/3} \tag{26}$$

then

$$VNAUT = DMA^{-1/2}L^{-3/2} \tag{27}$$

and

$$XJJ = \frac{XJ}{1 - XJ \cdot VNAUT}. \tag{28}$$

It should be noted that the above substitutions render the calculation of the effects of convection not only dimensionless, but also strictly in terms of simulation variables. Thus, by replacing the concentrations now in the center of each box with the concentrations at the distance calculated from Eq. (28), a new array will have been generated which will represent the effects of diffusion and convection normal to the electrode for a time $\Delta t$. [Since the value of XJJ calculated from Eq. (28) is not usually an integer, it usually designates a point which is not in the center of a box. The calculations could be made by using as the concentration at XJJ the concentration in the center of the box containing XJJ. However, it seems more reasonable to take as the value of the concentration at XJJ an interpolated value based upon the known concentrations on either side of XJJ and a linear concentration profile.]

In a similar manner, the effects of convection in the radial direction can be calculated. Beginning with the equation for the fluid velocity in the radial direction [18],

$$v_r = -0.51\omega^{3/2}\nu^{-1/2}xr \tag{29}$$

where r is the distance from the axis of rotation, one obtains

$$\frac{dr}{dt} = -0.51\omega^{3/2}\nu^{-1/2}xr. \tag{30}$$

Rearranging, integrating, and evaluating between $t_2$ and $t_1$, one obtains

$$\ln(r_2) - \ln(r_1) = -0.51\omega^{3/2}\nu^{-1/2}x(t_2 - t_1). \tag{31}$$

As before, $r_2$ and $r_1$ are the positions of a solution volume at times $t_2$ and $t_1$, respectively. Rearranging and substituting for x and $(t_2 - t_1)$,

$$\ln\frac{r_1}{r_2} = 0.51\omega^{3/2}\nu^{-1/2} \cdot XJ \cdot \Delta x\Delta t. \tag{32}$$

If we let

$$r_1 = RK \cdot \Delta r \tag{33}$$

and

$$r_2 = RKK \cdot \Delta r, \tag{34}$$

then substituting Eqs. (22), (32), and (33) into Eq. (31) yields

$$\ln\frac{RK \cdot \Delta r}{RKK \cdot \Delta r} = VNAUT \cdot XJ \tag{35}$$

or

$$RKK = \frac{RK}{\exp(VNAUT \cdot XJ)} \tag{36}$$

where

$$RK = K - 2 + IR1 \tag{37}$$

and is the radial distance of a solution volume in the center of the K-th box from the axis of rotation. The RKK is the distance from the axis of rotation at which the solution volume which will be at RK at the end of $\Delta t$ sec is to be found in the existing array of concentrations. Using Eq. (36) in the same manner as Eq. (28) was used in the case of convection normal to the electrode, an array of concentrations is generated which represents the effects of diffusion and convection for $\Delta t$ sec.

## VI. HOMOGENEOUS KINETICS

If the mechanism to be simulated involves a homogeneous chemical reaction, the effects of such a reaction upon the concentration profile would be calculated at this point. Consider as an example a first-order decomposition of the intermediate B to the electroinactive species C. The rate law for the disappearance of B from any box can be expressed as

$$\frac{dC_B(J, K)}{dt} = -kC_B(J, K) \tag{38}$$

where k is the first-order rate constant in $sec^{-1}$. If $\Delta t$ is small enough, the above equation can be approximated by

$$\frac{\Delta C_B(J, K)}{\Delta t} = -kC_B(J, K) \tag{39}$$

where $\Delta C_B(J, K)$ is the change in the concentration of B in the J, K-th box during the $\Delta t$ interval. Dividing both sides of the above equation by $C_A^o$ and rearranging, one obtains

$$\Delta FB(J, K) = -k\Delta t FB(J, K). \tag{40}$$

Substituting from Eq. (24) for $\Delta t$,

$$\Delta FB(J, K) = -kt_k FB(J, K)/L. \tag{41}$$

The product $kt_k$ is dimensionless and is designated as XKT. This parameter is read into the program and can be expressed as

$$XKT = (0.51)^{-2/3} k\omega^{-1} \nu^{1/3} D_A^{-1/3}. \tag{42}$$

Thus

$$\Delta FB(J, K) = XKT \cdot FB(J, K)/L, \tag{43}$$

and the new concentration array can be generated by the FORTRAN statement

$$FB(J, K) = FB(J, K) - XKT * FB(J, K)/L. \tag{44}$$

By similar reasoning, the concentration array for C would be generated by using the following equation:

$$FC(J, K) = FC(J, K) + XKT * FB(J, K)/L. \tag{45}$$

## VII. CURRENTS AND COLLECTION EFFICIENCY

At this point an array of concentrations has been generated which represents the concentration profile one would expect to observe under these conditions $\Delta t$ sec after the initiation of electrolysis. From this profile, one can calculate the currents one would expect to observe at this point in time.

The contribution to the current $i(K)$ from a given box K due to a species Z will be proportional to the number of moles of Z which diffuse to the electrode surface and are consumed in $\Delta t$ sec. The fractional flux of Z at the electrode surface is given by

$$FF_Z(K) = DMZ \cdot (FZ(2, K) - FZ(1, K)). \tag{46}$$

This fractional flux, when multiplied by the bulk concentration of the original species A, $C_A^o$, represents the number of moles per liter of Z which diffuse to the electrode surface in $\Delta t$ sec. This product, when multiplied by the volume of the K-th box, gives the number of moles of Z consumed in $\Delta t$ sec. The volume of the K-th box is $A(K)\Delta x$, where $A(K)$ is the area of the K-th box. Thus

$$i(K) = nF \cdot FF_Z(K)C_A^o A(K)\Delta x/\Delta t. \tag{47}$$

Rearranging Eq. (47) and substituting for $\Delta x$ and $\Delta t$, one obtains

$$i(K)/nFC_A^o = FF_Z(K)A(K)D_A^{1/2} DMA^{-1/2} L^{1/2} t_k^{-1/2}. \tag{48}$$

[The choice of $D_A$ and DMA is arbitrary. Since $D_A/DMA = D_B/DMB = \Delta t/(\Delta x)^2$ which is a constant for a given simulation, Eq. (48) could be written in terms of $D_B$ and DMB with equal validity.] Substituting for $t_k$ and rearranging,

$$\frac{i(K)}{(0.51)^{1/3} nFC_A^O D_A^{2/3} \omega^{1/2} \nu^{-1/6}} = FF_Z(K)A(K)L^{1/2}DMA^{-1/2}. \tag{49}$$

Dividing both sides of the above equation by the area of the disk electrode, $A_D$, renders both sides dimensionless and yields

$$\frac{i(K)}{(0.51)^{1/3} nFA_D C_A^O D_A^{2/3} \omega^{1/2} \nu^{-1/6}} = FF_Z(K)A(K)A_D^{-1} L^{1/2}DMA^{-1/2}. \tag{50}$$

Equation (50) can now be used to calculate the disk and ring current parameters. The current at the disk electrode due to species A can be represented by

$$\frac{i_d}{(0.51)^{1/3} nFA_D C_A^O D_A^{2/3} \omega^{1/2} \nu^{-1/6}} = FF_A(1)L^{1/2}DMA^{-1/2}. \tag{51}$$

In most cases, one is interested in simulating the limiting disk current. As mentioned above, the appropriate boundary condition for A at the disk electrode in that case is

$$FA(1, 1) = 0.0. \tag{52}$$

Thus the fractional flux of A to the disk electrode is

$$FF_A(1) = DMA \cdot FA(2, 1) \tag{53}$$

and Eq. (51) becomes

$$\frac{i_d}{(0.51)^{1/3} nFA_D C_A^O D_A^{2/3} \omega^{1/2} \nu^{-1/6}} = DMA^{1/2} L^{1/2} FA(2, 1). \tag{54}$$

Each side of Eq. (54) is dimensionless and is designated as ZD. Note that the lhs of Eq. (54) contains only terms dealing with the real system while the rhs contains only simulation parameters. Thus in terms of the real system,

$$ZD = \frac{i_d}{(0.51)^{1/3} nFA_D C_A^O D_A^{2/3} \omega^{1/2} \nu^{-1/6}} \tag{55}$$

while in terms of simulation variables

$$ZD = DMA^{1/2} L^{1/2} FA(2, 1). \tag{56}$$

In a similar manner parameters related to the limiting ring current can be calculated using Eq. (50). In general, the ring current will be given by

$$\frac{i_r}{(0.51)^{1/3} nFA_D C_A^O D_A^{2/3} \omega^{1/2} \nu^{-1/6}} = \sum_{K_r} FF_Z(K)A(K)A_D^{-1} L^{1/2}DMA^{-1/2} \tag{57}$$

where the summation is taken over all K boxes corresponding to the ring electrode. If all the ring current is due to species B and if the boundary condition for B at the ring electrode is

$$FB(1, K_r) = 0.0 \tag{58}$$

then Eq. (57) can be rewritten as

$$\frac{i_r}{(0.51)^{1/3} nFA_D C_A^o D_A^{2/3} \omega^{1/2} \nu^{-1/6}} = \sum_{K_r} DMB \cdot FB(2, K)A(K)A_D^{-1} L^{1/2} DMA^{-1/2}. \tag{59}$$

Each side of Eq. (59) is dimensionless and is designated ZR. Thus in terms of experimental parameters,

$$ZR = \frac{i_r}{(0.51)^{1/3} nFA_D C_A^o D_A \omega^{1/2} \nu^{-1/6}} \tag{60}$$

and in terms of simulation variables

$$ZR = \sum_{K_r} DMB \cdot FB(2, K)A(K)A_D^{-1} L^{1/2} DMA^{-1/2}. \tag{61}$$

Had the current at the ring or the disk been due to more than one chemical species, the parameters ZR and/or ZD would have contained additional but analogous terms including the flux of the other species.

Note that the ratio of ZR to ZD is simply

$$\frac{ZR}{ZD} = \frac{i_r}{i_d}. \tag{62}$$

This ratio is the important parameter for the ring-disk electrode and is called the collection efficiency N. Thus,

$$\frac{i_r}{i_d} = N = \frac{ZR}{ZD}. \tag{63}$$

Using the above approach, one can calculate the ring and disk current parameters and the collection efficiency one would observe $\Delta t$ sec after the initiation of electrolysis for any mechanism one might choose. Repeating the calculations using the existing concentration arrays (rather than the initial arrays) one will obtain the predicted current parameters and collection efficiency at $2\Delta t$ sec after the initiation of electrolysis. After I such iterations, the results will correspond to a time t in the real system where

$$t = I\Delta t. \tag{64}$$

Substituting for $\Delta t$ we find that

$$\frac{I}{L} = \frac{t}{t_k} = (0.51)^{2/3} \omega t D_A^{1/3} \nu^{-1/3}. \tag{65}$$

The output obtained from such a calculation is a relationship between ZD or ZR or ZR/ZD and the ratio I/L (Fig. 2). If the simulation has

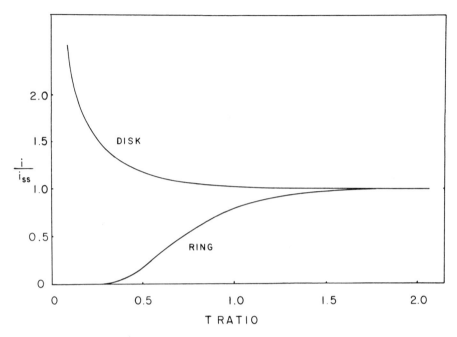

Fig. 2. Simulated limiting ring and disk currents normalized by their steady-state values as a function of TRATIO. TRATIO = $(0.51)^{2/3} \omega t D_A^{1/3} \nu^{-1/3}$, IR1 = 83, IR2 = 94, IR3 = 159.

correctly modeled the experimental system, then these graphs should also represent the relationships between the various experimental parameters, in particular between the currents or collection efficiency and time or rotation rate.

One of the advantages of the ring-disk electrode is that kinetic information can be obtained from steady-state measurements. This steady state is effectively reached by the time I/L = 3.0 (Fig. 2). Thus the ring and disk currents and collection efficiencies predicted when I/L = 3.0 should be those observed experimentally at steady state. By calculating steady-state collection efficiencies for various values of the rate parameter XKT, one can construct a working curve as shown in Fig. 3. This curve can then be used to obtain rate constants from the experimental results.

## VIII. PRACTICAL CONSIDERATIONS

Needless to say, these calculations are carried out using a digital computer. A typical program is presented in Appendix II. The reader will recall that one of the basic sssumptions of this technique is that Δt is small. This requires that L be large. It was found [12] that if L was taken to be 1000,

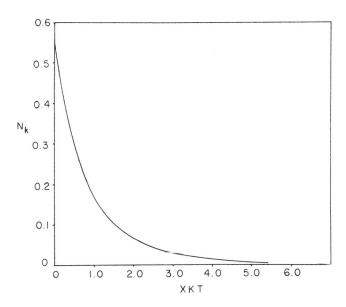

Fig. 3. Simulated collection efficiency working curve for first-order decay of the intermediate. $XKT = (0.51)^{-2/3}k\omega^{-1}\nu^{1/3}D_A^{-1/3}$, $IR1 = 83$, $IR2 = 94$, $IR3 = 159$.

the simulated results agreed very well with the existing theory for the ring-disk electrode. However, such a calculation required 45 min of computation time on a very fast CDC 6600 computer. Fortunately, however, it was found that the introduction of additional factors into the calculation of the effects of convection resulted in identical results being obtained for $L = 50$. These calculations required from 20 sec to 1 min depending upon the mechanism. For $L = 50$ and $DMA = 0.45$, Eqs. (28) and (36) become

$$XJJ = \frac{XJ}{1.0 - (1.11 \cdot XJ \cdot VNAUT)} \tag{66}$$

and

$$RKK = \frac{RK}{\exp(1.03 \cdot XJ \cdot VNAUT)}. \tag{67}$$

## IX. SUMMARY

The ring-disk electrode is a very attractive tool for studying fast chemical follow-up reactions. When coupled with the versatility of digital simulation, the ring-disk electrode becomes a very powerful tool for unraveling the complex reaction mechanisms with which Nature presents us.

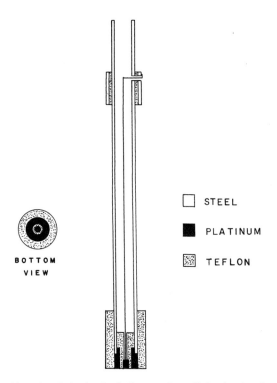

STEEL

PLATINUM

TEFLON

BOTTOM
VIEW

Fig. 4. A typical platinum ring-disk electrode.

APPENDIX I.  THE ROTATING RING-DISK ELECTRODE

A number of different rotating electrode configurations have been developed since the rotating disk electrode (RDE) was proposed by Levich [19] in 1942. Probably the most used rotating electrode other than the simple RDE is the rotating ring-disk electrode (RRDE) which was introduced by Frumkin and Nekrasov [1] in 1959. The RRDE consists of a disk and a concentric ring of electrode material separated from one another by a thin circular gap of inert insulating material. Both the ring and the disk are held coplanar in the same matrix (see Fig. 4). This electrode is placed in solution and rotated about the axis through the center of the disk perpendicular to the plane of the ring and disk. This rotation gives rise to a steady-state convective fluid flow with respect to the electrode surface.

The convective fluid flow to the surface of such a planar rotating electrode is independent of the manner in which active electrode material is distributed on the surface. Thus the theory developed by Levich [18] for the fluid flow to the RDE also applies to the RRDE. The pertinent results of this theory are that the fluid flow in the region of the electrode can be described to a good approximation by the following equations, which give the normal and radial components of the fluid velocity with respect to the electrode as a function of the normal distance from the electrode surface X, and the radial distance from the axis of rotation R.

$$V_X = -0.51\omega^{3/2}\nu^{-1/2}\chi^2 \tag{68}$$

$$V_R = -0.51\omega^{3/2}\nu^{-1/2}XR. \tag{69}$$

In the above equations, $\omega$ is the rotation rate of the electrode in rad/sec and $\nu$ is the kinematic viscosity of the solution in $cm^2$/sec. In particular, it should be noted that at the electrode surface X = 0, there is no convective fluid flow relative to the surface. This gives rise to a region near the electrode surface in which the primary mode of mass transport of electroactive species is diffusion.

If the potential of the disk electrode with respect to some reference half-cell is appropriately chosen, a chemical species A which is present in the bulk of the solution will be oxidized (or reduced) to a species B according to the following equation,

$$A \pm ne \rightarrow B. \tag{70}$$

The molecules of B which are produced at the disk electrode will then begin to diffuse outward from the electrode surface. As they diffuse away from the electrode, they will be caught in the radial convective flow and swept outward away from the disk and toward the ring where their presence may be detected by applying a potential to the ring electrode which is sufficient to cause the regeneration of species A,

$$B \pm ne \rightarrow A. \tag{71}$$

The current which is passed at the ring electrode is then a direct measurement of the amount of B reaching the ring electrode.

If B is a stable species, then that fraction of the B generated at the disk which will be detected at the ring will be a function only of electrode geometry. This fraction is called the collection efficiency N and is determined experimentally from the following equation:

$$N = i_r/i_d \tag{72}$$

where $i_r$ is the limiting current at the ring electrode due to the electrolysis of the B produced at the disk by the current $i_d$.

If N is not stable and instead undergoes a homogeneous chemical reaction such as

$$B + D \overset{k}{\rightarrow} E + F \tag{73}$$

then the fraction of B detected at the ring $N_k$ will be smaller and will be dependent not only upon electrode geometry, but also upon such parameters as the rate constant k, the rotation rate $\omega$, the concentrations of species A and D, the diffusion coefficients, and the kinematic viscosity $\nu$ of the solution. Thus the RRDE provides a steady-state method for both detecting short-lived intermediates and determining the kinetics of their disappearance.

## APPENDIX II. A TYPICAL PROGRAM

A listing of the FORTRAN computer program which was written to simulate the first-order mechanism described in this chapter is presented in this appendix. Although a more efficient program has been written, this one is presented in the interest of clarity. Since species C affects neither the ring nor the disk current, its concentrations are not calculated in the interest of minimizing computer time.

The first page of the program sets up the initial conditions. performs calculations which must be done only once, and calculates the effects of the first $\Delta t$ interval. The READ statements are self-explanatory except for the parameters LIMIT and LBJ. LIMIT is the total number of iterations which the program is to complete before termination. The value of this parameter is usually 150 for L = 50. If LBJ is any value other than zero, the program, upon completion of LIMIT iterations, will return to statement number 10 and begin reading a new set of input parameters.

JMAXM is a parameter that will later limit the number of boxes in the J direction which must be considered. VNAUT is a collection of terms used in the calculation of the effects of convection [see Eq. (27)].

In order to minimize edge effects, calculations are carried out on ten radial boxes beyond the outer edge of the ring electrode. To this end a parameter IR4 is defined as IR3 + 10. The parameters K2$\emptyset$, K3I, K3$\emptyset$ and so on are the radial box numbers of the outer edge of the gap, the inner edge of the ring, the outer edge of the ring, and so on. These parameters are used in designating regions in which certain boundary conditions apply.

AD is the area of the disk electrode. ZDNORM is a collection of parameters which, when multiplied by the fractional concentration of A in the (J = 2, K = 1) box, gives the disk current parameter ZD. ZRNORM is similar to ZDNORM in that the product of ZRNORM and the fractional concentration of B in a (J = 2) ring box yields the current density for that box. The product of the current density and the ratio of the area of that box to the area of the ring electrode yields the contribution of that box to the ring current parameter ZR.

The statement

$$FA(1, 1) = 0.0 \ \$ \ FB(1, 1) = 1.0 \tag{74}$$

sets up the disk boundary condition after one iteration. The statements in the 110 DO LOOP set up the boundary conditions for the rest of the (J = 1) boxes. The 150 DO LOOP sets up all the remaining boxes with this initial concentration.

The 999 DO LOOP begins near the bottom of the first page of the listing. This is the large time DO LOOP which must be completed LIMIT times before termination. The second statement after the DO 999 statement (top of the second page) calculates the effects of diffusion and the passage of current upon the concentration of B in the (J = 1, K = 1) box. Similarly, the 210 DO LOOP calculates the effects of diffusion and the passage of current upon the concentration of A in the (J = 1) ring boxes. The 220 and 230 DO LOOPS calculate the effects of diffusion upon the concentrations in the (J = 1) boxes corresponding to the gap and the outer insulating regions.

The 350 DO LOOP calculates the effects of diffusion upon all other boxes. Since the change in concentration in any one box depends upon the concentrations of the boxes on either side of it, a newly calculated concentration is placed in a variable DRIP and then in a variable DROP until the old concentration is no longer needed. Then the new concentration replaces the old value in the concentration array. Also calculated in this loop is the parameter JMAXM which reflects the largest value of J such that the concentration in that box is different from the bulk concentration.

The effects of convection are now calculated in the subroutines CONNORM and CONRAD. These will be discussed later. The 490 DO LOOP calculates the effects of the homogeneous reaction upon all boxes containing species B. The remaining statements in the main program calculate the disk current parameter ZD, the ring current parameter ZR, and the ratio of the two ZRATIO, and output this information.

Subroutine CONNORM uses Eq. (66) to calculate the effects of convection in the normal direction. The calculations are carried out from the surface of the electrode outward so that concentrations which will be needed in later calculations are not replaced. Since Eq. (66) usually yields a value of XJJ which is between two points of known concentrations, a linear interpolation is performed to yield the concentration at XJJ. The parameter JJ is the layer number such that the point XJJ from the electrode surface is between the centers of the JJ and (JJ + 1) layers. DJJ is the distance from the electrode surface to the center of the JJ-th layer in units of $\Delta x$ cm. Thus, assuming a linear concentration change from the center of the JJ layer to the center of the (JJ + 1) layer, the concentrations at XJJ can be calculated from the equations in the 101 DO LOOP.

The effects of convection in the radial direction are calculated from Eq. (67) in subroutine CONRAD. In this case the calculations are carried out from the outer radial box inward in order not to replace concentrations needed in later calculations. Again, if RKK calculated from Eq. (67) is between two points of known concentrations, then a linear interpolation is performed to obtain the concentration at RKK. In those cases where RKK is less than IR1, the concentration in the (K = 1) box is used.

A far less time-consuming program has been written in which the parameters JJ, XJJ, and DJJ are calculated just once as functions of J and stored for use each time the effects of convection normal to the electrode are to be calculated. In a similar manner KK, RKK, and DKK are calculated only once as functions of J and K and then stored. These modifications greatly reduce the time required to run the program but greatly increase the required storage space. The KK, RKK, and DKK arrays are typically 20 by 100 each. Thus, depending upon the storage capacity of the individual computer system, one may be forced to use the program as presented below at the expense of computer time. It has been found that storing these arrays in blocks on a disk, and recalling the blocks of information as needed, solved the storage problem and did not significantly increase the time required to run the program.

```
          PROGRAM RRDE
          COMMON    FA(18,90),FB(18,90)
C         THIS PROGRAM SIMULATES AN RRDE WITH FIRST-ORDER
C         DECOMPOSITION OF THE DISK PRODUCT, B, TO THE
C         ELECTRO-INACTIVE SPECIES, C.
C         THIS SIMULATION USES ONLY ONE BOX FOR THE DISK.
          READ 20,L,LIMIT
       20 FORMAT(8I10)
          READ 30,DMA,DMB
       30 FORMAT(8F10.6)
          READ 20, IR1,IR2,IR3
          XL=L $ ZA=SQRT(XL/DMA) $ TONE=1.0/XL
          VNAUT=1.0/(SQRT(DMA)*((SQRT(XL))**3))
          IR4=IR3+10
          K20=IR2-IR1+1 $ K31=K20+1 $ K30=IR3-IR1+1
          K41=K30+1 $ K40=IR4-IR1+1 $ K5=K40-1
          RI=IR1 $ AD=(RI-0.5)**2
          ZONORM=DMA*ZA
          ZRNORM=DMB*ZA/AD
       10 READ 30,XKT
          READ 20,LBJ
          XKTL=XKT/XL
          ZR=0.0 $ ZRATIO=0.0
          JMAXM=0
          PRINT 1010
     1010 FORMAT(1H1)
          PRINT 1020, IR1,IR2,IR3
     1020 FORMAT(14H RRDE WITH R1=,I5,5H, R2=,I5,5H, R3=,I5//)
          PRINT 1030, L,LIMIT
     1030 FORMAT(3H L=,I5,5X,6HLIMIT=,I5//)
          PRINT 1040,DMA,DMB
     1040 FORMAT(5X,4HDMA=,F10.6,10X,4HDMB=,F10.6)
          PRINT 1045,XKT
     1045 FORMAT(5H XKT=,F10.3//)
          PRINT 1050
     1050 FORMAT(5X,6HTRATIO,9X,2HZD,13X,2HZR,11X,6HZRA[10//)
          PRINT 1060,TONE,ZA,ZR,ZRATIO
     1060 FORMAT(F10.3,5X,F10.6,5X,F10.6,5X,F10.6)
          FA(1,1)=0.0 $ FB(1,1)=1.0
          DO 110 K=2,K40
          FA(1,K)=1.0
          FB(1,K)=0.0
      110 CONTINUE
          DO 150 J=2,18
          DO 140 K=1,K40
          FA(J,K)=1.0 $ FB(J,K)=0.0
      140 CONTINUE
      150 CONTINUE
          DO 999 I=2,LIMIT
          TI=I
```

```
      FB(1,1)=FB(1,1)+DMA*FA(2,1)-DMB*(FB(1,1)-FB(2,1))
      DO 210 K=K3I,K30
      FA(1,K)=FA(1,K)+DMB*FB(2,K)-(DMA*(FA(1,K)-FA(2,K)))
210   CONTINUE
      DO 220 K=2,K20
      FA(1,K)=FA(1,K)-DMA*(FA(1,K)-FA(2,K))
      FB(1,K)=FB(1,K)-DMB*(FB(1,K)-FB(2,K))
220   CONTINUE
      DO 230 K=K4I,K40
      FA(1,K)=FA(1,K)-DMA*(FA(1,K)-FA(2,K))
      FB(1,K)=FB(1,K)-DMB*(FB(1,K)-FB(2,K))
230   CONTINUE
      DO 350 K=1,K40
      DROPA=FA(2,K)+DMA*(FA(3,K)-2.0*FA(2,K)+FA(1,K))
      DROPB=FB(2,K)+DMB*(FB(3,K)-2.0*FB(2,K)+FB(1,K))
      JMAX=17
      DO 340 J=3,JMAX
      DRIPA=FA(J,K)+DMA*(FA(J+1,K)-2.0*FA(J,K)+FA(J-1,K))
      DRIPB=FB(J,K)+DMB*(FB(J+1,K)-2.0*FB(J,K)+FB(J-1,K))
      FA(J-1,K)=DROPA
      FB(J-1,K)=DROPB
      DROPA=DRIPA
      DROPB=DRIPB
      IF(DROPA-0.999999)340,330,310
310   JMAX=J
      IF(JMAX-JMAXM)340,340,320
320   JMAXM=JMAX
      GO TO 340
330   JMAX=J+1
340   CONTINUE
      FA(JMAX,K)=DROPA
      FB(JMAX,K)=DROPB
350   CONTINUE
      CALL CONNORM(FA,FB,JMAXM,VNAUT,K40)
      CALL CONRAD(FA,FB,JMAXM,VNAUI,K40,K5,IR1)
      DO 490 K=1,K40
      DO 480 J=1,JMAXM
      IF(FB(J,K))480,480,410
410   DEL=XKTL*FB(J,K)
      FB(J,K)=FB(J,K)-DEL
480   CONTINUE
490   CONTINUE
      ZD=FA(2,1)*ZDNORM
      ZR=0.0
      DO 510 K=K3I,K30
      RK=K-2+IR1
      ZR=ZR+(((RK+0.5)**2)-((RK-0.5)**2))*FB(2,K)*ZRNORM
510   CONTINUE
      ZRATIO=ZR/ZD
      THATIO=TI/XL
```

```
          PRINT 1510,TRATIO,ZD,ZR,ZRATIO
     1510 FORMAT(F10.3,5X,F10.6,5X,F10.6,5X,F10.6)
      999 CONTINUE
          IF(LBJ) 10,2000,10
     2000 CONTINUE
          END


          SUBROUTINE CONNORM(A,B,JMAXM,VNAUT,K40)
          DIMENSION A(18,90),B(18,90)
          DO 102 J=1,JMAXM
          XJ=J-1
          XJJ=XJ/(1.0-(1.11*VNAUT*XJ))
          JJ=XJJ+1.0
          DJJ=JJ-1
          DO 101 K=1,K40
          A(J,K)=A(JJ,K)+(A(JJ+1,K)-A(JJ,K))*(XJJ-DJJ)
          B(J,K)=B(JJ,K)+(B(JJ+1,K)-B(JJ,K))*(XJJ-DJJ)
      101 CONTINUE
      102 CONTINUE
          RETURN
          END


          SUBROUTINE CONRAD(A,B,JMAXM,VNAUT,K40,K5,IR1)
          DIMENSION A(18,90),B(18,90)
          DO 202 J=1,JMAXM
          DO 201 KINV=1,K5
          K=K40-KINV
          RK=K-2+IR1
          XJ=J-1
          RKK=RK/EXP(1.03*VNAUT*XJ)
          KK=RKK+1.0
          DKK=KK-1
          IF(KK-IR1) 100,105,110
      100 KK=1
          KKK=1
          GO TO 120
      105 KK=1
          KKK=2
          GO TO 120
      110 KK=KK-IR1+1
          KKK=KK+1
      120 A(J,K)=A(J,KK)+(A(J,KKK)-A(J,KK))*(RKK-DKK)
          B(J,K)=B(J,KK)+(B(J,KKK)-B(J,KK))*(RKK-DKK)
      201 CONTINUE
      202 CONTINUE
          RETURN
          END
              FINIS
```

## REFERENCES

[1].  A. N. Frumkin and L. N. Nekrasov, Dokl. Akad. Nauk SSSR, 126:115 (1959).

[2].  W. J. Albery, Trans. Faraday Soc., 62:1915 (1966).

[3].  W. J. Albery and S. Bruckenstein, Trans. Faraday Soc., 62:1920 (1966).

[4].  W. J. Albery, S. Bruckenstein, and D. T. Napp, Trans. Faraday Soc., 62:1932 (1066).

[5].  W. J. Albery, S. Bruckenstein, and D. C. Johnson, Trans. Faraday Soc., 62:1938 (1966).

[6].  W. J. Albery and S. Bruckenstein, Trans. Faraday Soc., 62:1946 (1966).

[7].  W. J. Albery and S. Bruckenstein, Trans. Faraday Soc., 62:2584 (1966).

[8].  W. J. Albery and S. Bruckenstein, Trans. Faraday Soc., 62:2596 (1966).

[9].  W. J. Albery, Trans. Faraday Soc., 63:1771 (1967).

[10]. W. J. Albery, M. L. Hitchman, and J. Ulstrup, Trans. Faraday Soc., 64:2831 (1968).

[11]. W. J. Albery, M. L. Hitchman, and J. Ulstrup, Trans. Faraday Soc., 65:1101 (1969).

[12]. K. B. Prater and A. J. Bard, J. Electrochem. Soc., 117:207 (1970).

[13]. K. B. Prater and A. J. Bard, J. Electrochem. Soc., 117:335 (1970).

[14]. K. B. Prater and A. J. Bard, J. Electrochem. Soc., 117:1517 (1970).

[15]. S. W. Feldberg and C. Auerbach, Anal. Chem., 36:505 (1964).

[16]. S. W. Feldberg, Electroanalytical Chemistry, Vol. 3 (A. J. Bard, ed.), Marcel Dekker, New York, 1969.

[17]. W. J. Albery and J. Ulstrup, Electrochim. Acta, 13:281 (1968).

[18]. V. G. Levich, Physiochemical Hydrodynamics, Prentice-Hall, Englewood Cliffs, N.J., 1962.

[19]. V. G. Levich, Acta Physicochim. USSR, 17:257 (1942).

Chapter 9

DIGITAL SIMULATIONS OF ELECTROGENERATED

CHEMILUMINESCENCE AT THE ROTATING RING-DISK ELECTRODE

J. T. Maloy

Department of Chemistry
West Virginia University
Morgantown, West Virginia

I. INTRODUCTION

Since its introduction by Frumkin and Nekrasov [1] in 1959, the rotating ring-disk electrode (RRDE) has become widely used in the study of the homogeneous kinetics of electrogenerated species. As shown in Fig. 1, the RRDE consists of concentric coplanar ring and disk electrodes mounted in a common inert matrix, coaxial with a shaft about which the entire assembly may be rotated, and separated from each other by an inert insulating material. Electrical contact is provided for each electrode so that the ring and disk may be independently connected to an external source with brush contacts.

When the RRDE is used conventionally, a species is electrogenerated at the disk electrode while the ring electrode is maintained at a potential at which this species will be electroactive; thus, the reverse of the disk reaction occurs at the ring and if A is the species present in the bulk of the solution and B is the species generated from A at the disk, the reactions

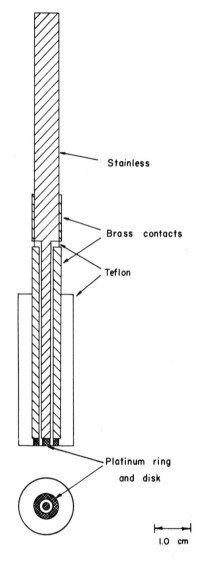

Fig. 1. The ring-disk electrode. (The ring and disk are not drawn to scale.)

occurring in conventional RRDE studies are

A → B (disk reaction)

B → A (ring reaction).
$$(1)$$

The ratio of the steady-state current at the ring to that at the disk under these conditions is called the collection efficiency (N) because it measures the fraction of the material generated at the disk which ultimately reaches the ring. If the species which is generated at the disk is stable, N is determined only by the geometry of the electrode and is, for any real electrode, calculable and less than unity [2]. If this species is unstable so that

$$B \xrightarrow{k_1} X \text{ (in solution)} \tag{2}$$

yielding an electroinactive product X, then N is an increasing function of the angular velocity ($\omega$) of the electrode which, at high $\omega$, approaches that value of N which would be obtained in the absence of reaction (2). This variation of N with $\omega$ has been treated theoretically by exact [3] and approximate [4] methods so that the lifetime of the unstable species may be determined experimentally. This is the conventional manner in which the RRDE has been used.

In conventional studies of electrogenerated chemiluminescence (ECL), both anion and cation radicals of an appropriate fluorescer are alternatively generated at a single electrode, thereby producing luminescence in the subsequent reaction of these ionic species within the diffusion layer about the electrode. One common method employs an unreferenced sinusoidal potential as a means of generating ionic species [5,6]; another uses a double (or multiple) potential step between known oxidative and reductive potentials [7,8]. While the former technique is probably the simplest means of observing the electroluminescent phenomenon, it is undesirable from an electrochemical standpoint because accurate potentials are not known; hence it is possible that dianions or dications are formed or that the decomposition of the solvent or supporting electrolyte at unknown potentials forms agents which inhibit the desired chemiluminescent reaction. Even if this difficulty can be eliminated by employing the latter technique, the current-time-intensity behavior obtained is observed to be quite complex because of the diffusional nature of the processes involved so that mathematical description of these processes is extremely difficult. In addition, large nonfaradaic double-layer charging effects inevitably result when a high-frequency alternating potential is applied to a cell having large uncompensated resistance. Most importantly, no alternating-potential process can achieve a true steady state; the intensity of chemiluminescent emission always varies with time.

These conventional ECL problems may be overcome through the unconventional use of the RRDE; instead of the ring electrode being used to collect the species made at the disk, it is used to generate a species other than that generated at the disk. Therefore, the least complicated set of electrode reactions in the RRDE-ECL experiment may be written

$A \rightarrow B$ (disk reaction)

$A \rightarrow C$                                                                                 (3)
        (ring reactions)
$B \rightarrow C$

where the latter ring reaction is thermodynamically unavoidable; in addition, the solution reaction of the anionic and cationic precursors of chemiluminescence occurs:

$$B + C \xrightarrow{k_2} \text{chemiluminescent products}$$                                  (4)

so that RRDE-ECL is really a new technique in which the disk-generated species is collected by a species generated at the ring rather than by the ring itself; the efficiency of this collection depends upon the rate of reaction (4) ensuing in solution.

The advantages of the RRDE in ECL studies are obvious. This type of electrode is particularly convenient for quantitative work because the solutions of the hydrodynamic equations which describe mass transfer as a function of rotation rate are known [9]. Since the steady state may be attained quickly at even moderate rotation rates, the potentials of both the ring and the disk may be set independently, thereby resulting in constant-current levels at both electrodes; this, in turn, results in the steady-state emission of chemiluminescence (unless the bulk concentration of the fluorescer is depleted or sufficient quantities of quenching agents are generated so as to deactivate the excited state of the chemiluminescent product). Because there is no alternation of potential at either electrode, charging currents are negligible once the steady state is attained. As an added advantage, the typically larger electrode area of the rotating ring-disk results in greater quantities of chemiluminescent emission.

Even though the RRDE is experimentally advantageous in the study of ECL phenomena, an exact theoretical treatment of the problem (which involves diffusion, convection, and at least one chemical reaction) is presently unobtainable. Therefore, it is necessary to treat the problem with digital simulation techniques so that the experimental advantages of the RRDE may be realized. As a result of these simulations and some pertinent experimental observations, it may be proposed that RRDE-ECL studies are an ideal way to determine $\Phi_{ECL}$, the efficiency of the current at producing luminescence. What follows, then, is a report of the initial theoretical treatment of ECL at the RRDE using digital simulation techniques.

## II. THE SIMULATION METHOD

Digital simulation techniques have been used to treat a variety of electrochemical [10], ECL [11, 12], and RRDE [4, 13] problems. The extension of the digital simulation method to the problem of RRDE-ECL is fairly straightforward. In fact, the computer programs written by Prater (see Chapter 8, this volume) need only slight modification to treat RRDE-ECL. Hence, only these modifications will be discussed herein.

In treating RRDE-ECL with digital simulation, one models the hydrodynamic layer of the solution by dividing it into a specified number of volume elements; the concentration of any species in solution is considered to be uniform within a given element. Due to the cylindrical symmetry of the RRDE, the solution is divided into several layers of thickness $\Delta x$, with each layer containing a cylindrical element of diameter $\Delta r$ which is centered on the axis of rotation; a series of concentric annular elements, each of width $\Delta r$, make up the remainder of each layer. Each element is numbered with its layer number J and its radial number K; hence, the relative concentration of any species (its real concentration divided by the bulk concentration of a species present initially) is specified as a function of two integers. Thus, in the nomenclature used by Prater, $F_B(J, K)$ is "the relative concentration of the B-th species in the J-th layer and the K-th ring."

The diffusion, convection, and chemical reactions (kinetics) of the species which occupy these elements are also functions of time. These temporal changes are modeled iteratively, each iteration of duration $\Delta t$ generating a new two-dimensional concentration array for each species present. The algorithms used to handle each of these changes are given by Prater. Normal diffusion is treated by a difference method; radial diffusion is neglected. Both normal and radial convection are simulated by using the laws of hydrodynamics to calculate the distance the solution in a given volume element travels during the interval $\Delta t$; a new array is generated by effecting these translations for each element. Chemical reactions occurring during $\Delta t$ are treated by using the difference equation form of the differential rate law to calculate new concentrations within each element following each iteration.

Variables of electrochemical interest are specified by the electrode boundary conditions that give rise to the concentration gradients in solution. Current is measured as the flux of electroactive material into the electrode under a given set of potential dependent electrode conditions. While it is easy to simulate any reversible electrode potential or any magnitude of constant current, the simulations reported herein all assume a potential such that the concentration of the electroactive species goes to zero at the electrode and the limiting current is achieved.

Obviously, the validity of any such simulation will depend upon the dimensions of $\Delta x$, $\Delta r$, and $\Delta t$; as these approach zero, the problem approaches differential form. Making these smaller, however, increases the computation time required to achieve a steady-state solution; therefore, some optimization is required. In practice these incremental quantities are defined at the outset of the simulation in terms of real time-distance parameters and dimensionless quantities which are specified within the program. For example, $\Delta x$ is defined in terms of $\Delta t$ and $D_A$, the diffusion coefficient of the species originally present in solution

$$\Delta x \equiv \left(\frac{D_A \Delta t}{DM_A}\right)^{1/2}. \tag{5}$$

Here $DM_A$ is the model diffusion coefficient of species A, and it is assigned a value less than 0.5 [10]. Similarly, $\Delta t$ is defined in terms of $\omega$ (the angular velocity), $D_A$, and $\nu$ (the kinematic viscosity of the solvent),

$$\Delta t \equiv \left(\frac{D_A}{\nu}\right)^{1/3} \frac{\omega^{-1}}{L} \tag{6}$$

where L is an assigned number of time iterations; it is found that after 3L iterations, steady-state conditions are achieved. It might be noted that the dimensionless coefficient $(D_A/\nu)^{1/3}$ appears as a result of the fact that hydrodynamic behavior depends upon solution viscosity. Finally, $\Delta r$ is defined by IR1, the number of elements representing the disk electrode,

$$\Delta r \equiv \frac{r_1}{IR1 - 0.5} \tag{7}$$

where $r_1$ is the actual radius of the disk electrode. [For the electrode used in the experimentation mentioned below, $r_1$ was 0.145 cm. The simulation variables IR2 and IR3 which respectively represent the inner radius ($r_2$ = 0.163 cm) and the outer radius ($r_3$ = 0.275 cm) of the ring electrode were in the same proportion to IR1 as $r_2$ and $r_3$ were to $r_1$.] Thus, it may be seen that the specification of the variables $DM_A$, L, and IR1 simultaneously specifies $\Delta x$, $\Delta t$, and $\Delta r$. In the simulation reported below, the following values were used for each of these quantities:

$$DM_A = DM_B = DM_C = 0.45, \quad L = 50, \quad IR1 = 81. \tag{8}$$

This set of input parameters is such that one steady-state simulation is obtained after 60 sec of computation with the Control Data Corporation Model 6600; in cases where comparisons with "exact" solutions are possible (e.g., the disk current in the absence of kinetic complications) the simulated results obtained in this computation time are found to agree within the error of electrochemical measurement (ca. 1%).

Since the basic RRDE program written by Prater calculates collection efficiencies, some modification of boundary conditions is required to treat RRDE-ECL according to Eq. (3). For the disk, these are unchanged by the introduction of C, a ring-generated species. They are

$$F_A'(1, \ 1) = F_A(1, \ 1) = 0$$

$$F_B'(1, \ 1) = F_B(1, \ 1) + DM_A[F_A(2, \ 1)] - DM_B[F_B(1, \ 1) - F_B(2, \ 1)] \quad (9)$$

$$F_C'(1, \ 1) = F_C(1, \ 1) = 0$$

where primed quantities exist at the start of the next iteration. Note that only one element is required to simulate the IR1 elements making up the disk electrode, and K = 1 in these equations; this is possible because the disk is a uniformly accessible surface so that the same conditions would appear for each element. The quantity $DM_A[F_A(2, \ 1)]$ represents the amount of species A which is converted to B at the disk; this is proportional to the disk current.

Since the ring is not uniformly accessible, each ring element must be treated separately. The boundary conditions for each ring element become

$$F_A'(1, \ K_R) = F_A(1, \ K_R) = 0$$

$$F_B'(1, \ K_R) = F_B(1, \ K_R) = 0$$

$$F_C'(1, \ K_R) = F_C(1, \ K_R) + DM_A[F_A(2, \ K_R)] + DM_B[F_B(2, \ K_R)]$$
$$- DM_C[F_C(1, \ K_R) - F_C(2, \ K_R)]$$

$$(10)$$

where $K_R$ is a radial element number of an element representing the ring. The current at the ring is proportional to the area-weighted sum of the currents of each ring element:

$$\sum_{\{K_R\}} \ \{DM_A[F_A(2, \ K)] + 2DM_B[F_B(2, \ K)]\} \cdot \left(\frac{A(K)}{A_D}\right)$$

where A(K) is the area of the K-th ring and $A_D$ is the area of the disk. Note that the total current is the sum of current contributions from both species A and species B.

Diffusion and convection are treated just as in previous RRDE simulations; therefore, the only remaining additional complication of the RRDE-ECL simulation is the treatment of Eq. (4), the reaction of B and C to produce light. Species B and C, of course, are the radical ions of A, the hydrocarbon present initially, and the reaction in question is an electron transfer reaction which may or may not produce light:

$$B + C \xrightarrow{k_2} {}^1A + {}^1A^*$$

$${}^1A^* \rightarrow {}^1A + h\nu$$

$$(11a)$$

or

$$B + C \rightarrow 2\,{}^1A.$$

$$(11b)$$

In any event, since the conversion of the excited singlet to ground-state singlet is quite rapid, the net material effect of each fruitful encounter of B and C is the production of 2A, and the number of quanta produced depends upon the fraction of fruitful encounters of B and C which proceed by (11a). Therefore, it is convenient, if not mandatory, that $\Phi_{ECL}$, the efficiency of the ECL process, be defined as the fraction of fruitful radical ion encounters which result in the emission of a photon.

With this background, one may rewrite the differential rate law

$$-\frac{d[B]}{dt} = k_2 [B] [C] \tag{12}$$

in difference form, using relative concentrations,

$$-\Delta F_B(J, K) = (k_2 C_A^0 \Delta t) \cdot F_B(J, K) \cdot F_C(J, K) \tag{13}$$

where $C_A^0$ is the bulk concentration of the A-th species. Combining Eqs. (6) and (13) yields the form actually used in the simulation,

$$-\Delta F_B(J, K) = \left[\left(\frac{D_A}{\nu}\right)^{1/3} \cdot \left(\frac{k_2 C_A^0}{\omega}\right)\right] \cdot F_B(J, K) \cdot \frac{F_C(J, K)}{L} \tag{14}$$

where the bracketed dimensionless coefficient is assigned the variable name XKTC in the simulation,

$$XKTC = \left(\frac{D_A}{\nu}\right)^{1/3} \cdot \left(\frac{k_2 C_A^0}{\omega}\right). \tag{15}$$

This is the final variable input parameter used in the simulation of RRDE-ECL; variations in XKTC determine the extent to which rotation competes with kinetics.

Equation (14) is used to determine the chemical changes due to radical ion reaction within each element and to quantify the chemiluminescence occurring as a result of these changes. The relative concentrations in each element are adjusted accordingly during each iteration:

$$F_A'(J, K) = F_A(J, K) - 2\Delta F_B(J, K)$$

$$F_B'(J, K) = F_B(J, K) + \Delta F_B(J, K) \tag{16}$$

$$F_C'(J, K) = F_C(J, K) + \Delta F_B(J, K).$$

### III. PARAMETRIC SUBSTITUTIONS

The specification of XKTC as an input parameter permits other time-dependent output parameters to be rendered dimensionless by the rate constant $k_2 C_A^0$ instead of $\omega^{-1}$. This parametric substitution results in no new information; however, the new dimensionless parameter which results from its application often expresses a desired relationship more clearly than that which was obtained initially. For example, consider the development of a dimensionless steady-state RRDE-ECL intensity parameter which follows.

The number of photons emitted from any element during $\Delta t$ may be written

$$-\Delta F_B(J, K) \cdot \Phi_{ECL} \cdot C_A^o \cdot [\Delta x \cdot A(K)]$$

so that I, the total ECL intensity (Quanta/sec), is the sum of the emission from all the elements divided by $\Delta t$:

$$I = - \sum_{J, K} \frac{\Delta F_B(J, K) \cdot \Phi_{ECL} \cdot C_A^o \cdot [\Delta x \cdot A(K)]}{\Delta t}. \tag{17}$$

Substitution of Eqs. (5) and (6) and division of both members by $A_D$ yields, after rearrangement,

$$\frac{I}{\Phi_{ECL} A_D C_A^o D_A^{1/3} \nu^{1/6} \omega^{1/2}} = -\left(\frac{L}{DM_A}\right)^{1/2} \sum_{J, K} \Delta F_B(J, K) \cdot \frac{A(K)}{A_D} \tag{18}$$

where all those quantities on the right-hand side (rhs) are simulation variables, while the left-hand side (lhs) is a valid dimensionless representation of ECL intensity. Unfortunately, however, in this representation I is rendered dimensionless by $\omega$, an experimental variable; hence, any use of this parameter in a comparison of real and simulated results requires that each experimentally observed intensity be divided by $\omega^{1/2}$ prior to any comparison.

This requirement may be eliminated by dividing both sides of Eq. (18) by $(XKTC)^{1/2}$ and by combining the resulting expression with Eq. (15). This results in

$$\frac{I}{\Phi_{ECL} A_D [D_A (C_A^o)^3 k_2]^{1/2}} = -\left(\frac{L}{XKTC \cdot DM_A}\right)^{1/2} \cdot \sum_{J, K} \Delta F_B(J, K) \cdot \frac{A(K)}{A_D} \tag{19}$$

where the lhs is a new dimensionless RRDE-ECL intensity parameter which contains no $\omega$ dependence like that given in Eq. (18); since the rhs of Eq. (19) is merely the rhs of Eq. (18) divided by $(XKTC)^{1/2}$, it may readily be calculated from simulation variables.

To model the steady-state intensity-rotation-rate dependence with digital simulation, one would vary XKTC, the dimensionless representation of $\omega$ in the program described in the previous section, thereby obtaining the dimensionless intensity parameter of Eq. (19) as a steady-state function of XKTC.

### IV. SIMULATION RESULTS

The program described above was used to model the intensity-rotation-rate behavior for RRDE-ECL at an electrode of the geometry given above. The results of this simulation are displayed in Fig. 2 which shows the variation of the dimensionless ECL parameter of Eq. (19) with the square root of the dimensionless $\omega$ parameter XKTC. Actually, Fig. 2 consists of two graphs, each having the same ordinate and lying adjacent at a common point; the abscissa of the graph on the right, however, is the reciprocal of that on the left plotted in reverse. Hence, the total abscissa is drawn with $\omega$ increasing in a nonlinear fashion from left to right, and all possible values of $\omega$ are shown. One may also observe that the denominator of the ordinate variable contains only factors which are constants, given a set of experimental conditions; similarly, the abscissa variables can change only when $\omega$ changes in a real experiment. Thus, Fig. 2 is the dimensionless representation of the I vs. $\omega$ behavior in RRDE-ECL.

The simulation, then, predicts that RRDE-ECL intensity will increase with increasing rotation rate, go through a maximum, and decrease to zero thereafter. In fact, the maximum (point b on the graph) is predicted when

$$\left(\frac{D_A}{\nu}\right)^{1/6} \left(\frac{\omega}{k_2 C_A^0}\right)^{1/2} = 0.4 \tag{20}$$

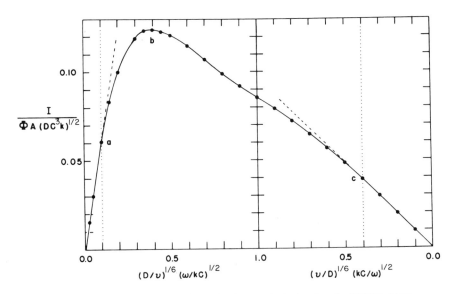

Fig. 2.  Dimensionless-intensity-rotation-rate behavior in RRDE-ECL. Each point required an independent simulation.

or, since $(\nu/D_A) \simeq 10^3$ for systems of electrochemical interest, when

$$\omega \simeq 1.6 \, k_2 c_A^{\,0}. \tag{21}$$

Therefore, since the maximum rotation rate possible with typical instrumentation is $\sim 10^3$ $sec^{-1}$ and the minimum practical concentration of electroactive species is $\sim 10^{-3}$ $\underline{M}$, the ECL maximum will be reached only if $k_2$ is less than $10^6$ $\underline{M}^{-1}$. This, it will be suggested, is highly unlikely.

In addition to point b, two additional points have been marked on Fig. 2, and vertical lines passing through these points divide the graph into three regions. Interior to point a, the simulation predicts linear behavior with the straight line passing through the left origin; exterior to point c, a straight line passing through the right origin is predicted. This indicates that as I is increasing, it increases with $\omega^{1/2}$, but as it decreases, it decreases with $\omega^{-1/2}$. More importantly, it may be noted that, in theory, three distinctly different types of ECL behavior may be observed, depending upon whether the reaction kinetics are fast, slow, or intermediate with respect to the rotation rate.

Concentration contours at these three rotation rates (a, b, and c) are shown for species A and species C in Fig. 3. These graphs depict lines of equal concentration within the region bounded by a radius of the electrode and the hydrodynamic layer to a thickness of $2\delta$ where

$$\delta = 1.8 D_A^{1/3} \nu^{1/6} \omega^{-1/2}. \tag{22}$$

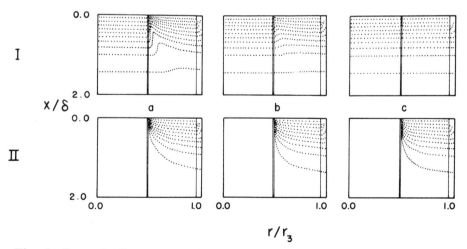

Fig. 3. Concentration contours for the parent hydrocarbon (I) and the ring-generated species (II) in RRDE-ECL. The points a, b, and c refer to those shown in Fig. 2.

Three vertical lines partition the hydrodynamic layer (which is drawn here increasing in a negative direction) into areas below the disk, gap, ring, and shield of the electrode. The lines of equal concentration are drawn every 0.10 unit of relative concentration. Figure 3, I depicts species A so that the outermost (lowest) line is the line of 0.99 relative concentration. In Fig. 3, II, species C is depicted with the outermost line of relative concentration 0.01. These differences result, of course, because A is moving toward the electrode whereas C is moving away from it. In all, Fig. 3 is not particularly edifying, however, and one need note only two facts:

(1) When the reaction is fast (Fig. 3, I, a), a high concentration of A forms at the inner edge of the ring electrode.

(2) At any rotation rate there is very little difference in the contour of species C. The species generated at the ring always surrounds the disk with a "doughnut" of reactive material.

Figure 4 is more informative; in Fig. 4, I the concentration contours of species B are shown with the outermost line at a relative concentration of 0.01. These contours really illustrate the effectiveness of the "doughnut" of C in blocking the penetration of B. In Fig. 4, I, a (fast kinetics) it is quite effective; in Fig. 4, I, c much of the disk-generated species escapes into the bulk of the solution.

The effect of this on the ECL produced by the reaction of B and C is shown in Fig. 4, II where $\rho$, the density of ECL emission (normal to the

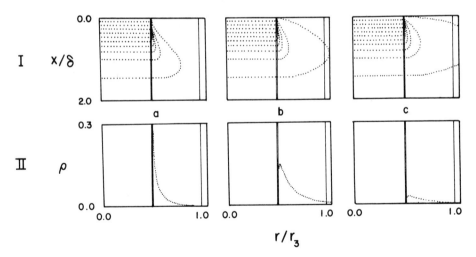

Fig. 4. Concentration contours for the disk-generated species (I) and ECL density (II) in RRDE-ECL. The rotation rates are specified by a, b, and c in Fig. 2.

electrode), is shown as a function of electrode radius. At a, the ECL would be seen as a sharp ring of light close to the inner edge of the ring. At b, the emission maximum would be displaced from the inner edge of the ring and the emission would be spread out, falling to about one-third of the maximum at midring. Finally, at c, the chemiluminescence would almost appear to cover the ring uniformly. Therefore, the appearance of the ECL emission is indicative of the rate of the process relative to the rotation rate.

If the process is rapid, ECL will be observed as a bright ring at the innermost edge of the ring. This region of fast kinetics, as noted above, is a region where I increases with $\omega^{1/2}$. In fact, one may write the equation of the straight line passing through point a to show this explicitly:

$$\frac{I}{\Phi_{ECL} A_D [D_A (C_A^0)^3 k_2]^{1/2}} = 0.62 \left(\frac{D_A}{\nu}\right)^{1/6} \left(\frac{\omega}{k_2 C_A^0}\right)^{1/2} \tag{23}$$

or, upon rearrangement,

$$I = 0.62 \Phi_{ECL} A_D C_A^0 D_A^{2/3} \nu^{-1/6} \omega^{1/2}. \tag{24}$$

This is quite similar to the Levich equation [9] for $i_d$, the current at the disk electrode,

$$i_d = 0.62 nF A_D C_A^0 D_A^{2/3} \nu^{-1/6} \omega^{1/2} \tag{25}$$

where n is the molar ratio of electrons to electroactive species, and F is the Faraday. Since the current at the disk is independent of conditions at the ring, Eq. (25) is valid even in RRDE-ECL, and combining the simulation result of Eq. (24) with the Levich equation yields

$$I = \Phi_{ECL} \cdot \frac{i_d}{nF} \tag{26}$$

which is valid in the region of fast kinetics. [By taking point a

$$\left(\frac{D_A}{\nu}\right)^{1/6} \left(\frac{\omega}{k_2 C_A^0}\right)^{1/2} = 0.1 \tag{27}$$

as the upper boundary of the region of fast kinetics, one obtains

$$\omega = 0.1 k_2 C_A^0 \tag{28}$$

as the rotation rate at this limit. This implies that if $k_2$ is greater than $10^7 \ M^{-1} \ sec^{-1}$, ordinary instrumentation is incapable of rotation rates outside this limit, and Eq. (26) is always valid.]

This consequence of a fast reaction between B and C is quite reasonable if one considers what can happen to a species B as it is swept across a ring which is generating species C: (1) it can react with C; (2) it can be

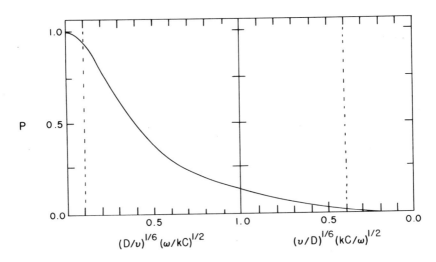

Fig. 5. The probability of fruitful encounter of ring- and disk-generated species in RRDE-ECL.

converted to C at the ring electrode; and (3) it can escape into the bulk of the solution. The result of the simulation in the limit of fast kinetics predicts that reaction with the ring-generated species is the most probable of these three events. In Fig. 5 this probability of fruitful encounter, P, is shown as a function of the dimensionless representation of $\omega$ in the manner of Fig. 2. Note that in the region where Eq. (26) is valid,

$$P \simeq 1.0 \tag{29}$$

so that the disk current is an effective measure of the rate of a reaction occurring in the hydrodynamic layer.

## V. THE DETERMINATION OF ECL EFFICIENCY

The results of these simulations suggest that RRDE-ECL is the ideal method for determining $\Phi_{ECL}$, the efficiency of the ECL process, so long as the reactions producing the ECL are rapid. In addition, the simulation gives the criteria by which this rapidity may be judged. Of course, since the reaction in question is the electron transfer between the anion radical and the cation radical of a hydrocarbon, there is every reason to suspect that it will be rapid indeed. It has been suggested that this redox reaction occurs at diffusion-controlled rates [11], and if this rate is calculated with the equation of Osborne and Porter [14]

$$k_d = \frac{8RT}{1000\eta} \tag{30}$$

T. V. Atkinson

Fig. 6. The ECL from DPA at the RRDE. The actual diameter of the inner bright ring is ~3.3 mm.

with $\eta$ = 0.79 cp for N, N-dimethylformamide (DMF), a solvent commonly used in ECL studies, a rate constant of $1.3 \times 10^{10}$ $M^{-1}$ is obtained. This is three orders of magnitude greater than the minimum rate constant required for Eq. (26) to be valid.

That the ECL emission is localized at the inner edge of the ring may be seen from Fig. 6, a photographic reproduction of the ECL emission that results from the simultaneous oxidation and reduction of 9, 10-diphenyl-anthracene (DPA) in dimethylformamide solution at the RRDE. Even though much clarity has been lost through the enlargement and reproduction of this print, it shows that the ECL emission is concentrated at the inner edge of the ring electrode, a condition that would be predicted if the reaction kinetics were rapid with respect to the rotation rate (~150 $sec^{-1}$). In addition, it may be observed experimentally that ECL intensity is directly proportional to disk current (and thereby proportional to $\omega^{1/2}$) just as predicted by the fast kinetics region of Fig. 2.

The facts, then, all indicate that the reactions producing ECL are quite rapid so that Eq. (26) is valid. Accordingly, $\Phi_{ECL}$ may be determined directly from the steady-state ratio of the total ECL intensity to the current at the disk

$$\Phi_{ECL} = nF \cdot \frac{I}{i_d} . \tag{31}$$

Therefore, only the precise measurement of the absolute ECL intensity and the disk current are required to determine the efficiency of the process.

## VI. SIMULATING THE EFFECT OF RADICAL INSTABILITY

Since the ion radicals of many aromatic hydrocarbons are unstable in solvents such as DMF, it is possible that this instability can have an effect on the ECL which results from the reaction of these radicals. For example, the DPA cation is known to be unstable in DMF yielding an unknown product X, which is electroinactive at potentials somewhat more negative than the oxidation of DPA [12]. While the cation probably reacts with the solvent, the rate law for the process is usually written in "pseudo-first-order" form

$$DPA^+ \overset{k_1}{\rightarrow} X. \tag{32}$$

At the RRDE the cation can be generated at either the ring or the disk so that Eq. (32) can imply that either

$$B \overset{k_1}{\rightarrow} X \tag{33}$$

or

$$C \overset{k_1}{\rightarrow} X \tag{34}$$

is a possible kinetic complication to be treated with digital simulation.

As an example of how simulated results can be combined with experimental results, the effect of the first-order decay of the disk-generated species [Eq. (33)] on the collection efficiency is shown in Fig. 7. Here the simulated collection efficiency is plotted as a function of a dimensionless rotation rate parameter; the filled circles are results obtained from the simulation. As expected, the collection efficiency is predicted to rise with increasing $\omega$, ultimately approaching the theoretical collection efficiency of the electrode in the absence of kinetic complications. The open circles are points obtained from collection efficiency measurements made on the disk-generated DPA cation; the experimental curve was fitted to the simulated curve by setting

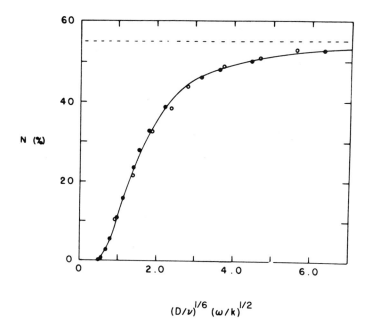

Fig. 7. The collection-efficiency-rotation-rate behavior for DPA$^+$. Filled circles show simulated points; open circles represent experimental data fitted to the simulated curve.

$$k_1 = 4.1 \ sec^{-1} \tag{35}$$

in Eq. (32). Hence, the lifetime of the DPA cation in DMF was found to be 0.25 sec by this study. (This lifetime is considerably longer than that reported previously [12]; the two measurements were made using different techniques and the solutions were prepared from solvents purified in a different manner. This increase in cation lifetime probably reflects improvements made in the solvent purification process rather than an error made in either determination.)

One may question how this cation decay effects RRDE-ECL. Clearly, this will depend upon whether the electroinactive product of the cation decay will react with DPA anions. It is quite possible that X is capable of oxidizing DPA$^-$ even though it cannot undergo electrode reduction at potentials just negative of DPA oxidation. If this is true, the problem depends on the relative chemical potentials of DPA$^+$, X, and DPA$^-$ and not upon digital simulation.

On the other hand, if X is inactive toward DPA$^-$, the ECL behavior can be predicted by simulation techniques. This may be accomplished by modifying the basic RRDE-ECL program to include the difference form of the differential rate law for Eq. (33) or (34). This results in an appropriate decrease in the quantity representing DPA$^+$ in each volume element. For example, if the generation of DPA$^+$ at the disk is to be simulated, one would write the rate law for Eq. (33) in relative concentration form,

$$-\Delta F_B(J, K) = (k_1 \Delta t) F_B(J, K). \tag{36}$$

When combined with Eq. (6) this becomes

$$-\Delta F_B(J, K) = \left(\frac{D_A}{\nu}\right)^{1/3} \frac{k_1}{\omega} \frac{F_B(J, K)}{L} \tag{37}$$

However, the total change in $F_B(J, K)$ is the sum of that which is depleted through cation decay, Eq. (37), and that which is used in Eq. (14), the radical redox reaction producing ECL. The addition of these two equations yields

$$-\Delta F_B(J, K) = \left(\frac{D_A}{\nu}\right)^{1/3} \frac{F_B(J, K)}{\omega L}[k_2 C_A^o F_C(J, K) + k_1]. \tag{38}$$

This may be simplified by setting

$$\frac{k_1}{k_2 C_A^o} = R \tag{39}$$

so that Eq. (38) becomes

$$-F_B(J, K) = \left[\left(\frac{D_A}{\nu}\right)^{1/3} \frac{k_2 C_A^o}{\omega}\right] \cdot \frac{F_B(J, K)}{L}[F_C(J, K) + R] \tag{40}$$

where the initial bracketed factor is XKTC. Hence, the problem may be treated by specifying two variables, XKTC which is $\omega$-dependent, and R which depends only upon the relative magnitudes of the rate constants and is $\omega$-independent.

The results of these simulations appear in Fig. 8 where curve a shows the effect of an unstable ring-generated species [Eq. (34)] on the dimensionless I vs. $\omega^{1/2}$ behavior in RRDE-ECL, while curve b shows the effect of generating the unstable species at the disk [Eq. (33)]. In either case, the reaction of B and C was considered fast with respect to the first-order decay ($R = 10^6$) and I and $\omega$ were rendered dimensionless by $k_1$ rather than by $k_2 C_A^o$.

Curve A, of course, corresponds to the generation of DPA$^+$ at the ring while b corresponds to the collection efficiency experiment where DPA$^+$ is generated at the disk. In both cases ECL is predicted to increase with increasing $\omega$, but in the latter case it is quite obvious that the effect of cation decay must be overcome before I is apparently linear with $\omega^{1/2}$. Thus, if I vs. $\omega^{1/2}$ is apparently linear but with a negative ordinate intercept, this can be attributed to the generation of an unstable species at the disk. On the

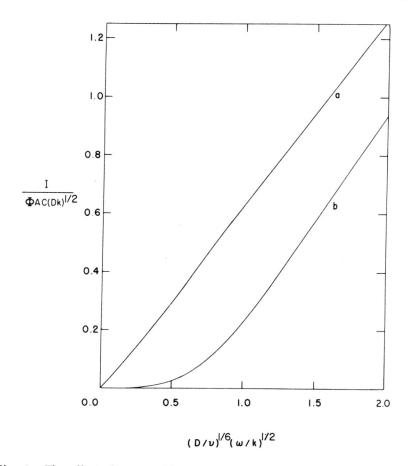

Fig. 8. The effect of an unstable species on intensity-rotation rate behavior. Curve a shows the effect of generating the unstable species at the ring, while b shows the effect of generation at the disk. In either case, $k_1/k_2 C_A^o$ was $10^{-6}$.

other hand, the generation of the unstable species at the ring is much less likely to show this variation (although curve a is nonlinear). The important result to observe here is that at rotation rates where the disk generation of the unstable species causes an apparently negative intercept in the I vs. $\omega^{1/2}$ curve, the ring generation of that species will produce apparent I vs. $\omega^{1/2}$ behavior which extrapolates through the origin. This is a necessary condition if either one (but only one) of the ionic precursors of ECL undergoes reaction to form a product which is inactive to the other.

## REFERENCES

[1].    A. N. Frumkin and L. N. Nekrasov, Dolk. Akad. Nauk SSSR, 126:115 (1959).

[2].    W. J. Albery and S. Bruckenstein, Trans. Faraday Soc., 62:1920 (1966).

[3].    V. J. Albery and S. Bruckenstein, Trans. Faraday Soc., 62:1946 (1966).

[4].    K. B. Prater and A. J. Bard, J. Electrochem. Soc., 117:335 (1970).

[5].    A. J. Bard, K. S. V. Santhanam, S. A. Cruser, and L. R. Faulkber, in Fluorescence (G. G. Gilbault, ed.), Marcel Dekker, New York, 1967.

[6].    C. A. Parker and G. D. Short, Trans. Faraday Soc., 63:2618 (1967).

[7].    D. M. Hercules, R. C. Lansbury, and D. K. Roe, J. Am. Chem. Soc., 88:4578 (1966).

[8].    A. Zweig, A. K. Hoffmann, D. L. Maricle, and A. H. Maurer, J. Am. Chem. Soc., 90:261 (1968).

[9].    V. G. Levich, Physicochemical Hydrodynamics, Prentice-Hall, Englewood Cliffs, New Jersey, 1962.

[10].   S. W. Feldberg, in Electroanalytical Chemistry, Vol. 3 (A. J. Bard, ed.), Marcel Dekker, New York, 1969.

[11].   S. W. Feldberg, J. Am. Chem. Soc., 88:390 (1966).

[12].   S. A. Cruser and A. J. Bard, J. Am. Chem. Soc., 91:267 (1969).

[13].   K. B. Prater and A. J. Bard, J. Electrochem. Soc., 117:208 (1970).

[14].   A. D. Osborne and G. Porter, Proc. Roy. Soc., Ser. A, 284 (1965).

Part III

INSTRUMENTATION


A. Analog Instrumentation

Chapter 10

OPERATIONAL AMPLIFIER INSTRUMENTS

FOR ELECTROCHEMISTRY

Ronald R. Schroeder

Department of Chemistry
Wayne State University
Detroit, Michigan 48202

## I. INTRODUCTION

The main components of an analog computer are its operational amplifiers. When each of these amplifiers is provided with an appropriate input and feedback network, it can perform a mathematical operation on one or more of the computer voltage signals. In the analog computer, these signals represent the mathematical variables and constants of a normal differential equation.

Analog computation has itself had limited applications to electroanalytical chemistry. However, the computing modules, both the circuits which perform the unit operations and the amplifiers on which these are based, have been widely used in electroanalytical instruments.

Operational amplifiers were first used in electrochemical instrumentation little more than a decade ago. Since then, in this country at least, the operational amplifier has emerged as the single most important electronic device for the construction of instruments for electrochemistry. Well over a hundred publications which show operational amplifier circuits have appeared and the investigations in which operational amplifiers were used would probably number in the thousands. Operational amplifiers and the circuits in which they are used have become the basis for the communication of instrumentation ideas among electroanalytical chemists. Students of electroanalytical chemistry generally find operational amplifiers one of their main topics of study.

To date, to this author's knowledge, no extensive review of operational amplifier instrumentation has appeared. Reviews of operational amplifier circuitry in electroanalytical chemistry covering limited periods or specific electrochemical techniques have appeared. Books and monographs covering the general application of operational circuitry are also available. However, no unified treatment of operational amplifier circuits specifically for analytical chemists has appeared.

Progress in operational amplifier circuitry and its application to electrochemistry have come from many sources and in a variety of forms. The types of operational amplifiers available and the interests of many electroanalytical chemists have changed during this last decade. Several of the "classic" instruments reported during the early portion of this operational amplifier age were composed of vacuum-tube amplifiers and, though still useful, those instruments have become outmoded in the general conversion to solid-state circuits.

The technological advance from vacuum tube to transistor to integrated-circuit operational amplifiers has been accompanied by a wealth of newly reported instruments; but most of these involve only new amplifiers in old circuits. Several of the major problems which confronted the users of vacuum-tube circuits are of little concern to users of solid-state amplifiers.

Thus, a very timely advance in a once critical area of operational-amplifier circuitry may have little bearing on the design of more modern circuits, even though this advance appeared only a decade ago.

A review of the application of operational amplifiers in electroanalytical chemistry must also involve a review of operational amplifier circuits in general. Circuits found useful by electroanalytical chemists have come from sources outside the field. In fact, much of the knowledge now possessed by many electroanalytical chemists comes from sources to which no primary reference can be given. For these and other reasons it has been necessary to arrange this review in a somewhat unusual manner.

Section II deals with common operational circuits, without regard to the characteristics of real amplifiers, and considers the principles of both operational and control circuits. Consideration is then given to the effects of real amplifiers on these circuits. This section is intended not only to define circuits and useful terms for those who are familiar with operational amplifiers, but to provide some necessary background for those who are not. Instruments constructed from operational amplifiers and developed by chemists for electrochemical applications are described and discussed. There then follows an extensive discussion of operational amplifier potentiostats. The advances in potentiostat design and in the development of the principles of their operation and design are given.

No attempt has been made to list all the operational amplifier circuits or all the publications concerned with them. Circuits given may be considered as "typical," although several represent advances of some importance. To simplify the discussion, the drawing, and the locating of important parts, many of the circuits presented have been reduced to the skeleton, but no important components have been left out. The original circuit is usually more complete and where interest merits, the reader is urged to consult the original reference.

## II. BASIC OPERATIONAL AMPLIFIER CIRCUITS

### A. The Ideal Operational Amplifier

An operational amplifier is generally an inverting (or differential), direct-coupled, high-gain, wide-band amplifier designed specifically for use in circuits with negative feedback. Some qualification of this definition is required since not all amplifiers fitting this description are called operational amplifiers, and not all operational amplifiers meet the above specifications. We are most interested in the operational amplifiers which are used in analog computers and these do meet the above requirements.

To begin a discussion of operational amplifiers by considering real amplifiers would create unnecessary complexity so that much of the simplicity of operational amplifiers would go unnoticed. We begin, therefore, by

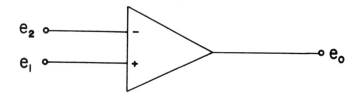

Fig. 1. The schematic representation of an operational amplifier showing two inputs and a single output. The negative sign denotes the inverting input (generally the upper input), and the positive sign the noninverting input (often omitted when connected to ground). The power supply common serves as the signal reference point, however power connections are not shown.

defining and using an ideal operational amplifier and will return later to the real and more practical aspects of operational amplifier circuitry. Beginning a discussion by assuming ideal behavior of the operational amplifier is not without precedent or purpose, and most circuits are probably designed from this view. For the sake of simplicity, we also consider only one type of amplifier, the differential input operational amplifier. This ideal amplifier, symbolized in Fig. 1, has the following characteristics:

(1) infinite input impedance, meaning it draws no current from the external circuit;

(2) very high gain (nearly infinite), which allows measurable output voltage from the amplifier with an extremely small input voltage;

(3) an output voltage given exactly by the relation $e_o = -g(e_2 - e_1)$, with the consequence that shorting the inputs gives zero output voltage;

(4) large output voltage (but finite) and high output current, to drive the most demanding load;

(5) infinite bandwidth, able to operate on a signal no matter how great the frequency or to operate even on a discontinuous signal without delay or loss of accuracy.

B. Circuits Employing Ideal Amplifiers

### 1. Operational Circuits

From this ideal amplifier we can begin to construct idealized circuits. One point worth mentioning here is that although the amplifier we have defined has many useful characteristics, by itself it is of little practical use. The gain of the ideal amplifier is so high that an immeasurably small input signal will produce a large output signal, probably to the finite output voltage limit of the amplifier. To make practical use of the amplifier it will be

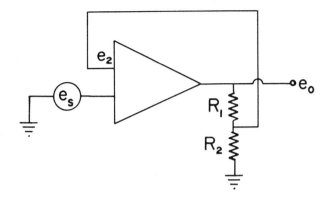

Fig. 2. A noninverting amplifier circuit employing a differential operational amplifier with negative feedback. A fraction, $\beta$, of the output signal is returned to the inverting input. The positive input acts as the signal input for the circuit. The circuit operation is given by $e_0/e_s = 1/\beta$ where $\beta = R_2/(R_1 + R_2)$.

necessary to sacrifice much of its inherent gain.

The circuit shown in Fig. 2 offers a possible use for the amplifier [1]. Here much of the amplifier's gain is sacrificed by the return of a portion of the output signal to the inverting input — an example of negative feedback. An analysis of the circuit follows from the definition of the amplifier output signal

$$e_0 = -g(e_2 - e_1). \tag{1}$$

Defining $e_2$ and $e_1$ as

$$e_2 = \frac{R_2}{R_1 + R_2} e_0 \tag{2}$$

$$e_1 = e_s \tag{3}$$

and substituting in (1) gives

$$e_0 = -g\left(\frac{R_2}{R_1 + R_2} e_0 - e_s\right). \tag{4}$$

Solving for $e_0/e_s$, the circuit operation gives

$$\frac{e_0}{e_s} = \frac{1}{\dfrac{1}{g} + \dfrac{R_2}{R_1 + R_2}} \tag{5}$$

and since $g \to \infty$, $1/g \to 0$; for any values of $R_1$ and $R_2$,

$$\frac{e_o}{e_s} = \frac{R_1 + R_2}{R_2} = 1 + \frac{R_1}{R_2}. \tag{6}$$

The circuit operation is to amplify the input signal, $e_s$, by the factor, $(R_1 + R_2)/R_2$, or to give a finite gain. Provided that g, the amplifier gain, is large it does not appear in the final equation. The significance of many of the ideal amplifier characteristics can be seen in this single circuit. Thus, the high gain allows an amplification factor independent of the amplifier gain or any variations in the amplifier gain. The high amplifier input impedance prevents any current drain from the feedback circuit to the negative input or from the source to the positive input. High current output and wide bandwidth allow any load and any signal frequency to be employed with the assurance that the circuit gain will still be that of Eq. (6). Only the finite output voltage limits the variety of resistor ratios which can be used.

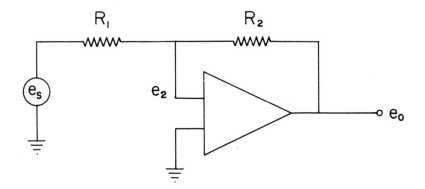

Fig. 3. An inverting amplifier circuit. $e_2$ is held at virtual ground by the amplification, and feedback. $e_o / e_s = -R_2 / R_1$.

An alternative means of producing a limited gain amplifier is shown in the circuit of Fig. 3. An analysis of this circuit, again starting with the definition of $e_o$, gives, since $e_1$ is held at zero,

$$e_o = -g(e_2). \tag{7}$$

Considering the currents in $R_1$ and $R_2$ and the infinite input impedance of the amplifier, it follows that

$$\frac{e_s - e_2}{R_1} = \frac{e_2 - e_o}{R_2}; \tag{8}$$

substituting for $e_2$ from (7) into (8) gives

$$\frac{e_s - e_o/g}{R_1} = \frac{e_o\left(\frac{1}{g} - 1\right)}{R_2}. \tag{9}$$

Terms with g in the denominator are again negligible and Eq. (9) reduces to

$$\frac{e_o}{e_s} = -\frac{R_2}{R_1}. \tag{10}$$

Again, negative feedback has reduced the over-all gain to a value independent of the amplifier characteristics. For this case an inversion of the input signal is obtained in addition to amplification.

An interesting fact about operational amplifier circuits can be learned by directing our attention to the actual magnitude of the $e_2$ signal for both of the preceding circuits. Combining Eqs. (2), (3) and (6) gives $e_2 = e_1$ for the circuit of Fig. 2. Inspection of Eq. (7) shows that for the circuit of Fig. 3

$$e_2 = \frac{e_o}{g} \approx 0. \tag{11}$$

Remembering that $e_1$ was held at ground, for this circuit $e_2$ has again been made equal to $e_1$ and $e_2$ is thus held at virtual ground. It can be shown from a generalized approach that, when a negative feedback path is provided, an operational amplifier maintains its input signals nearly equal. In circuits where the positive input is ground, the negative input is always held at virtual ground. This general fact will prove very helpful in the design of operational amplifier circuits and will further allow one to easily describe the output-input relation for any circuit.

Other examples of operational amplifier circuits can now be readily written down and analyzed. A useful operational amplifier circuit can be made from the circuit of Fig. 2 by setting $R_1 = 0$ and $R_2 = \infty$ as shown in Fig. 4.

Substituting these values in Eq. (4) and solving gives

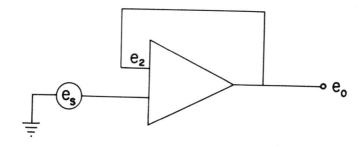

Fig. 4. An operational amplifier voltage follower, where $e_o/e_s = 1$.

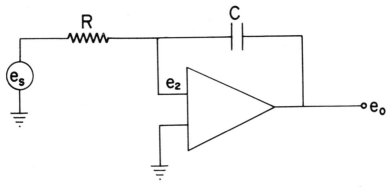

Fig. 5. An operational amplifier integrator, where $e_o = (-1/RC) \int_0^t e_s \, dt$.

$$\frac{e_o}{e_s} = \frac{g}{1 + g} \tag{12}$$

and when g is large, as considered here, $e_o = e_s$. This unity gain circuit, called a voltage follower, performs a trivial mathematical operation: it can transmit a signal from a high impedance source to a low impedance circuit without altering the voltage value.

In the circuit of Fig. 5, we place a capacitor in the feedback path, rather than a resistor (see Fig. 3). Equating the currents in the input and feedback impedances and remembering that $e_2$ is held at virtual ground, gives

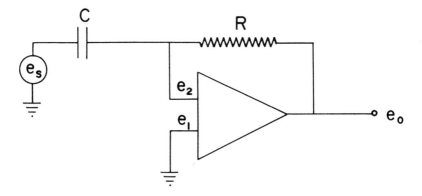

Fig. 6. A differentiating circuit, where $e_o = -RC(de_s/dt)$.

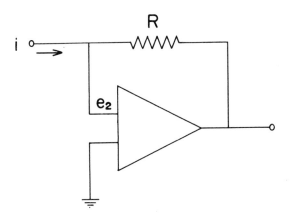

Fig. 7. A current follower in which $e_2$, the signal at the input, is maintained at virtual ground by feedback, where $e_0 = -iR$.

$$e_0 = -\int_0^t \frac{e_s}{RC} \, dt. \tag{13}$$

The circuit integrates the input voltage signal with the term $1/RC$ as the scale factor.

Reversing the location of resistor and capacitor, as shown in Fig. 6, and following a similar analysis gives

$$e_0 = -RC \frac{de_s}{dt}. \tag{14}$$

The circuit of Fig. 7 presents a new facet of operational amplifier circuits, that of operating on the input current directly. A current, i, at the input produces a corresponding output voltage, $e_0$, where $e_0 = -iR$. Since $e_2$ is at virtual ground, this current follower can be introduced in a grounded current-carrying circuit without concurrently introducing any impedance to disrupt the circuit parameters. Replacing the resistor of Fig. 7 with a capacitor would produce a current integrator with a virtually grounded input.

The inverting amplifier of Fig. 3 can be modified by the introduction of additional input resistors, such as in Fig. 8. Summing the input currents and equating these to the feedback loop current gives, with $e_2 = 0$,

$$e_0 = -R\left(\frac{e_A}{R_A} + \frac{e_B}{R_B} + \frac{e_C}{R_C}\right). \tag{15}$$

With all resistors equal, a precise voltage-summing network is produced. In any summing inverter the total current flowing into the point common to

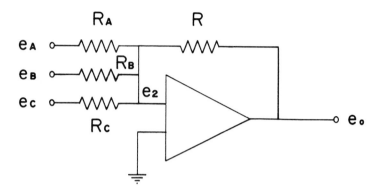

Fig. 8. A summing amplifier, which is similar to the inverter of Fig. 3, except in having multiple inputs. $e_2$ is held at virtual ground by negative feedback, where $e_0 = -[(R/R_A)e_A + (R/R_B)e_B + (R/R_C)e_C]$.

all input resistors — called the summing point — will be zero. It follows that if all resistors connected to the summing point are equal, the sum of the voltages fed to the amplifier input, including feedback signal, must also be zero.

In general, for a circuit of the type shown in Fig. 9 with only a resistor as the input impedance, the circuit character is determined by the nature of the feedback impedance, and the specific frequency characteristics of the

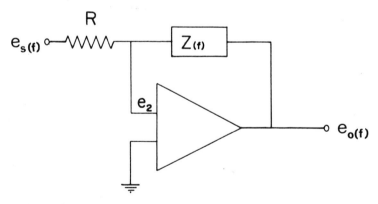

Fig. 9. A generalized amplifier circuit whose gain operation is $e_{o(f)}/e_{s(f)} = -Z(f)/R$. The nature of the gain for a particular frequency, f, is determined primarily by the impedance of Z(f) at the same frequency.

amplifier circuit are identical to those of the feedback impedance. Feedback impedance need not be restricted to resistors and capacitors or combinations of these. Nonlinear impedances can produce circuits which give a log, exponential, square, square root, and other functions of the input signal.

With some fear of being self-contradictory, it should be mentioned that circuits of ideal operational amplifiers which do not use negative feedback can also be useful. Equation (1) indicates that the operational amplifier output is very sensitive to input signal variations. Any finite voltage difference between the positive and negative inputs will produce the full amplifier output voltage. Although increasing the difference has no effect, reversing the sign of the voltage difference between inputs will reverse the output, giving the opposite voltage limit. This makes operational amplifiers well suited for level-detecting circuits. In Fig. 10 such a circuit is shown. With $e_R$, the reference signal, connected to the positive input and $e_x$, the unknown signal, connected to the inverting input, $e_x > e_R$ produces the full negative output and $e_x < e_R$ the full positive output. Only at $e_x = e_R$ is the output voltage zero. Circuits such as that of Fig. 10 have many uses which can include decision-making operations in logic circuits, square-wave signal generation, and on-off type controlling circuits operation.

## 2.   Control Circuits

The latter use of the level-sensing circuit brings us to a very valuable application of circuits which include negative feedback — the control operation. It has been indicated that, when negative feedback is employed, the action of an operational amplifier is to maintain equal input signals. Considering the circuit of Fig. 11 (identical to the circuit in Fig. 2) and directing our attention away from the amplifier output toward the inputs, we see that the amplifier is performing a control operation.

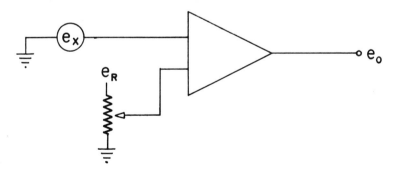

Fig. 10. A level sensing or comparator circuit. The output signal $e_0$ assumes only its limiting values ($\pm E_L$) since no feedback is present, where $e_0 = -E_L$ when $e_x > e_R$ and $e_0 = E_L$ when $e_x < e_R$.

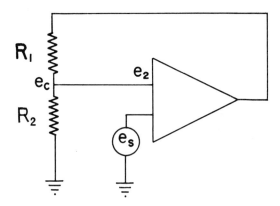

Fig. 11. A simple controller circuit. Negative feedback maintains signal $e_2$ equal to $e_s$.

For this operation it is most helpful to consider the amplifier action as dependent on an error signal. The desired operation in this circuit is the control of the $e_2$ signal, in this case to make the voltage at $e_2$ identical to the source signal $e_s$ (or $e_1$). Any deviation from this condition results in a finite error signal $e_e$, where $e_e = e_2 - e_s$. The direction and amount of amplifier response is determined by the sign and magnitude of the error signal. A positive error signal, brought about by $e_2 > e_s$, causes the amplifier output signal to decrease in accord with Eq. (1). This action decreases $e_2$ and continues until $e_e$ is zero. A negative error signal ($e_e < 0$, $e_2 < e_s$) causes an increase in the amplifier output until $e_e$ is again zero. In this control circuit, a finite error signal can arise from (1) a variation of $e_s$ or (2) a change in a circuit component, $R_1$ or $R_2$. For this application our interest is in the forced response of the controlled variable, $e_c$. Our concern is then with the accuracy of this control operation. Using an approach similar to that employed in analyzing previous circuits involves, in this case, solving for the ratio $e_2/e_s$. Starting from a relation similar to Eq. (1),

$$e_0 = -K(e_2 - e_1) \tag{16}$$

where $-K$ is the amplifier gain. Defining $e_1$ and $e_0$ from the circuit

$$e_1 = e_s \tag{17}$$

$$e_0 = \frac{R_2 + R_1}{R_2} \cdot e_2 \tag{18}$$

and substituting and solving for $e_2/e_s$ gives

$$\frac{e_2}{e_s} = \frac{K\beta}{1 - K\beta} \tag{19}$$

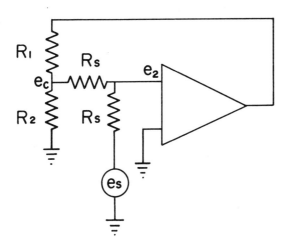

Fig. 12. A controller circuit based on a summing amplifier. Negative feedback and the summing operation maintain $e_c = e_s = 0$.

where

$$\beta = \frac{e_2}{e_0} = \frac{R_2}{r_1 + R_2}. \tag{20}$$

Rearranging Eq. (19) to give

$$\frac{e_2}{e_s} = 1 - \frac{1}{1 - K\beta} \tag{21}$$

shows that the error in the control operation is given by the term $1/(1 - K\beta)$. For the case at hand, K is very large making $K\beta \gg 1$; this gives $e_2/e_s = 1$ with a high degree of accuracy.

Other operational amplifier circuits can be used for control applications. In Fig. 12 an amplifier of the summing-inverter type serves as the central control amplifier. With equal summing resistors, as we have seen, the sum of the input signals $e_c + e_s$, must equal zero. Therefore, $e_c$ is controlled to equal $-e_s$. Unlike Fig. 11, where no current is drawn from the feedback circuit, here finite current flows into the summing resistor, and unless $R_s \gg R_1$, the currents through $R_1$ and $R_2$ will differ. This difference does not affect the control accuracy.

Control is also obtained with a circuit like that in Fig. 13. This circuit is not suited for analog computer applications, but serves nicely for a control system. The virtual grounding of $e_2$ forces the sum, $e_c + e_s$, equal to zero, again controlling $e_c$ to equal $-e_s$.

We have now established much of the nature of operational amplifier

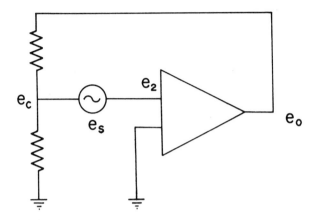

Fig. 13. A controller circuit in which the controlled and reference signal appear in series with the amplifier negative input. Negative feedback maintains $e_2$ at virtual ground and forces $e_c + e_s = 0$.

circuitry. For ideal amplifiers, the circuit function is wholly determined by the external circuit components and all operations are performed with perfect accuracy.

### C. Circuits Employing Real Operational Amplifiers

When we use actual operational amplifiers in circuits, we can start with the ideal cases as first approximations. A comparison of the characteristics of real amplifiers to those of ideal amplifiers will allow us to determine if and by how much our approximations fail. Real operational amplifiers come with a large variety of characteristics; some approximate the ideal amplifier types more than others. To be perfectly correct one would have to say that operational amplifiers possess none of the ideal characteristics listed earlier. However, to obtain near ideal behavior from most circuits, the ideal characteristics are not always necessary.

Thus, let us consider amplifier gain as an example. In the ideal case, the gain is infinite; in the real amplifier, the gain is very high but obviously not infinite. Yet an inspection of most of the circuits given earlier would show that, if the amplifier gain is large, a near ideal circuit characteristic is still obtained. The actual magnitude of the gain affects only the accuracy of the operation performed by a real circuit. Perfect accuracy is not achievable or, in fact, necessary with ideal amplifiers. Real external circuit components have themselves only a limited accuracy. Therefore, attempting to improve circuit operation beyond the accuracy limits set by the external components would be costly and fruitless. To be sure, the gain of a real amplifier should be as large as is practical but that it is less than infinite is not restricting.

The first notable restriction introduced by a real amplifier is in the type of components which can be used in the external circuit. The maximum output current of most operational amplifiers in general use is between 2 and 100 mA. Voltage signals are commonly in the 1-50-V range, and resistors in the 1-k$\Omega$ to 10-M$\Omega$ range are most often used. Larger resistance values are not useful since the input impedance of most amplifiers falls in the several-megohm region. The finite currents drawn by the amplifier must remain very small compared to external circuit current. The lack of ideality in the input impedance and in the output current characteristics restricts only the type of amplifier circuits; however, within these limitations nearly ideal behavior is observed.

Less apparent but more important and far more troublesome are the nonideal gain and gain frequency characteristics of available operational amplifiers. The effect of the finite gain has been discussed earlier and it was seen that reasonably large gains were entirely satisfactory for nearly ideal circuit operation. The dc gains of commercially available amplifiers range from $10^4$ to about $10^8$ and with some restrictions such gains are quite acceptable. For signals other than dc the gain of an operational varies, generally decreasing with frequency. At high frequencies the capacitive component of the amplifier impedance and lead capacitances, both within the amplifier and in the external circuit, become important causing this decrease. The effectiveness of the amplifier, as seen in its ability to perform operations, diminishes when the gain becomes so small that the $1/g$ terms in Eqs. (5), (9), and (12) are not negligible. At frequencies beyond this point, the amplifier gain and the circuit effectiveness continue to decrease. The frequency at which the amplifier has a gain of unity can range from about 100 kHz up to 100 MHz in commercial operational amplifiers.

The only characteristic left to consider is the definition of the amplifier output. Unfortunately, here again some difficulty is encountered since the defining equation for the real amplifier differs from that given earlier. A more realistic equation would be

$$e_0 = -g(e_2 - e_1 + e_n + e_c + e_d) + g'(e_1 + e_2) \tag{22}$$

where $e_0$, $g$, $e_2$, and $e_1$ are as previously defined, $e_c$ represents the constant output in the absence of input signals, $e_n$ the noise voltage, $e_d$ the very low frequency noise (drift), and $g'$ the common mode gain of the amplifier. In well designed amplifiers the common mode gain is reasonably small, giving a common mode rejection ratio, $g/g'$, which is quite large. A constant voltage signal from a high impedance source can be introduced to offset the value of $e_c$. The high frequency noise signal, $e_n$, is normally small in well designed circuits and represents only a small portion of the output signal. Signal $e_d$ represents a small but very low frequency signal, one for which the amplifier has its largest gain value. Special circuits have been devised to counteract this drift at the amplifier input.

The magnitude and importance of each of the parameters in Eq. (22) depends to a great extent on the operational amplifier type, cost, and intended application. In vacuum tube differential amplifiers, the terms $e_n$, $e_c$, and $e_d$ are often quite large and in circuits using these amplifiers special wiring and extra circuitry are often necessary.

In circuits where high dc gain was desired, drift due to the $e_d$ term posed a particularly knotty problem. Such circuits can be drift-stabilized by using two amplifiers wired in a specific manner (Fig. 14). The inverting input of a normal dc amplifier is ac-coupled to the summing point through a capacitor. A low frequency ac amplifier with a chopper-modulated input and output stage is directly coupled to the summing point and its output is wired to the positive input of the differential amplifier. This tandem action increases the dc gain of the amplifier combination since the chopper amplifier output is further amplified in the differential amplifier. Also, any slow drift of the summing point signal is counteracted by the chopper amplifier action. The amplifiers wired in this way act as a single "stabilized amplifier" with superior drift and gain characteristics. Generally, in circuits like these the second input, which is now the chopper reference, is connected to ground. Thus the resultant amplifier has only one signal input.

In electron tube operational amplifiers, chopper stabilization was ad-

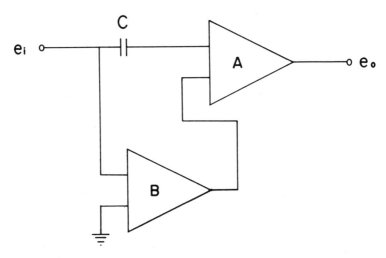

Fig. 14. A stabilized amplifier consisting of an ac coupled differential amplifier (A) and a dc coupled chopper amplifier (B). This single input amplifier has the broadband, high gain character of the differential amplifier and the low drift character of the chopper operated type.

vantageous for almost all applications and an absolute necessity in many.
As far as electron tube amplifiers are concerned, we will consider only the
stabilized type in our discussion except in limit-to-limit applications, such
as in signal generators.

Operational amplifiers constructed from transistors or as integrated
circuits generally possess more favorable characteristics than the electron
tube varieties, except perhaps for their lower input impedance (a transistor
base as opposed to the open grid of a tube) and lower output voltage (not
really a disadvantage) of transistorized amplifiers. Noise ($e_n$), offset ($e_c$),
and drift ($e_d$) are generally much less a problem and, even with the need for
an additional offset current correction, solid state amplifiers are superior
and far more convenient. For critical applications solid state amplifiers
which are chopper stabilized are available.

To continue with an evaluation of real amplifier circuits, we return to
the circuits previously discussed using ideal amplifiers but substituting a
typical amplifier: gain, dc, $10^5$; unity gain frequency $10^6$ Hz; input impedance, between inputs, 4 M$\Omega$; input to common, 20 M$\Omega$; maximum output
voltage, ±10 V; current, ±10 mA. These characteristics are chosen as
those of an average transistorized differential operational amplifier; however, comparisons of the circuit properties to those obtained with typical
electron tube amplifiers and with more specialized amplifiers can be easily
made.

In the circuit of Fig. 2 (a noninverting amplifier) our typical amplifier
introduces several restrictions on the circuit characteristics and range of
operation. Most limiting are the finite gain, the gain frequency character,
and the output current and voltage limits of the amplifier. These restrict
the useful over-all gain of the circuit to 10 or so depending on the size of
the input signal. The accuracy of the circuit operation is poor at frequencies
beyond which the amplifier gain is less than 100 $(R_1 + R_2)/R_2$ — the condition
for 1% accuracy. Assuming a typical gain-frequency character as shown in
Fig. 15, the 1% accuracy limit would be at 1 kHz for a circuit gain of 10 and
10 kHz for a gain of unity.

Tube-type differential amplifiers are not well suited for this circuit nor
can single-ended amplifiers, stabilized or unstabilized, be used. Improved
accuracy and frequency response can be obtained with amplifiers possessing
large gains or higher unity-gain crossover frequencies.

The circuit of Fig. 3 is similarly, but generally less, affected by
specific amplifier characteristics than is the noninverting amplifier circuit.
The input impedance of the circuit is determined by the input resistor rather
than by the amplifier input to ground impedance. The circuit gain and accuracy are restricted in a similar way however. All types of amplifiers,
differential and single ended, both stabilized and unstabilized, are suited to
this circuit. Improved accuracy and frequency response again require

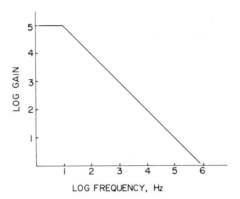

Fig. 15.  The gain vs. frequency characteristic of a typical operational amplifier.

improved operational amplifier characteristics.

The characteristics of the voltage follower circuit of Fig. 4 are strongly dependent on certain amplifier characteristics. Using the typical amplifier under consideration, the input impedance would be about 20 M$\Omega$ at dc (the input to ground impedance of the amplifier). The voltage follower provides accurate operation at higher frequencies than of previously considered circuits and provides 1% accuracy even at an amplifier gain of 100. This would be 10 kHz for the case at hand. Since the only application of voltage followers is in impedance matching, high input impedance and maximum frequency response are usually desired.

For the integrator circuit of Fig. 5 two particular aspects of normal operational amplifiers present problems: the constant voltage, $e_c$, and the drift signal, $e_d$. In the absence of any input signal, the high dc circuit gain amplifies these low frequency signals and produces a steadily increasing output signal due to capacitor charging. Fine offset voltage control is necessary to eliminate the effect of the constant input voltage. Since even a relatively small drift in the amplifier input produces a large output signal, chopper stabilization is essential to proper integrator functioning. Even high quality solid state amplifiers have enough drift at the input to make them unsuitable for integrator applications without stabilization. The accuracy of the integration function is good over the entire frequency range provided that the integrator constant 1/RC is much smaller than the unity gain crossover frequency expressed in rad/sec.

The differentiator circuit of Fig. 6 is little affected by extraneous low frequency signals. Instead, since the circuit gain of the differentiator increases with frequency ($e_o \propto de/dt$), high frequency noise becomes a limitation and low noise amplifiers are required. A second problem with

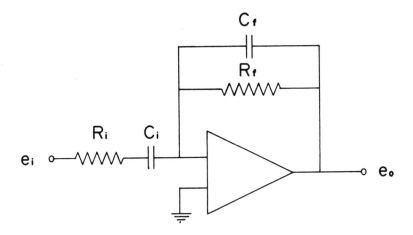

Fig. 16. A stabilized differentiator. The gain frequency character of the differentiator of Fig. 6 has been trimmed to allow stable and relatively noise-free operation.

differentiator circuits is their tendency to be unstable. This problem can be averted by wiring the differentiator as shown in Fig. 16. This modification limits the frequency response of the circuits and reduces the high frequency noise. Since in this circuit low frequency signals do not interfere, stabilization is not necessary. Differentiators have good operational accuracy over a fairly large frequency band.

Generally, operational amplifier circuits designed for specific frequencies resemble one of the preceding circuits and their characteristics can be reasoned from the previous discussions.

In limit-to-limit applications several amplifier characteristics affect the circuit performance. First, the finite amplifier gain causes level sensing circuits to show hysteresis in switching potentials. To somewhat offset this, positive feedback can be employed as shown in Fig. 17. This regenerative feedback increases the rate of change of the amplifier. Second, in spite of positive feedback which acts on the amplifier input, the maximum rate of change of the output is limited. This characteristic, called the slewing rate (units V/sec or V/nsec), limits the quality and maximum frequency of square wave outputs.

Most of the effects that real amplifiers have on control circuits are related to the K term in Eq. (19). The effect of the $K\beta$ product on control accuracy has been discussed in the section on ideal amplifiers and is further discussed later in this chapter. The gain-frequency characteristic is a more difficult point to consider. The characteristics shown in Fig. 15 embody the features essential for our consideration. The action of the amplifier, at low

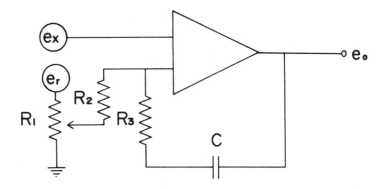

Fig. 17. A comparator circuit similar in operation to the circuit of Fig. 10. Positive feedback through C and $R_3$ increases the rate of change from limit to limit.

frequencies, is to both amplify and invert (for symmetrical ac signals, inversion can be designated by a 180° phase shift) the input signal. This is represented on the diagram as the constant gain segment. At higher frequencies the gain decreases as shown by the sloping of the gain-frequency plot. Such a change in the gain also indicates an additional 90° shift of the phase angle between output and input signals. This brings the total phase shift to 270°. For the circuit of Fig. 11 and this gain-frequency character-istic, the feedback signal, $e_f$, is no longer in phase with the source voltage, $e_s$; however, the signals are not out of phase either and their ratio is most easily defined using complex numbers. If we look at the $\beta$ term of the cir-cuits, since $\beta$ relates two voltages it can also often be defined as a complex number. The product of two complex numbers, $K\beta$, can then assume values positive, complex, and negative (Fig. 15). This presents no great complication except in the region where $K\beta$ crosses the unity gain axis. For if it has a positive (360°) sign at the unity gain axis then Eq. (19) be-comes indeterminate. This generally results in the oscillation of the ampli-fier output and the circuit is termed unstable. For control, this circuit is no longer of value. Obviously, avoiding this condition is necessary in control circuitry. The gain-frequency characteristic shown in Fig. 15 is typical of many operational amplifiers. Thus, over most of the frequency region a 270° phase shift between input and output signals is observed. Avoiding the onset of circuit oscillation then requires that the phase shift in the external circuit does not amount to the additional 90° at the frequency where $K\beta$ is unity.

The response rate of a control system involving a real amplifier is also closely related to the gain-frequency characteristic of the circuit. Frequency-dependent and time-dependent characteristics of the circuits are related by integral transformations. Fortunately, general conclusions regarding these relations can be drawn without actually employing the transformations. The general appearance of the $K\beta$ vs. frequency plot or Bode diagram for the circuit indicates the type of response expected, while the unity gain crossover frequency determines the response rate. For a Bode diagram showing a decade change in $K\beta$ per decade of frequency change, a smooth response to a change in signal $e_s$ is expected.

The time required for ~99% response to an ideal step input signal is $1/2\pi f_0$ where $f_0$ is the frequency at which $K\beta$ is unity. For larger slopes the response time improves but the system overshoots the desired control point, eventually returning to this value. For slopes less than one decade change in $K\beta$ per decade frequency increase, a somewhat sluggish but smooth response is obtained. Further consideration is given to the response time, stability, and accuracy of control circuits, with a more extensive discussion of feedback principles, in another section of this chapter.

## III. AMPLIFIER CIRCUITS AND INSTRUMENTS

### FOR ELECTROCHEMISTRY

### A. Basic Circuits

One of the most valuable features of operational amplifiers is that even those who have only a limited background in electronics can use them to design and build useful electronic circuits. Moreover, the circuits obtained are of high quality often suitable for even the most critical research applications. These facts have been amply demonstrated in electroanalytical instrumentation especially with regard to the generating, detecting, and conditioning of electrical input and output signals. The experimental electroanalytical chemist can, with little difficulty, devise circuits to perform whatever operations are necessary for the technique which interests him. This has allowed him to proceed with the development of new methods with reasonable assurance that suitable instrumentation can be constructed. He is no longer obliged to wait for others to develop the circuitry for his needs or to be content with the limitations of commercial circuits.

Operational amplifier circuits can be used to generate a wide variety of voltage signals. These generators often consist of a basic circuit which produces a sine wave, square wave, step or pulse function and may contain other circuits for performing various mathematical operations on these basic signals. For more complex waveforms the various signals can be combined in an adder or adder-subtracter to produce the desired signal output. From just a few basic circuits a wide variety of signals can be produced, signals which are suited to almost any electrochemical technique.

Signal detectors must sample the output signals from an electrochemical experiment and provide these signals in a suitable form for conditioning or measurement. These signals are usually just the cell current or voltage. This sampling must be done with a minimum perturbation of the electrolysis process, and requires either a very high impedance voltage detector or a nearly zero impedance current sampler. Where these are not practical, the measuring device introduced will cause variations in the electrolysis process unless some compensating signal is provided to correct the experimental system for this error.

Signal conditioning involves either performing a specific mathematical operation on the output signals or some form of sorting and isolation of the desired portion of the signal. These three signal operations, generation, conditioning, and detection, will be discussed together. For most instruments these are designed as a unit and the generators, detectors, and conditioners are usually matched to one another. Some circuits are, however, quite general in the application and can be used in a variety of techniques; these will be discussed first.

### 1.   Voltage Follower

Not the first developed, but probably the most important circuit in these classes is the voltage follower shown in Fig. 4, and discussed earlier with ideal amplifiers. A full utilization of operational amplifier circuits often requires that impedance matching devices be placed between the various signal sources, especially the reference electrode, and the controlling amplifier input. Differential amplifiers possess the two active inputs necessary for this circuit but are subject to an appreciable drift error over long periods. Stabilized amplifiers generally need one grounded input, and the proper feedback and input arrangement was not practical. DeFord [2] solved this problem and used a direct-coupled and chopper amplifier combination in creating the stabilized voltage follower shown in Fig. 18.   In this circuit the chopper reference input was not grounded in the

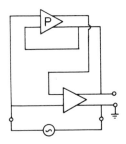

Fig. 18.   The DeFord stabilized voltage follower.   P is a chopper modulated amplifier.   (Reprinted [21, p. 1773] by courtesy of the American Chemical Society.)

usual manner, but was connected to the direct-coupled amplifier output. The chopper amplifier was therefore located in the feedback loop of the direct-coupled amplifier. The chopper amplifier input signal was effectively the difference between the circuit input and output signals. The combined action of the two amplifiers minimized this difference signal, creating the unity gain operation. The importance of DeFord's circuit cannot be over-stated. It made possible many new and useful electroanalytical circuits and vastly increased the number of areas in which operational amplifier circuits could be used. DeFord's circuit was effective, but did require a chopper amplifier that was separate or separable from the direct-coupled amplifier so that the indicated wiring modification could be introduced. This unfortu-nately prevented the use of many of the high quality single-ended circuits available at that time.

A circuit with somewhat better characteristics is that given by Booman and Holbrook [3]. They showed that by altering the normal means of con-necting single-dnded chopper stabilized amplifiers, a high quality unit gain amplifier could be obtained. In this circuit, shown in Fig. 19, the normal input is used; however, the circuit output is obtained from the power supply common (floated supply) and the amplifier output is grounded through a resistor. The amplifier output acts to balance the input signal and chopper reference (now the output) signal by raising or lowering its potential relative to ground.

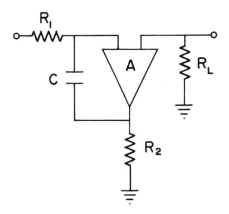

Fig. 19.   The grounded output voltage follower. The output (across $R_L$) is taken from the chopper reference (power supply common). (Reprinted [3, p. 1801] by courtesy of the American Chemical Society.)
For the circuit $R_1$ = 10 K, $R_2$ = 1 K, C = 50 pf, and amplifier A is a USA-3M3 chopper stabilized operational amplifier (Philbrick Researches Inc., Dedham, Massachusetts).

Both of these follower circuits are valuable when used with vacuum tube amplifiers or in cases where long term dc accuracy is of the utmost importance. The much smaller drift characteristics of high quality solid state differential amplifiers permits the normal differential form of the follower circuit to be used. With field-effect transistor input such amplifier circuits [4] possess an extremely high dc input impedance and give accurate and drift-free operation.

Voltage followers have found application in numerous circuits where impedance matching is necessary or desirable. They are especially suited to cell signal measurements where high impedance or easily polarized reference electrodes are used. Their small input currents prevent any polarization of the reference electrode or the appearance of a significant iR drop error in the cell voltage measurement.

The frequency response and accuracy of the voltage follower is determined by the type of amplifier used. The circuit loop gain is essentially the amplifier open loop gain. The follower is usually one of the best components of a multiamplifier circuit and generally the one least likely to hinder the circuit performance. The capacitive component of the amplifier input impedance does degrade the follower input impedance in the high frequency region. This has been reported as a significant problem in only a few of the numerous cases where follower circuits have been used [3].

## 2. Circuits for Current Measurement and Load Resistor Compensation

Current amplifiers, or current followers, have been used in electrochemistry as long as operational amplifiers themselves have been in use. Booman [5] utilized a boosted current follower with his first instrument and very few variations on this original circuit have been made (Fig. 20). Kelly et al. [6] did use a somewhat different form of the circuit (Fig. 21) but its operation is essentially the same. The Booman type circuit has become an almost standard feature of electrochemical controlled potential circuits.

The inclusion of the current follower as part of the electrolysis-cell, current-carrying loop provides the virtual grounding of the cell-working electrode. This feature simplifies the measurement of the cell potential in control applications and allows the use of single-input controllers. The cell-reference electrode potential can be measured and controlled relative to ground with only an insignificant error. The single-ended output of the current follower also eases the demands placed on the voltage readout device, which can have one input at ground potential. A drawback to the current follower type circuit is that the amplifier used must pass the same current as the controller, requiring that the output characteristics of the current following amplifier must be as good as those of the controller. Thus in large scale electrolysis apparatus, a high current amplifier is required and the use of a booster in the current follower is often necessary. In polarography

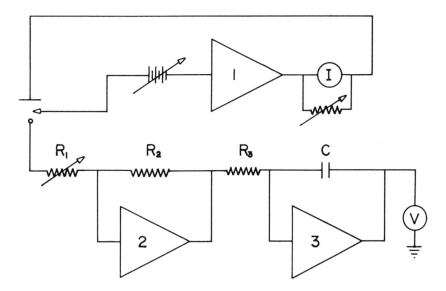

Fig. 20. Booman's controlled-potential coulometer. Amplifier 1 (the controller), amplifier 2 (the current follower), and amplifier 3 (the integrator) were chopper-stabilized operational amplifiers (Philbrick K2-X, K2-P) R $R_2$ = 10 $\Omega$ to 50 K, $R_3$ = 1.33 M, C = 30 $\mu$F. (Reprinted in part [5, p. 215] by courtesy of the Americal Chemical Society.)

and other low current techniques very little demand is placed on this amplifier and in circuits for these, the current follower behaves almost ideally.

The current follower has an extremely large voltage gain which combined with its wide bandpass, can produce large amounts of noise. It is common in current followers to trim the frequency response by placing a capacitor in parallel with the feedback resistor. This generally reduces both noise and ringing on the output, and causes only a slight to moderate loss in the frequency response.

Placing the load resistor in the portion of the circuit compensated by the controller requires a differential or floating measuring device. Evans and Perone [7] averted this problem by using a differential input operational amplifier adder-subtracter (Fig. 22). Placing the two circuit inputs across the load resistor, they obtained a single-ended signal proportional to the load resistor iR drop. It is, of course, necessary to minimize the current flow into the adder-subtracter circuit by using input resistors which are much larger than the load resistor. The high frequency accuracy of this measuring technique is questionable but the circuit is adequate for dc and low frequency measurements.

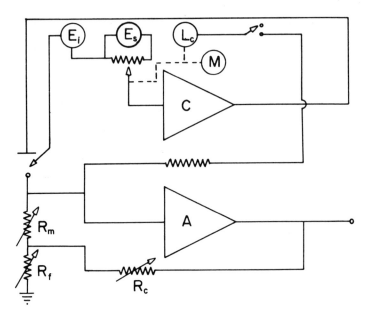

Fig. 21. Portion of the controlled potential and derivative polarograph circuit of Kelley et al. [6]. Controller C and current amplifier A are Philbrick USA-3 operational amplifiers. (Reprinted [6, p. 1477] by courtesy of the American Chemical Society.)

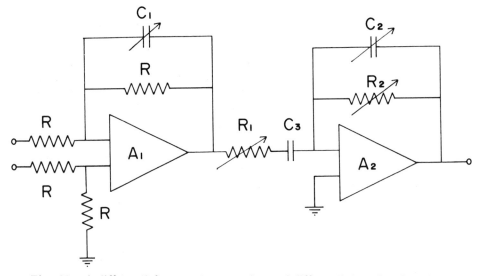

Fig. 22. A differential current measuring and differentiator circuit. (Reprinted [7, p. 310] by courtesy of the American Chemical Society.)

Keeping the load resistor in series with the working electrode requires some form of compensation [3] or the use of a differential controller [8]. Compensation for the resistor iR drop would involve subtracting a signal equal to this voltage from the controlled potential signal.

Booman and Holbrook [3] provided such a compensation by using an inverter to provide the necessary signal. This subtracts the iR drop which the controller sees as part of the cell potential. The inverter input resistor must be much larger than the load resistor. The voltage proportional to the cell current can be measured at a low impedance point at the follower or inverter outputs or with a high input impedance device at the load resistor itself. The inverter form of current measurement and compensation lessens the required current output of this amplifier since it is not in the cell current-carrying loop. A quality amplifier with good frequency response is still required if inaccurate control through poor compensation is to be avoided.

These basic detectors are useful in almost all electrochemical circuits since current or potential is the measured output from most techniques. In many cases, however, a function of the current or only a portion of the current is the desired output. Mathematical functions of the current or voltage can be obtained from recordings of these by graphical or other computational procedures. The use of operational amplifiers to obtain these functions is generally more convenient and makes the function available to act as a controllable variable, to perform other instrumental functions, or for further mathematical operations. Besides, the accuracy obtainable from the operations of these amplifier circuits is as good as, if not better than, those obtained by first converting the data to numerical form and performing the operations by hand. Sorting and signal isolation can be done only on the instantaneous current or voltage output.

Numerous output signal detectors have come into general use. Among these, integrators, differentiators, phase and frequency selective circuits, and algebraic compensators are the most commonly used.

### B.  Circuits for Controlled Potential Electrolysis,

### Coulometry, and Coulometric Titrations

Controlled potential electrolysis was the first electrochemical technique in which an operational amplifier instrument was employed. The types of circuits and amplifiers that are necessary are dictated by the characteristics of the technique itself; and suitable instrumentation, aside from the controller, is relatively simple.

Booman's first controlled potential electrolysis instrument [5] (Fig. 20) consisted basically of a constant voltage source; a controller; a current detector, a current follower; and a single signal conditioner, an integrator.

The current follower converted the cell current into a proportional voltage signal which acted as the input to the integrator. The integrator provided an output voltage proportional to the number of coulombs passed; an adjustable scale factor, the integrator time constant, converted the readout in volts to a more convenient quantity. In Booman's experiments the integrator output in volts was made equal to the number of milligrams of uranium in the sample.

The amplifiers Booman used in both the current follower and the integrator were stabilized electron tube amplifiers (USA-3). The current follower needed extra current capacity which was provided by a multitube booster.

The controlled potential coulometer of Kelley et al. [8] (Fig. 23) also provided an integrated current readout. The output circuit was essentially the same as Booman's except that their differential controller did not require a current follower and the integrator was connected directly to the current measuring load resistor.

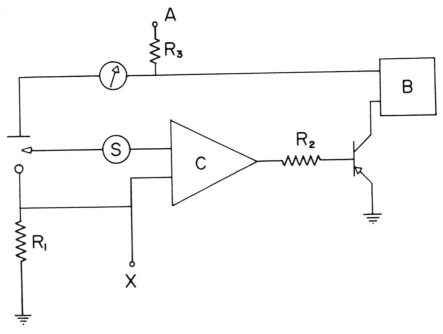

Fig. 23. A portion of the controlled potential coulometer of Kelley et al. [8] where A is a -300-V power supply, B is the cell power supply, C is the controller, a chopper stabilized difference amplifier (modified Philbrick USA-3M3), and X is the connection to the integrator. (Reprinted, in part [8, p. 489], by courtesy of the Americal Chemical Society.)

In 1963 Booman and Holbrook [3] reported a coulometer circuit which used an integrated readout circuit similar to the Kelley et al. type. They also discussed integrator performance with regard to the characteristics of the amplifier, the quality of the capacitor, and the size of the input resistance. They pointed out the effect of offset correction in producing a long-term integrator drift and concluded that either an extremely low offset amplifier must be used or the offset signal should be derived from very high impedance sources. They further considered high dc gain and high input impedance in the amplifier as extremely important to good integrator performance. They indicated that the dielectric material of the integrating capacitor determines the long-term accuracy of the integrator and that polystyrene capacitors had proven to be the most suitable for integrator applications. They claimed that operational amplifier integrators, when properly designed, could function with a precision of 0.01%. In a comparison of operational amplifier circuits to other types of integrators they concluded that only elegant digital integrators possessed a greater accuracy.

Other circuits useful in controlled potential electrolyses have appeared. In general these circuits have been identical to the Booman or the Kelley et al. types with some modification in the external circuitry or with solid state amplifiers.

Instruments which scan the potential during the controlled potential electrolysis have been employed by several workers. Such a scanning coulometer constructed from operational amplifiers has been given by Propst [9]. His circuit consisted of a low current potentiostat of the adder type and a linear scan signal generator controlled by the electrolysis cell current. The scan signal generator was an integrator whose input signal varied as the inverse of the current. The input signal was supplied by a novel adder circuit which functioned as a divider. This circuit is shown in Fig. 24. A current, proportional to the electrolysis current, was obtained from an inverter-current follower combination and passed through a diode network, generating a voltage, across the diodes, proportional to the log of the current. This voltage was added to a constant signal in the adder, producing an output proportional to log k (the constant) minus log i. The voltage output from the adder passed through a second diode network to a resistor, producing a current through, hence a voltage across, this resistor proportional to k/i or inversely proportional to the current. This signal then became the scan integrator input and produced a scan rate inversely proportional to the cell current. The detector was a simple current follower integrator combination producing an output proportional to the total charge passed. The recording of this signal versus the cell potential on an x-y recorder provided a coulomb versus potential readout. Various other circuits for background current correction were also included. Stabilized amplifiers were used throughout the instrument.

In addition to the coulometer circuits already discussed, other

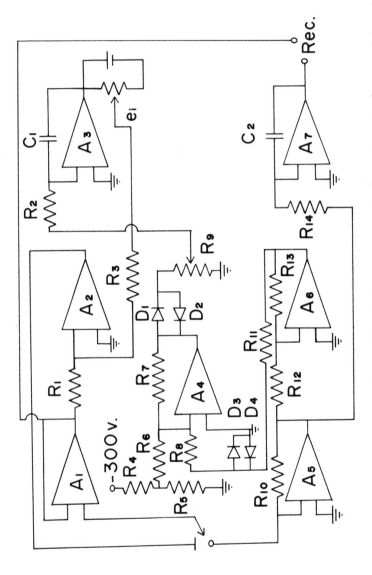

Fig. 24. A voltage scanning coulometer: $A_1$, voltage follower; $A_2$, controller; $A_3$, signal source integrator; $A_4$, signal-conditioning circuit; $A_5$, current follower; $A_6$, an inverter; and $A_7$, output integrator. (Reprinted [9, p. 959] by courtesy of the American Chemical Society.)

instruments for coulometry and coulometric titrations have been reported by Stephens and Harrar [10], Buck [11], Stromatt and Connally [12], Jones et al. [13], Booman and Holbrook [3], and Harrar and Behrin [14].

## C. Circuits for Linear Scan Techniques, Polarography,

### and Related Methods

The methods in electroanalytical chemistry which have received the most benefit from operational amplifier instrumentation are the controlled potential, linear sweep methods. Operational amplifiers have been important in the development of suitable control amplifiers, accurate signal generators, and sensitive detectors for these methods. Such circuits have made stationary electrode polarography, cyclic voltammetry, and the variations on these, very useful and practical techniques. Polarography, although it is theoretically a constant potential technique, is instrumentally a linear scan method; and polarographic instrumentation is considered in this section as well.

The first operational amplifier instrument for polarography was the controlled potential and derivative polarograph of Kelley et al. [6]. In addition to the controller and current follower circuits previously mentioned, this instrument included two signal conditioning circuits: a peak current reader and a dc current differentiator.

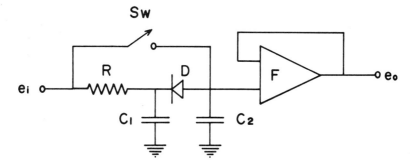

Fig. 25. A peak reader. This circuit is connected through an RC filter to the output of current amplifier A in Fig. 21. (Reprinted [6, p. 1477] by courtesy of the American Chemical Society.)

In the peak reader circuit (Fig. 25), the output of the current follower was filtered and passed through a diode to a capacitor. The diode allowed capacitor charging in one direction only. The capacitor voltage was monitored by an operational amplifier voltage follower. The high input impedance of the amplifier and the high reverse impedance of the diode prevented any discharging of the capacitor. Thus, at any time during a polarographic scan, the output of this circuit corresponded to the maximum current level. The output of this circuit corresponds to the polarographic peak current provided that a maximum does not appear or that the diffusion current does not decrease due to changes in the droptime. Instead of passing into the peak follower, the output of the current follower could be filtered and fed to a differentiator circuit. This circuit provides a first derivative with respect to time. If the cell potential varies linearly with time, as was the case here, the derivative, $di/de$, of the polarographic wave is obtained. This circuit output was inverted, amplified or attenuated, then filtered and recorded. The conditioning circuits contained in the controlled-potential and derivative polarograph were constructed with unstabilized, electron-tube type amplifiers.

A convenient signal generator for polarography was also developed by Kelley et al. [15]. They used an operational amplifier integrator with a constant input voltage to obtain a ramp signal. The integrator output was attenuated and used as the controlled-potential polarograph input signal. The sweep rate could be changed by altering the integrator input resistor. A shorting switch reset the capacitor after a sweep and held the integrator output at zero between experiments. The circuit employed a nonstabilized, electron-tube amplifier (K2-X) which proved sufficient for the signal quality desired. This sweep generator also provided a compensating signal to the current amplifier which, when added to the polarographic current, produced an essentially zero residual current.

Based on these early circuits and on the suggestions of DeFord [16], several workers have built operational amplifier instruments for polarography and for linear scan methods.

Two such circuits, one for polarography ane one for stationary electrode polarography, employed steplike signals rather than the normal linear forms.

Auerbach et al. [17] used a staircase signal in an incremental approach to derivative polarography. At a specific time during each growing mercury drop the potential was increased by a fixed amount. The currents before and after the potential step were compared and their difference recorded.

The staircase signal was formed by discharging a small capacitor into an operational amplifier integrator. The voltage steps were synchronized with, but delayed from, the fall of the mercury drops. The steps were, thus, applied at constant drop age. The difference between the currents flowing before and after the potential step was detected in a unique memory circuit, shown in Fig. 26, that consisted of a current follower whose output passed to a low pass filter followed by a current integrator with a capacitor input. The capacitor input to the integrator, called the memory capacitor, was charged, by reason of virtual summing point ground, to the current

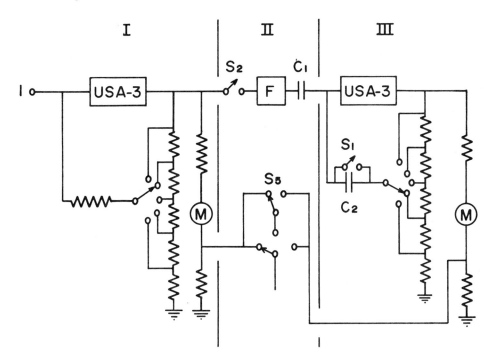

Fig. 26. A "memory" circuit for recording the current increase following
a stepwise change in the potential of a polarographic cell. The current
follower circuit (I) charges the memory capacitor, $C_1$, according to the
value of i, when $S_2$ is closed. $S_2$ is opened and the potential is increased.
$S_2$ is then closed and $S_1$, previously closed, is open. Capacitor $C_2$ in the
integrator (III) is charged to the difference between the $C_1$ value before and
after the potential change. F — voltage follower; $S_1$ and $S_2$, relay switches.
Current follower and integrator employ USA-3 stabilized amplifiers (Philbrick Researches Inc.) (Reprinted [17, p. 1481] by courtesy of the
American Chemical Society.)

follower output. Prior to the potential step, a relay opened the circuit at
the current follower output, and a second relay shorted the feedback capacitor of the current integrator. After the potential step the relay shorting the
current integrator was opened. Then the second relay again connected the
circuit to the current follower output. The current flow into the current
integrator represented the difference between the current follower output
and the original potential on the memory capacitor. The difference between
them was stored on the current integrator and a voltage proportional to this
difference could be read directly. One difference signal was obtained for
each mercury drop and the resulting output appeared like the first derivative

of a polarogram.

Mann's staircase circuit [18] involved the generation of a more rapid staircase signal. The main timing and signal-generating unit was a unijunction transistor oscillator. Control of the steptime was accomplished by varying the charging rate of a capacitor in the emitter section of the circuit. The pulses obtained at the base of the transistor were used as the input to a current source amplifier which charged a high quality capacitor. The staircase signal across the capacitor was amplified and fed to the cell. The pulsed charging produced the necessary staircase signal. The size could be varied by attenuating the output signal to the cell.

The bulk of the circuits for polarography and for linear scan methods followed more closely the suggestions of DeFord [16] and the integrator circuit of Kelley et al. [15].

In 1963 Underkofler and Shain [19] reported a number of useful circuits which were available from a single multipurpose instrument. A variety of operational amplifier circuits were used to produce, measure, and condition the voltage signals for several different experimental techniques. Their ramp generator for polarography and for stationary electrode polarography

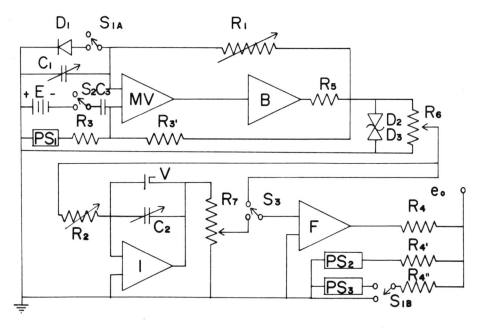

Fig. 27. A precise square and triangular waveform generator for single-scan and multicycle operation (I). (Reprinted [19, p. 1781] by courtesy of the American Chemical Society.)

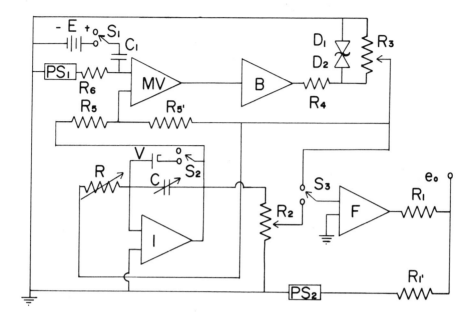

Fig. 28. A precise square and triangular waveform generator for single-scan and multicycle operation (II). (Reprinted [19, p. 1782] by courtesy of the American Chemical Society.)

was similar in concept to the earlier circuit of Kelley et al. [15], but was more sophisticated and included a stabilized operational amplifier (K2-X, K2-P) which produced very accurate and reproducible signal forms. A wide variety of sweep rates could be obtained since the input resistance of the integrator could be varied from 1000 $\Omega$ to 100 M$\Omega$ and different feedback capacitors, 0.1, 1, and 10 $\mu$f could be employed.

For cyclic triangular wave voltammetry and step function controlled potential electrolysis, two new signal generating circuits were given by Underkofler and Shain. Both circuits (Figs. 27 and 28) contained an operational amplifier multivibrator for the generation of square waves, an integrator to convert the square wave to a trangular signal, and a voltage follower to transmit either signal form to the controller.

In the first circuit, the multivibrator could act in either an astable or monostable fashion. A single-cycle or a repetitive square waveform could be obtained from the multivibrator output. These signals could be integrated to produce a single-cycle or multicycle triangular wave. A diode in the integrator feedback loop selected the direction of the scan. In single-cycle operation, a diode in the input circuit of the multivibrator dictated the stable state and the required polarity of the trigger pulse. The signal frequency (or period) was determined by the multivibrator input, where a delay circuit controlled the switching time. This RC circuit time constant could be varied to produce a range of scan rates: the input signal to the integrator and the integrator time constant could also be varied to change both the scan rate and scan amplitude. This latter feature was a drawback of this generator circuit in that changing the integrator input or feedback impedance always changed the signal amplitude as well as the scan rate.

In the second circuit, the scan rate could be varied without necessarily changing the signal amplitude. Although the basic circuit components are the same for this second circuit as for the first, the operation was quite different. In this circuit the multivibrator was bistable. Its output state was determined by the sign and magnitude of the signals at its input. One of the signals was a dc reference voltage. The other signal was obtained at the midpoint of a resistor network connecting the integrator output and the attenuated output of the multivibrator. When the multivibrator state was reversed by a trigger signal or a sign reversal at its inputs, this midpoint signal acted as a regenerative feedback signal. The change in the sign of the integrator input signal caused the integrator output to reverse direction. When the magnitude of the integrator output signal exceeded the attenuated output of the multivibrator the multivibrator again changed state.

In multicycle operation, this cycling process was repeated at both the positive and negative limits of the triangular signal. A diode in parallel with the integrating capacitor restricted the integrator to either a positive or negative triangular signal. This diode limiting also placed one trigger point out of the reach of the integrator output swing, and switching action stopped after one half-cycle of the triangular wave. Only a trigger signal could initiate a multivibrator change and produce another half-cycle. The integrator time constant determined the signal period, and the integrator diode determined the scan direction and the trigger signal polarity. For the second circuit, the repetitive triangular wave contained both positive and negative voltages.

The second circuit was the most suited for single-cycle operation, while the first produced the more desired multicycle operation.

A circuit embodying the convenience of a conventional polarograph and the three electrode capabilities of operational amplifiers was reported by Annino and Hagler [20]. A conventional recording polarograph (Sargent Model XV) acted as the signal source and current detector while a stabilized operational amplifier was used as the control amplifier. Two different circuits were given (Fig. 29). In one the load resistor iR drop was left uncompensated, but no modification of the polarograph was required. In the other circuit the load resistor appeared in the compensated portion of the circuit, but a simple modification of the polarograph was necessary. A single-ended chopper stabilized amplifier was used for both circuits. The configurations correspond to circuits i and c in Fig. 46 [21].

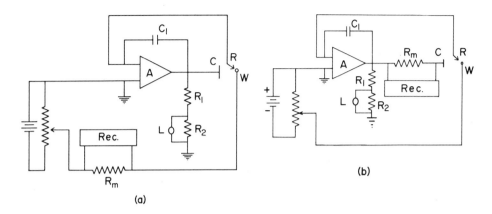

Fig. 29. Two circuits for utilizing a two-terminal Sargent polarograph in a three-electrode circuit: (a) current measuring resistor uncompensated; (b) current measuring resistor compensated, polarograph circuit modified. (Reprinted [20, p. 1558] by courtesy of the American Chemical Society.)

More recently Schroeder (22), using a differential solid state operational amplifier controller, has given a similar circuit but with the load resistor iR drop uncompensated. This circuit is similar to that of Fig. 46d. Floating the polarograph recorder or the operational amplifier power supply could provide the more ideal case, such as the circuit of Fig. 46b.

Circuits giving first-, second-, and third-derivative readout of the current in polarography and linear-scan voltammetric techniques have been employed by numerous workers [7, 23, 24]. These techniques generally require one or more operational amplifier differentiator circuits operating on the filtered output from a current follower. These circuits are the same as, or similar to, the Kelley et al. [6] derivative polarograph circuit; most employ differentiator circuits like that shown in Fig. 16; many use solid state amplifiers [24].

Other interesting signal generating and conditioning circuits appear regularly. Almost all of them employ solid-state or integrated-circuit operational amplifiers; many include other relatively new solid state devices, such as integrated circuits for gating, switching, and digital operations. Such a circuit for triangular-wave generation has been made by Myers and Shain [25].

An example of the application of these new modules in operational amplifier circuits is the triangle-wave circuit of Huntington and Davis [26]. As shown in Fig. 30, this generator incorporates a silicon-controlled rectifier (SCR) gate which, when triggered, begins the signal generation cycle. Before the gating pulse, the undefined input of the flip-flop amplifier causes output

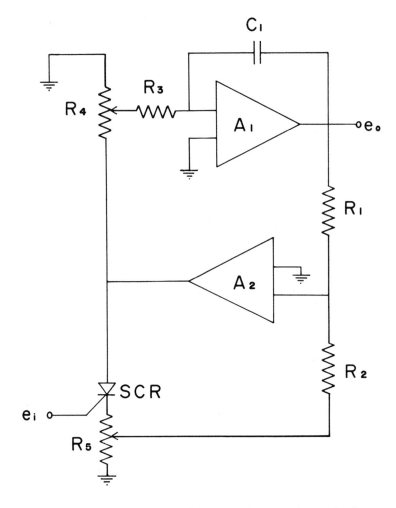

Fig. 30. A signal source for triangular waveforms [26].

oscillation which produces no output from the integrator. A pulse causes the
rectifier to conduct and the flip-flop amplifier is driven to its positive limit
by the positive feedback through the SCR, $R_5$, and $R_2$. The integrator, operating on a fraction of that positive signal, gives a negative ramp output.
Eventually the integrator output becomes sufficiently negative to reverse the
sign at the flip-flop amplifier positive input. The flip-flop output reverses,
causing the integrator output to reverse direction. The SCR, which was held
on by the positive output signal, is now shut off by the negative one. When
the integrator output reaches its initial value, the circuit returns to its

original state. Reversing the SCR produces a triangular wave of opposite sign. This circuit, which uses solid state circuitry throughout, is simple in operation, compact in size, and relatively inexpensive to construct.

In addition to the circuits discussed, numerous instruments for polaro-graphy and for linear scan voltammetry have been reported. These include a polarograph by Durst et al. [27], a circuit for cyclic voltammetry by Alden et al. [28], a Kalousek polarograph by Kinard et al. [29], and a solid state controlled potential polarograph by Jones et al. [24].

## D. Circuits for Alternating-Current Methods

Operational amplifiers have frequently been used in instruments for ac polarography and related techniques. The controllers and followers are the same as those used in other controlled potential methods. In ac methods, circuits for the isolation and amplification of a specific frequency signal are quite important.

The twin-T network forms the basis for the most useful of the specific

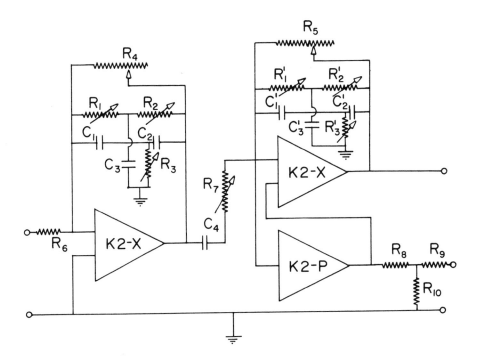

Fig. 31. A twin-T amplifier for application in ac polarography. (Reprinted [31, p. 1813] by courtesy of the American Chemical Society.)

frequency amplifier circuits. This circuit has been used extensively in ac circuits since the work of Smith and Reinmuth [30].

The twin-T amplifier is merely an inverter where the feedback impedance is the twin-T network and a limiting resistor (Fig. 31). The frequency of maximum gain is that where the impedance of the twin-T network is a maximum. The limiting resistor, in parallel with the twin-T network, determines the maximum value of the feedback impedance and hence sets the amplifier gain. This resistor also prevents instability and oscillation. The twin-T oscillator shown in Fig. 32 is basically a high gain amplifier circuit made unstable by positive feedback. The maximum gain, and hence the maximum oscillator output, is at the frequency of maximum impedance for the twin-T network.

Somewhat later, Smith [31] reported using several operational amplifier circuits in an instrument for various forms of phase-selective, second-harmonic, and phase-angle-recording ac polarography.

Some of the basic circuits used included a rectifying amplifier (Fig. 33), a zero crossing detector (Fig. 34), and an adder, as well as the twin-T

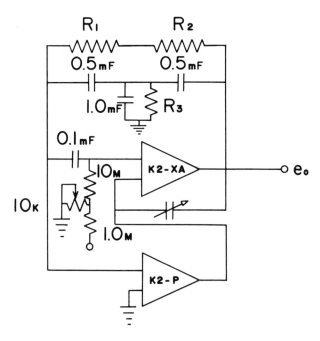

Fig. 32. A twin-T oscillator for generating a sinusoidal waveform. (Reprinted [31, p. 1813] by courtesy of the American Chemical Society.)

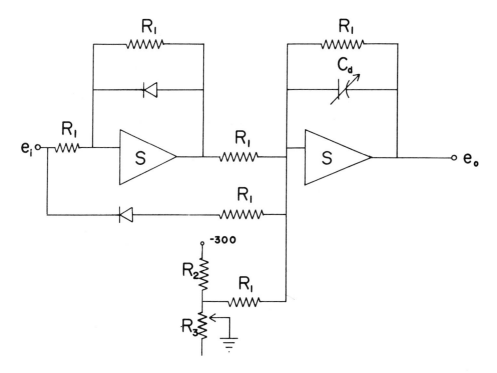

Fig. 33. A full wave rectifier-amplifier circuit for rectification and frequency doubling. (Reprinted [31, p. 1817] by courtesy of the American
Chemical Society.)

amplifier. From several of these units, Smith constructed a variety of
circuits including a second-harmonic ac polarograph (Fig. 35), a second-
harmonic phase angle recording polarograph (Fig. 36), and others.

Operational amplifier circuits for pure ac signals offer only a slight
advantage over normal tube or transistor circuits. Operational amplifiers
can be used over a broad frequency region, however, and do offer simplicity
in the design of ac amplifier circuits.

Other circuits for alternating-current techniques include a phase-
selective polarograph for stripping analysis by Underkofler and Shain [32],
an admittance-recording ac polarograph by Hayes and Reilley [33], an ac
polarograph for stripping analysis by Eisner et al. [34], a phase-selective
ac polarograph by Evilia and Diefenderfer [35], an instrument for admittance
recording by DeLevie and Husovsky [36], and a circuit for square-wave
voltammetry by Krause and Ramaley [37].

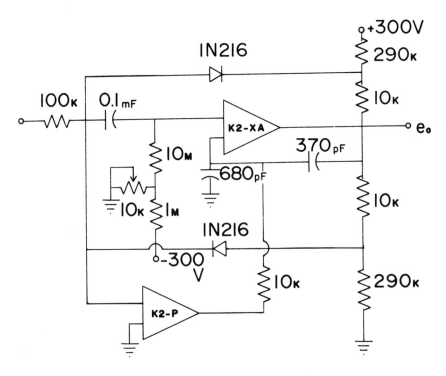

Fig. 34. A zero crossing detector for generating a square wave signal of fixed amplitude. The square wave output is in phase with and of the same frequency as the input signal. (Reprinted [31, p. 1817] by courtesy of the American Chemical Society.)

## E. Circuits for Controlled-Current Methods

Operational amplifiers have been used in circuits for controlled current methods. In controlled current circuits, they play a less critical and less demanding role than in controlled potential applications.

### 1. Circuits for Controlling the Cell Current

Basically, three different types of galvanostats have been reported. The first was reported by Reilley [1] and others. In this circuit the cell replaces the resistor in the feedback loop of an inverter (Fig. 37). Since the current through the feedback loop is the same as the current through the input resistor, the cell current is directly related to the input signal. With this arrangement one side of the cell is at virtual ground. The cell potential can be measured relative to ground and single-ended measuring devices can be used.

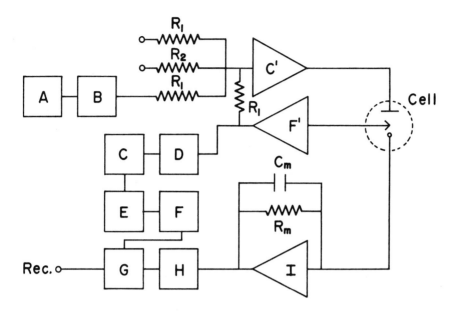

Fig. 35. The block diagram of an instrument for phase selective second harmonic ac polarography: $C'$, controller; $F'$, voltage follower; I, current follower; A, twin-T oscillator (see Fig. 32); B, D, E, H, tuned amplifiers (see Fig. 31); C, full wave rectifier (see Fig. 33); F, phase shifter; and G, phase detector. (Reprinted [31, p. 1916] by courtesy of the American Chemical Society.)

The second type of galvanostat circuit makes direct use of the potentiostats used in controlled potential applications. The potential that is controlled is the iR drop across a resistor in series with the cell. Potentiostats can be used directly in this manner by placing the cell in the potentiostat feedback loop and the current-determining resistor between the reference input and ground. Generally, a differential or floatable voltage measuring device is required since all cell leads are removed from ground or virtual ground.

Schwarz and Shain [21] systematically evaluated the performance of galvanostat circuits of this type. They determined how the arrangement of the circuit components affected the accuracy of current control. Using criteria similar to those employed in evaluating potentiostats, they found several circuits which provided accurate control and reasonably convenient operation. They investigated single-amplifier and multiamplifier galvanostat circuits with two types of system grounding. Their type-A grounding represents the normal case, where the power supply common is at ground. Type-B grounding required a floating supply and a grounding of the normal amplifier output. The dominant problem found in the eight possible arrangements with

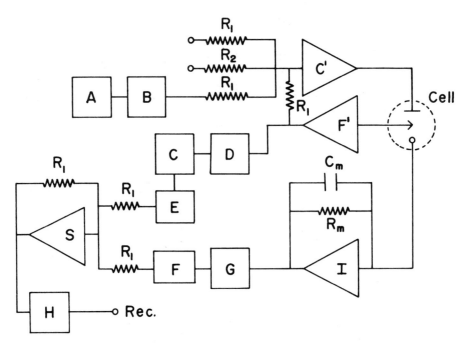

Fig. 36. The block diagram of an instrument for phase angle recording:
C′, controller; F′, voltage follower; I, current follower; S, adder; A,
twin-T oscillator (Fig. 32); B, D, G, tuned amplifier (Fig. 31); C, phase
shifter; E, F, trigger (see Fig. 34); H, full wave rectivier (Fig. 33).
(Reprinted [31, p. 1817] by courtesy of the American Chemical Society.)

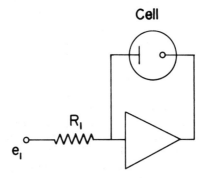

Fig. 37. A simple controlled current circuit. (Reprinted [1, p. A862] by
courtesy of the American Chemical Society.)

type-A grounding was that the voltage measuring device could seldom be used with one end at ground, whereas in the eight circuits using type-B grounding the signal source could not have one end at ground. Most circuits could use stabilized amplifiers, but in very few cases was signal generator iR drop error absent.

In simple experiments where easily floated battery sources can be used, type-B grounded circuits would be satisfactory. The circuit g in Fig. 38, with the upper output grounded, possesses the best characteristics of the circuits in this class.

The utilization of type A grounded circuits appears from Table 1 to be disadvantageous. However, circuit b of Fig. 38 is identical with the one

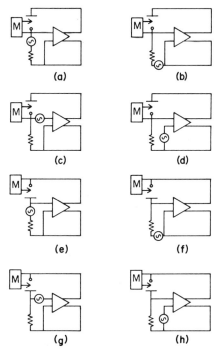

Fig. 38. Circuit configurations for a single-amplifier galvanostat. Circuits are shown without ground. (a-d) Type A grounding — circuit is as shown, lower output (power supply common) at ground. (e-f) Type B grounding — upper output at ground, lower output (power supply common) floating, signal source is connected to alternate input and cell current is drawn from the floating output. (Reprinted [21, p. 1776] by courtesy of the American Chemical Society.)

TABLE 1

Interrelations of Characteristics of Single-Amplifier Galvanostats[a]

| Characteristics[b] | Circuit configuration (Fig. 38) | | | | | | | | | | | | | | | |
| --- | --- | --- | --- | --- | --- | --- | --- | --- | --- | --- | --- | --- | --- | --- | --- | --- |
| | Type A grounding | | | | | | | | Type B grounding | | | | | | | |
| | a | b | c | d | e | f | g | h | a | b | c | d | e | f | g | h |
| 1. No SG interaction | - | - | + | + | - | - | + | + | - | - | + | + | - | - | + | + |
| 2. Stabilization | + | + | + | - | + | + | + | - | + | + | + | - | + | + | + | - |
| 3. SG grounding | - | + | - | + | - | + | - | + | - | - | - | - | - | - | - | - |
| 4. M grounding | + | + | - | - | - | - | - | - | - | - | - | - | + | + | + | + |

[a]Reprinted [21, p. 1777] by courtesy of the American Chemical Society.

[b]Notes: (+) indicates that:
   1. signal generator (SG) is not located in the current carrying loop;
   2. circuit can be stabilized;
   3. signal generator (SG) can be referred to ground;
   4. readout device (M) can be referred to ground.

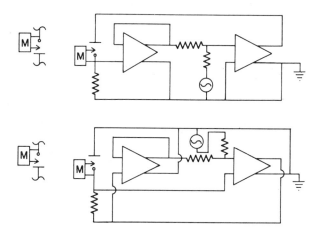

Fig. 39. Circuit configurations for adder-type galvanostats. Both types A and B grounding are shown. (Reprinted [21, p. 1777] by courtesy of the American Chemical Society.)

given by Reilley [1]. One might see that signal generator iR drop error is not really a problem when the signal source does have one end at ground. Thus circuit b is quite acceptable because the signal source is separated from the amplifier output by the virtual ground at the inverting input. The signal source must supply current but this is part of its normal function. Little error is present if the resistor in series is greater than the output impedance of the source.

For galvanostats with adder-controllers the signal generator is attached to one of the adder inputs and one end of the load resistor is connected through a follower to a second input. This is the only practical circuit arrangement. However, four different configurations (Fig. 38) are possible when types A and B grounding and the two alternate ways of connecting the cell are considered. Only with type A grounding can stabilized amplifiers be used, but only one circuit, a B grounding type, allows the voltage measuring devices to have one terminal grounded.

A third type of circuit for controlling the cell current has been given by Ewing and Brayden [38] (Fig. 40). This circuit uses two amplifiers in inverter circuits. One inverter is the source of the load current. The second inverter detects the voltage across the load and transmits the inverse of this signal to the main inverter. This feedback action increases the apparent input signal to the main inverter by the iR drop in the load. This results, essentially, in the control of the voltage across resistor R which is in series with the load. Provided that the current into the input resistor of the detecting inverter is negligible, the current through the load is constant.

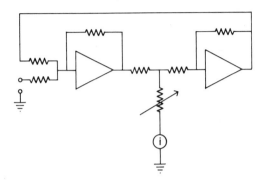

Fig. 40. The circuit for a constant current source. (Reprinted [38, p. 1827] by courtesy of the American Chemical Society.)

## 2.  Other Circuits for Controlled Current Techniques

In 1963 Murray [39] reported a circuit for a controlled current technique in which the current through the cell was varied as a power of time. This circuit consisted of a controller similar to the one shown in Fig. 37 and an interesting circuit for generating the controller input signal (Fig. 41). The signal generator consisted of four operational amplifier circuits; one summing amplifier, two inverters, and an inverter with a nonlinear input impedance.

In operation, a linearly increasing voltage, $e_s$, at the input of the summing amplifier is imposed at the sliding contact of the variable voltage divider by the feedback action of the upper inverter. This requires that voltages at the ends of the voltage divider, $e_1$ and $e_2$, have values such that

$$Ne_1 + (1 - N)e_2 = e_s \tag{23}$$

where

$$N = R_2/R_1 + R_2. \tag{24}$$

The quadratron input impedance of inverting amplifier Z causes the output voltage, $e_1$, of this inverter to equal 0.01 times the square of its input voltage, $e_o$. The output voltage of the lower inverter, $e_2$, is just the inverse of its input signal, $e_o$. Thus, $e_1 = 0.01e_2^2$ and this relation can be used to solve Eq. (23) for either variable in terms of $e_s$. Approximate values for these relations are

$$e_2 = k_2 e_s^r \tag{25}$$

and

$$e_1 = k_1 e_s^r \tag{26}$$

where r may assume values between 0.5 and 1 in the first equation and

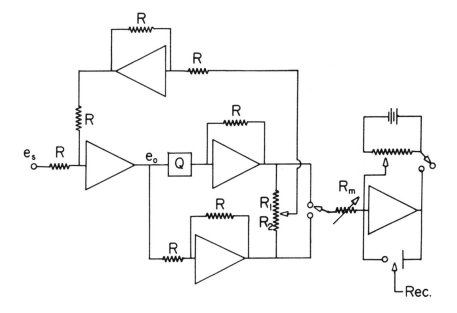

Fig. 41. A circuit for power-of-time chronopotentiometry. (Reprinted [39, p. 1784] by courtesy of the American Chemical Society.)

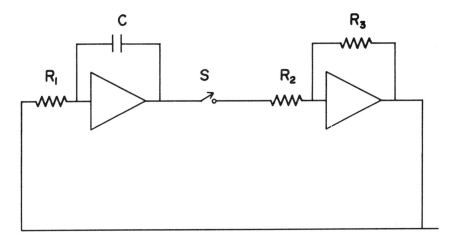

$$e_0 = \propto e^{\beta t} R_m$$

Fig. 42. An exponential time-current generator, connects to the $R_m$ in the circuit of Fig. 41. (Reprinted [39, p. 1787] by courtesy of the American Chemical Society.)

between 1 and 2 in the second. Since $e_s$ varies linearly with time, two power-of-time signals result.

Murray also reported a circuit for obtaining signals which varied exponentially with time. This circuit, shown in Fig. 42, is a straightforward use of analog computing, and represents the general differential equation

$$\frac{dx}{dt} - x = 0. \tag{27}$$

The solution, obtained at the inverter output, is

$$e_o = B \exp(-\alpha t) \tag{28}$$

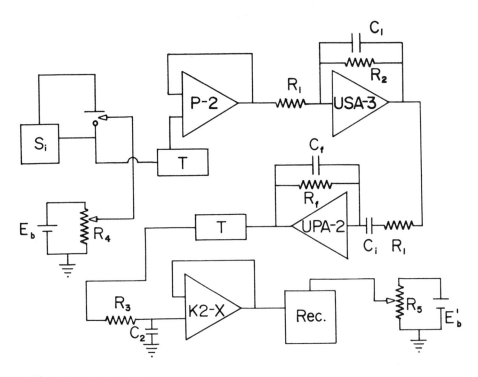

Fig. 43. A circuit for derivative chronopotentiometry. Circuit components are a controlled current source, $S_i$; low pass filters, T; a voltage follower employing a P-2 carrier amplifier; an inverter employing a USA-3 chopper stabilized amplifier; a differentiator employing a UPA-2; a second voltage follower using a K2-X (all amplifiers, Philbrick Researches Inc.); a cell bias signal from $E_b$ and $R_4$ and a recorder bias from $E_b'$ and $R_5$. (Reprinted [40, p. 531] by courtesy of the American Chemical Society.)

where the constants are determined by the resistor and capacitor values in the circuit.

Peters and Burden (40) have devised an instrument for derivative chronopotentiometry, where $dE/dt$ is the desired experimental output. Their circuit, which employed a conventional controller and differentiating circuit, is shown in Fig. 43.

Other circuits for controlled current techniques include circuits for chronopotentiometry by Deron and Laitinen [41], for chronopotentiometry with compensation for extraneous current by Shults et al. [42], and for cyclic chronopotentiometry by Herman and Bard [43].

### F. Other Applications of Operational Amplifiers

An interesting application of operational amplifiers is the circuit used by Napp et al. [44] for the simultaneous control of two working electrodes. This circuit, shown in Fig. 44, uses only one reference electrode in the control operation but allows the two working electrodes to be controlled independently and their currents to be measured simultaneously.

The circuit contains two controller functions, each operating on one of

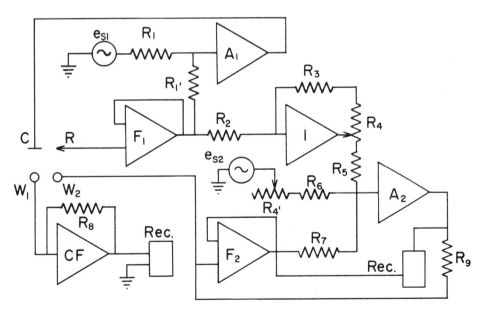

Fig. 44. A potentiostat for the simultaneous control of two working electrodes. (Reprinted [44, p. 481] by courtesy of the American Chemical Society.)

the working electrodes. The first controller sets the potential of working electrode $W_1$ and is composed of adder $A_1$, voltage follower $F_1$, and current follower CF, and is of conventional design. The second control circuit, operating on electrode $W_2$, is composed of adder $A_2$, followers $F_1$, and inverter I. The first controller imposes potential $e_{s1}$ between electrode $W_1$ and the reference electrode. The second controller imposes potential $e_{s2}$ between electrode $W_2$ and the reference electrode.

The inverter I between follower $F_1$ and adder $A_2$ serves to maintain the control of electrode $W_2$ independent of potential $e_{s1}$. With $e_{s2}$ zero, for any $e_{s1}$ value adder $A_2$ maintains the potential of electrode $W_2$ equal to the reference electrode potential.

In practice, Napp et al. adjusted potentiometer $R_4$ to obtain this condition. With the circuit so adjusted, any $e_{s2}$ value will cause adder $A_2$ to bring follower $F_2$, and hence electrode $W_2$, to a potential of $-e_{s2}$ versus the reference electrode. Potentiometer $R_4$ is adjusted to bring about this condition while $e_{s1}$ is zero. When both adjustments are completed, the independent control of the two electrodes is achieved.

Other reported circuits for special applications include a four-terminal potentiostat by Matsen and Linford [45], instruments for chronocoulometry by Christie et al. [46] and by O'Dom and Murray [47], a device for measuring the slopes of reaction rate curves by Pardue and Dahl [48], an instrument for recording electrocapillary curves by Bard and Herman [49], a device for drop-time measurements by Hayes et al. [50], a detector for use in a continuous titrator by Blaedel and Laessig [51], and circuits for impedance and galvanostatic measurements on ion-selective electrodes by Brand and Rechnitz [52, 53].

### G. Multipurpose Instruments

Several instruments which perform a variety of techniques have been reported. These are generally centered about a potentiostat system which can be altered to perform as a galvanostat. Generally a variety of signals can be provided to the controller from either internal or external signal sources. These circuits include the multipurpose instruments of Underkofler and Shain [19], Buck and Eldridge [54], Lauer et al. [55], Morrison [56], Oglesby et al. [57], Goolsby and Sawyer [58], Mueller and Jones [59], and Bezman and McKinney [60].

### IV. OPERATIONAL AMPLIFIER POTENTIOSTATS

### A. Early Development

One could view the development of operational amplifier circuits for electroanalytical chemistry as occurring in two stages. The first stage

began with the introduction of operational amplifiers by Booman [5] and
continued with the circuits developed by Kelly et al. [6, 8], and the improvements and developments introduced by DeFord [16].  At the end of this
period most of the basic operational amplifier circuits had been introduced
and successfully used in electrochemical instruments.  The value of operational amplifiers had been established.  The period following DeFord's work
has seen an expansion in the use of operational amplifier circuits and
instruments, the conversion of operational amplifiers from vacuum tube to
solid state devices, and the steady improvement in the science of operational amplifier instrumentation.

In 1957, Booman [5] first reported the use of operational amplifiers in
an instrument for controller potential coulometry.  In this instrument (Fig.
20) an operational amplifier was used in the controller, an operational
amplifier current follower was used for the detection and conditioning of the
electrolysis cell current, and an operational amplifier was used as an integrator to obtain direct readout of total charge passed.

The control circuit was relatively simple and consisted of an operational
amplifier and a high current output stage.  The controller circuit was similar
in concept to those developed earlier by several workers [61-63].  The current follower actually played an essential part in the control circuit operation;
however, in the ideal case it served only to maintain the cell cathode at
virtual ground.  The signal appearing at the control amplifier input was the
sum of the cell potential and the battery source voltage signals.  The feedback path through the cell allowed the controller to maintain its input signal
at virtual ground.  This forced the cell potential to equal the battery source
voltage.

With a constant voltage source the controller maintained the cell potential constant during the electrolysis.  The output stage provided the large
current necessary for a mass scale electrolysis.  Aside from this function
the current booster did not markedly affect the nature of the control operation.  This was true because the booster had a gain near unity and did not
introduce significant phase shift or invert the signal from the operational
amplifier.

A chopper stabilized amplifier was used in the controller to avoid any
long-term drift error and to provide maximum control accuracy.  The signal
source could not be grounded but, except for long-term stability, little
demand is placed on this component.

The current follower operation was characterized by two main features;
it maintained its input near ground potential, and provided an output signal
proportional to its input current.  A stabilized amplifier was also used in
this circuit.  The feedback resistor was a variable precision wirewound type
and its value was determined by the expected range of electrolysis currents.

The second reported use of operational amplifiers in electrochemistry was the controlled potential device of Kelley et al. [8] (Fig. 23). Their circuit was similar in its function to the controller devised by Booman but was composed of a single amplifier with a transistorized current booster. The simplicity of their control circuit was due to a modification of the stabilized amplifier which allowed it to be used as a differential amplifier. Their controller operated directly on the difference between the working and reference electrode potentials. The virtual grounding action of a current follower was not necessary. The cell current could be measured across the resistor connecting the working electrode to ground. The voltage source characteristics and the current booster action were similar to those in the Booman coulometer.

The controlled potential and derivative polarograph developed by Kelley [6] (Fig. 21) included several new uses of operational amplifiers. It included an operational amplifier controller, a current follower, a peak follower, and a differentiator. The controller was essentially the same as Booman's original circuit but did not employ a current booster. The current follower was somewhat different from Booman's but operated on the same principle.

Many of the early circuits using operational amplifiers were more fully developed by DeFord [16]. He reported a number of useful circuits including a stabilized voltage follower, a versatile controller, a galvanostatic controller, and operational amplifier signal sources.

The voltage follower (Fig. 18) he developed was the first to employ a chopper amplifier-differential amplifier combination. This follower circuit possessed superior drift characteristics and high-input impedance. These made it especially suited for impedance matching applications in electrochemical circuits where precise and accurate potential control are critical.

DeFord's follower opened the way for the use of lower-input impedance controllers, such as the adder type controller circuit (Fig. 45). This controller circuit allowed the simultaneous use of several input signals and thus increased the number of techniques to which operational amplifier instruments could be applied.

Following the works of Booman [5], Kelley et al. [6, 8], and DeFord [16], numerous operational amplifier circuits and several operational amplifier-based instruments were reported. Most of the circuits devised later were based on the ideas developed by these early workers and included many of the same basic circuits. In addition to these applications and developments, several workers reported on studies which were concerned solely with improving the nature or the functioning of operational amplifier circuits.

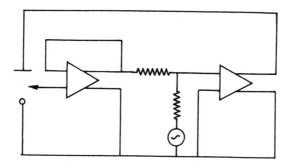

Fig. 45. A generalized adder controller with a voltage follower for impedance matching. (Reprinted [21, p. 1775] by courtesy of the American
Chemical Society.)

## B. Development of Potentiostat Circuits

The circuits constructed by Booman, by Kelley et al., and by DeFord
were each designed for a specific electroanalytical technique. The circuits
reported in the works to follow were also designed for a specific purpose,
and tended to become even more specialized. The controller has remained
the main component of most instruments. Yet it has undergone only minor
changes since the first controlled designs were reported. These early
designs for controller potential applications have been the subject of several
studies, which have resulted in the improved stability, accuracy, and over-
all performance of these circuits.

### 1. Design of Potentiostat Circuits

The primary goal in the design of a controller circuit commonly called
a potentiostat is that it maintain the electrolysis cell potential equal to some
desired value, generally that of a reference source. The reference signal
may be a constant voltage or a varying signal of any desired form. Several
secondary characteristics of the control circuit must also be considered.
These include insuring that the available signal sources, necessary current
measuring devices, and basic electrical components of the circuit are compatible with the controller design.

A study of the arrangements of the necessary components in a controlled
potential circuit was made by Schwarz and Shain [21]. They adopted a few
standards which an optimum circuit should meet. First, the circuit should
provide accurate potential control. Second, a design allowing the use of a
stabilized amplifier is preferred over one requiring a differential amplifier.
Third, since most auxiliary devices operate best when one output or input is
at ground potential, the most advantageous circuit design would allow the
reference signal source and the current measuring device (load resistor)

each to have one terminal grounded.

In a systematic fashion these authors devised several possible circuit arrangements. They considered single-amplifier potentiostats and two-amplifier potentiostats (adder with voltage follower), with and without auxiliary devices.

Schwarz and Shain considered that a generalized single-amplifier potentiostat contains two separate loops and performs two separate functions. One loop is the potential comparison portion of the circuit which contains the two amplifier inputs and the reference and working electrodes. This loop must also contain the signal source. The second loop, the current carrying loop, contains the two amplifier outputs (one output is the power supply common, not grounded), the counter, and working electrodes, and must also contain the current measuring device or load resistor. With these considerations the possible locations of the signal source and load resistor produce nine possible circuit arrangements (Fig. 46).

In the one-amplifier case an acceptable circuit should meet two

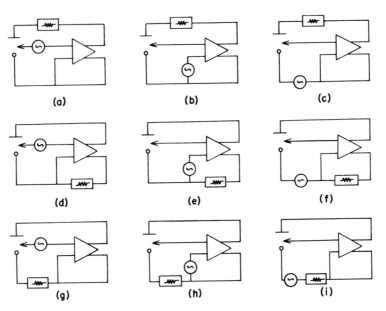

Fig. 46. Circuit configurations for single-amplifier potentiostats. (a-c) Type A grounding — circuit is as shown, lower output (power supply common) at ground. (d-i) Type B grounding — upper output at ground, lower output (power supply common) floating; (g-i) signal source is connected to alternate input and cell current is drawn from the floating output. (Reprinted [21, p. 1771] by courtesy of the American Chemical Society.)

TABLE 2

Interrelations of Characteristics of Single-Amplifier Potentiostats[a]

| Characteristics[b] | Circuit Configurations (Fig. 46) | | | | | | | | | | | | | | | | | |
| --- | --- | --- | --- | --- | --- | --- | --- | --- | --- | --- | --- | --- | --- | --- | --- | --- | --- | --- |
| | Type A grounding | | | | | | | | | Type B grounding | | | | | | | | |
| | a | b | c | d | e | f | g | h | i | a | b | c | d | e | f | g | h | i |
| 1. No SG interaction | + | + | − | + | + | − | + | + | − | + | + | − | + | + | − | + | + | − |
| 2. No $R_L$ interaction | + | + | + | + | − | + | − | − | − | + | + | + | + | + | + | − | − | − |
| 3. SG grounding | − | + | − | − | + | − | − | + | +[c] | − | − | − | − | − | − | − | − | − |
| 4. $R_L$ grounding | − | − | + | + | − | + | + | + | +[c] | + | + | + | − | − | − | − | − | − |
| 5. Stabilization | + | − | + | − | − | − | + | − | + | + | − | + | − | − | + | + | − | + |

[a]Reprinted [21, p. 1772] by courtesy of the American Chemical Society.

[b]Notes: (+) indicates that:

1. signal generator (SG) is not located in the current carrying loop;
2. current measuring device ($R_L$) is not located in the comparison loop;
3. signal generator can be referred to ground;
4. current measuring device (and readout device) can be referred to ground;
5. circuit can be stabilized.

[c]In circuit i (type A), the signal generator and load resistor cannot both be grounded simultaneously.

additional standards: first, that the signal generator not appear in the current carrying loop and, second, that the current measuring device not be contained in the potential comparison loop.

Of the nine possible arrangements not one meets all of the standards established by Schwarz and Shain (see Table 2). In fact, the adoption of any one of these circuit arrangements requires the relaxation of two or more of these standards. In general, the importance of these criteria varies with the intended application. In short duration experiments, for example, a stabilized amplifier might not be necessary. In low current techniques, the inclusion of the current measuring device in the potential comparison loop may not be harmful. Battery sources are easily used without grounding one terminal. High quality differential measuring devices remove the need for grounding the load resistor. Thus several of these circuits may be useful in particular applications.

The use of an auxiliary device to compensate for the iR drop in the current measuring resistor improves several of these circuits. Schwarz and Shain also evaluated adder type potentiostats by applying essentially the same criteria. With this circuit, the connection of the signal generator to an input resistor fixes its location outside of the current carrying loop and

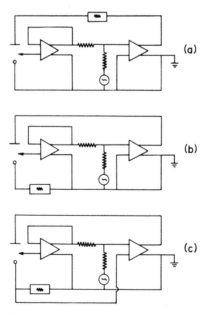

Fig. 47.   Three configurations for adder-type potentiostats with type A grounding. (Reprinted [21, p. 1775] by courtesy and the American Chemical Society.)

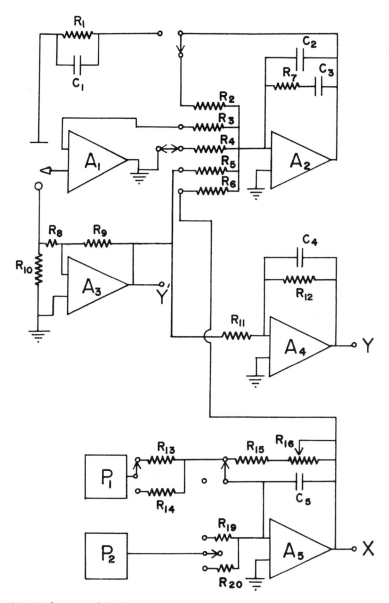

Fig. 48. Polarographic circuit. $P_1$, $P_2$ — low voltage dc power supplies; $A_1$ — a grounded output voltage follower; $A_2$ — an adder type controller (shown in standby mode) with stabilization network $R_7$, $C_2$, and $C_3$; $A_3$ — inverter to compensate for the load resistor iR drop; $A_4$ — inverter for current output; $A_5$ — ramp generator; Y — locations of current output signal (for recorder Y axis); X — ramp output signal (for recorder X axis). (Reprinted [65, p. 796] by courtesy of the American Chemical Society.)

allows one terminal to be grounded. The three locations for the current measuring resistor produce three different circuits (Fig. 47). These circuits do not meet the standards unless a circuit modification is made or an auxiliary device is included.

In the first circuit, a differential measuring device is required for satisfactory operation. In the second, the load resistor iR drop error can be eliminated from the potential comparison loop with a current follower. In the third circuit a differential amplifier would be required to obtain accurate control (see ref. [64] for a correction of this circuit).

An alternate configuration for each of the three circuits (shown incorrectly in original) is obtained by grounding the normal output of the control amplifier and floating the power supply common. This modification does not increase the utility of any of these circuits.

A few changes in operational amplifier potentiostat design have appeared since the study by Schwarz and Shain. In 1963, Booman and Holbrook [65] reported a modified form of the adder-controller circuit. Using a circuit like that shown in Fig. 47b they compensated for the load resistor using an inverter instead of a more conventional current follower (Fig. 48). The inverter signal was fed to an adder input and, in effect, subtracted the voltage across the load resistor (iR) from the voltage follower output. The controlled variable in this circuit is the sum of the voltage follower ($E_c$ + iR) and inverter (-iR) outputs which is just the cell potential, $E_c$.

Another basic controller circuit form was used in 1966 by Lauer and Osteryoung [66] (Fig. 49). This arrangement required a differential amplifier for satisfactory operation and embodies many of the advantages of both adder type and differential circuits.

## 2. Factors Affecting Potentiostat Accuracy

Schwarz and Shain considered the accuracy of potential control circuits under circumstances where the operational amplifiers behaved ideally. They were concerned with finding the optimum arrangement of the components in the external circuit and not with studying amplifier performance. Obviously, such a view is valid and valuable so long as work is conducted under conditions where ideal amplifier performance is obtained. Less than ideal amplifier response degrades a circuit's performance and the ideal case represents an optimum situation. Real amplifiers, of course, do not behave ideally. The purely inverting nature and very high gain associated with operational amplifiers are dc or low frequency characteristics. At the higher frequencies within the bandpass of operational amplifiers, the amplifiers show an additional phase shift between their output and input signals and a decreasing gain. The effect of these nonideal properties on the performance of operational amplifier potentiostats was first considered and studied by Booman and Holbrook [3]. Using the principles of feedback and

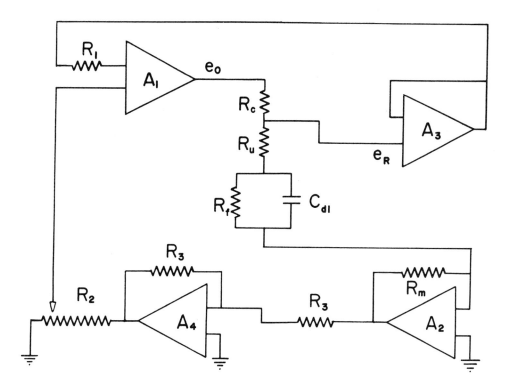

Fig. 49. A potentiostat employing positive feedback for compensation of the cell iR drop. $A_1$ — adder type controller (some inputs omitted) with the positive input active; $A_2$ — current follower; $A_3$ — voltage follower; $A_4$ — inverter which provides positive feedback to the noninverting input of $A_1$; $R_c$, $R_u$, $R_f$, and $C_{d1}$ — simulated electrolysis cell. (Reprinted [66, p. 1107] by the courtesy of the American Chemical Society.)

control theory they showed how these properties limited the accuracy and response of potentiostat circuits and also applied these principles in the design and evaluation of an instrument for controlled potential electrolysis.

A second important point when considering potentiostat accuracy is that, even with ideal amplifier performance and with the best possible arrangement, the accuracy of potential control is limited by the electrolysis cell. In electrochemistry, the major interest is the potential between the center of the working electrode and a point just outside the diffuse double layer. This is, essentially, the potential across the double layer at the working electrode surface. It is most often the case that the controlled potential differs from the potential across the double layer capacitance. This is not unusual since these potentials are each derived from a different

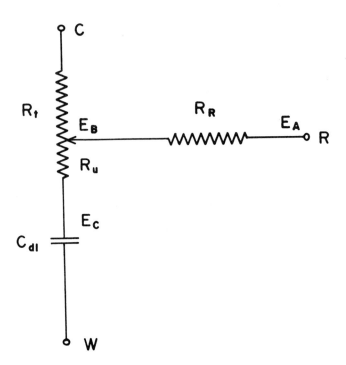

Fig. 50. A cell equivalent circuit. $R_T$ — total solution resistance between counter and working electrodes; $R_u$ — uncompensated resistance; $R_R$ — reference electrode resistance; $C_{dl}$ — double layer capacitance; $E_A$ — the voltage at the reference electrode lead; $E_B$ — the voltage at the tip of the reference electrode capillary; and $E_C$ — the potential at a point just outside of the diffuse double layer of the working electrode. (All voltages referred to ground.)

point. An analog of the electrolysis cell (Fig. 50) will help to clarify this problem.

For common potentiostatic circuits the controlled variable is really the signal transmitted to the input of the controller for comparison to the reference signal. In the equivalent circuit representation this is the potential difference between the reference electrode and the working electrode terminals, $E_A$. With proper impedance matching this signal can be made nearly identical to the potential between the tip of the reference probe and the working electrode terminal, $E_B$. The controlled signal is then the total voltage between point B and ground, and is given by

$$E_B = E_C + i(R_u).$$                                    (29)

This of course differs from the true working electrode potential $E_C$ which is the first term on the right-hand side of Eq. (29). It is not possible to directly measure $E_C$ by, for example, placing the reference electrode close enough to the working electrode so that $R_u$ equals zero. Even minimizing $R_u$ might disrupt the electrochemical processes of the cell. Thus for this case the actual controlled potential must differ from the electrochemical potential, over which direct control is desired.

One must further conclude that a difference exists between the controlled system accuracy and the actual electrochemical accuracy. Perfect control may be obtained from a system, yet the experiment may be only a partial success, in the electrochemical view. In poorly designed cells, the electrochemical accuracy may be extremely poor. Obviously, if the control system accuracy is poor this will be reflected in an even worse electrochemical accuracy.

Although control system accuracy must be satisfactory, the electrochemical accuracy cannot be sacrificed to obtain this or, obviously, the usefulness of the electrochemical experiment will be lost. In other words, the electrolysis cell should be designed to suit the needs of the experiment and to provide an acceptable electrochemical accuracy. The control system must be chosen to provide the system accuracy required.

### C. Dynamic Performance of Potentiostat Circuits

DC accuracy is essential in potential control circuits. However, other characteristics of the circuit are equally important and also play a vital role in the potentiostat performance.

The preceding discussion dealt with the accuracy of potentiostat circuits but assumed that the performance of the operational amplifier is ideal. As indicated earlier, such performance is obtained at low frequencies. However, all the frequencies within the bandpass of the amplifiers must be considered in determining the over-all performance of a circuit. Over a large portion of their frequency range, operational amplifiers behave in a less than ideal fashion.

Booman and Holbrook first investigated the over-all performance of operational amplifiers in feedback control circuits. They considered the factors of stability, accuracy, and response rate of potentiostats, using the principles of automatic control theory.

The principles of feedback control have been discussed extensively by Booman and Holbrook [3, 65] and by Schroeder and Shain [67]. In the design and evaluation of feedback systems the stability, accuracy, and response rate are important considerations since these characteristics determine the effectiveness of the system in performing its intended operation. A high degree of accuracy and a rapid response rate are generally desired in control systems designed for electrochemical applications. Maximum accuracy is obtained when high amplifier gain is available, and rapid response requires this high gain over a broad frequency region. Yet high gain over a broad frequency range often leads to unstable operation. Gain can be reduced to obtain stability but this affects the accuracy and response rate of a system. Booman and Holbrook used feedback theory as an aid in the design and evaluation of their circuit to obtain maximum accuracy and response with stable operation.

The properties of a feedback system can be determined by investigating the gain-frequency characteristics of the system or the components of the system. These characteristics can then be judged according to certain criteria to determine the operational stability of the system. Various modifications in the circuit can be used to affect the gain-frequency characteristics and, thus, to insure stability. Accuracy and response rate also can be determined from the gain-frequency properties of the system. These properties can be used as design aids in modifying a control circuit to obtain the desired performance.

## 1.  Feedback Circuits

Feedback control systems are composed of a few essential units: the controller which performs the primary function, and the load which contains the variable to be controlled. A reference signal source which prescribes the desired value of the controlled variable may also be included. In a controlled potential electrochemical experiment these units would be the operational amplifier potentiostat, the electrolysis cell, and a reference signal source, respectively. The potentiostat operation consists of comparing the cell and reference signals and increasing or decreasing the cell current as necessary to maintain the cell potential equal to the reference signal. These operations might be termed error detection and error correction. The basis of feedback control is in the action of the potentiostat to reduce any difference between the cell and reference signals. If the cell signal is greater than the reference, the potentiostat output decreases; if the cell signal is too small the potentiostat output increases. This opposing action is characteristic of negative feedback control systems.

In the simplest operational amplifier potentiostat, one employing a single amplifier, negative feedback control is obtained by connecting the reference source to the noninverting input, the reference electrode to the inverting input, and the auxiliary or counter electrode to the output of the amplifier. The cell working electrode is connected to the amplifier ground (power supply common). In simple terms, as long as the inverting action is maintained between the amplifier negative input and its output, control will be realized. If this relation reverses, the amplifier will increase the error signal, instead of diminishing it, and control will be lost. This latter situation, termed positive feedback, must be avoided if control is to be maintained.

These cases are really oversimplifications of the actual situation. But, to consider real systems, a mathematical approach is most suitable. The equation given in an earlier section describes the operation of a feedback control system:

$$\frac{e_c}{e_r} = \frac{K\beta}{1 - K\beta}. \tag{30}$$

For the case at hand K is the amplifier gain and $\beta$ is the ratio of the cell (feedback) signal to the amplifier output signal.

Negative feedback results when the product $K\beta$ is negative, and positive feedback results when $K\beta$ is positive. Actually positive feedback, unless it occurs at one critical point within the frequency range of the system, does not destroy the control operation although it does diminish the quality of control. A positive $K\beta$ value equal to unity produces an undefined control ratio and totally destroys the control operation. It is exactly this situation which produces unstable control systems and which must be avoided.

Avoiding the case where $K\beta$ is unity is somewhat more complicated than it might appear. Both K and $\beta$ represent signal ratios. These values must represent not only the ratio of the magnitudes but also the phase angle between the two signals. This is conveniently done by representing K and $\beta$, when necessary, as complex numbers. The real and imaginary parts represent the magnitude and phase angle, respectively, of these ratios. The $K\beta$ product can therefore be positive, negative, or complex. When both K and $\beta$ are complex quantities, avoiding a positive $K\beta$ product becomes a more difficult task.

## 2. Stability Criteria

The essential feature of unstable circuits is that $K\beta$ can assume the value of unity. For stable circuits this value must not occur or be approached too closely. Several methods for evaluating stability are in use and each involves a different way of representing the characteristics of K, $\beta$, and $K\beta$. For each method the point $K\beta = +1$ would be represented in a different way and thus for each method a different criterion for stability can be established.

a. Direct Calculation of System Parameters. In order to evaluate the stability of a circuit it is not necessary to calculate the control operation ratio $e_c/e_r$. From Eq. (30) and the preceding discussion it can be seen that the evaluation of $K\beta$ alone provides sufficient information for an evaluation of the system stability. The control ratio $e_c/e_r$ is a closed loop function of the control system while it is in operation. The term $K\beta$, however, refers to the system open loop characteristics. The $K\beta$ term is the gain of the system about the control loop while the control loop is broken at some point. The term $K\beta$ is referred to as the open loop gain of the control system or simply the loop gain.

The most obvious approach to determining stability involves the rigorous calculation of $K\beta$ from known K and $\beta$. The sign of $K\beta$ when $|K\beta|$ is unity can be determined directly. This has been done for a simple control system by Schroeder and Shain [67] (Fig. 51 and Table 3). This approach tends to be lengthy when several frequencies are considered. However, unless results at several frequencies are obtained this procedure will not be helpful in correcting any difficulties discovered in the analysis. Shorter and less tedious, but fully rigorous, ways of obtaining the desired $K\beta$ information have been developed and have been reported in numerous places [68-70]. However, it is not necessary to use rigorous calculation procedures. Convenient graphical procedures have been devised for the evaluation of control system stability, and these have proven to be quite helpful in systems analysis.

b. Nyquist Method.    A Nyquist diagram for a system can be constructed if the magnitude and phase angle of $K\beta$ are known or can be determined. These values are then plotted on a polar coordinate graph, generally as the log of the gain versus phase angle. The Nyquist plot for the circuit of Fig. 51 is shown in Fig. 52.

The normal inversion operation of the operational amplifier is equivalent, for a single frequency sine wave, to a $180^\circ$ phase lag between output and input signals. Additional phase lag that is introduced by the K or $\beta$ term is added to this. Positive feedback is a $360^\circ$ phase lag and thus occurs when an additional $180^\circ$ of phase lag is introduced. Avoiding positive feedback becomes a

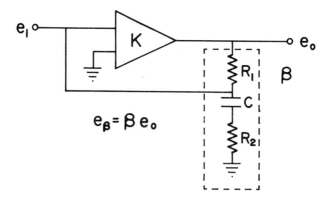

Fig. 51.  A control system composed of a controller and a reactive load. K and $\beta$ values are given in Table 3 for $R_1$ = 900 k$\Omega$, $R_2$ = 100 k$\Omega$, and C = 0.001 $\mu$F [67].

TABLE 3

Gain and Phase-Shift Characteristics for the Circuit of Fig. 51 [67]

| Frequency Hz | rad/sec | $|K|$ | $|\beta|$ | $|K\beta|$ | K, deg | $\beta$, deg | $K\beta$, deg |
|---|---|---|---|---|---|---|---|
| 1 | 6.28 | $-1.00 \times 10^4$ | 1.00 | $-1.00 \times 10^4$ | -180.6 | -0.3 | -180.9 |
| 10 | 62.8 | $9.95 \times 10^3$ | 0.998 | $9.93 \times 10^3$ | 185.7 | 3.2 | 188.9 |
| 50 | 314 | $8.95 \times 10^3$ | 0.959 | $8.58 \times 10^3$ | 206.6 | 15.6 | 222.2 |
| 100 | 628 | $7.07 \times 10^3$ | 0.848 | $6.00 \times 10^3$ | 225.0 | 28.6 | 253.6 |
| 200 | $1.26 \times 10^3$ | $-4.27 \times 10^3$ | 0.627 | $-2.68 \times 10^3$ | -243.6 | -44.3 | -287.9 |
| 500 | $3.14 \times 10^3$ | $1.96 \times 10^3$ | 0.318 | $6.24 \times 10^2$ | 258.7 | 54.9 | 313.6 |
| $1.00 \times 10^3$ | $6.28 \times 10^3$ | $9.95 \times 10^2$ | 0.186 | $1.85 \times 10^2$ | 264.3 | 48.8 | 313.1 |
| $2.00 \times 10^3$ | $1.26 \times 10^4$ | $4.97 \times 10^2$ | 0.127 | 63.3 | 267.2 | 34.0 | 301.2 |
| $5.00 \times 10^3$ | $3.12 \times 10^4$ | -200 | 0.105 | -21.0 | -268.9 | -15.8 | -284.7 |
| $1.00 \times 10^4$ | $6.28 \times 10^4$ | 100 | 0.101 | 10.1 | 269.4 | 8.1 | 277.5 |
| $5.00 \times 10^4$ | $3.14 \times 10^5$ | 20.0 | 0.100 | 2.00 | 269.9 | 1.6 | 271.5 |
| $1.00 \times 10^5$ | $6.28 \times 10^5$ | 10.0 | 0.100 | 1.00 | 270.0 | 0.8 | 270.8 |
| $5.00 \times 10^5$ | $3.14 \times 10^6$ | -2.00 | 0.100 | -0.200 | -270.0 | -0.2 | -270.2 |
| $1.00 \times 10^6$ | $6.28 \times 10^6$ | 1.00 | 0.100 | 0.100 | 270.0 | 0.1 | 270.1 |
| $5.00 \times 10^6$ | $3.14 \times 10^7$ | 0.200 | 0.100 | 0.0200 | 270.0 | 0 | 270.0 |
| $1.00 \times 10^7$ | $6.28 \times 10^7$ | 0.100 | 0.100 | 0.0100 | 270.0 | 0 | 270.0 |

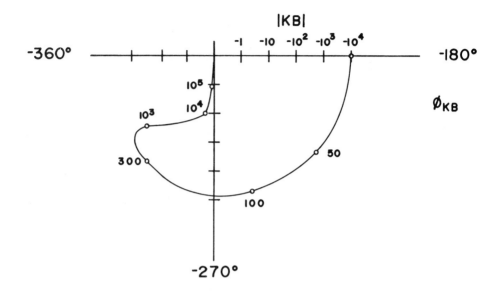

Fig. 52. The Nyquist diagram for the circuit of Fig. 51 using the data from Table 3. The point (+1, 360°) is not enclosed or intersected. The circuit is unconditionally stable [67].

problem, then, of keeping the additional phase lag within the system to less than 180°. As indicated earlier, positive feedback is not totally destructive to the control operation unless it occurs when $K\beta$ has a unit value. The stability criterion, in these terms, is that a stable system will result if, at the point where the magnitude of $K\beta$ is unity, the phase shift is not 360°. Practical considerations also prohibit phase shifts greater than 360° at this point. The circuit of Fig. 51 will be stable, since the phase angle of $K\beta$ at $|K\beta| = 1$ is significantly different from 360°. This criterion, stated in terms of the Nyquist diagram, would be that in the plot of the magnitude versus the phase shift of $K\beta$, the point ($|K\beta| = 1$, $\theta = -360°$) be neither enclosed nor intersected.

For practical reasons, the system phase shift at $|K\beta| = 1$ should be less than 360° by about 35°. Providing this stability margin gives a reasonable assurance that stable operation will result. The Nyquist plot again shows that the circuit of Fig. 51 will be stable. Circuits where the 360° total phase shift is not approached are termed unconditionally stable. For Nyquist plots where the 360° axis is crossed but the point (1, -360°) is not enclosed or intersected, the circuit is termed conditionally stable. For these circuits, positive feedback is present over the range of frequencies where $\theta$ is greater than, equal to, or nearly equal to 360°.

The Nyquist plot and stability criterion are quite effective in predicting stability and also provide some idea about how the circuit might be altered to obtain acceptable closed loop operation. The information required to construct the plot is somewhat difficult to obtain.

c. Bode Method. A more convenient method which is applicable to most electrochemical circuits is available. In circuits where a minimum number of reactive elements produce the observed phase shift it is possible to develop a simpler approach to stability evaluation. For such circuits, a Bode diagram and the Bode criterion serve to determine the stability or instability of a system. The Bode diagram involves essentially a plot of the log of the loop gain versus the log of the frequency. This plot is also called the transfer function of the system. Similar log-log plots of K and $\beta$ give the transfer functions of these operations. It is common to denote the gain in terms of decibels (dB) which is related to the log of the gain by

$$dB = 20 \log \frac{e_o}{e_i}$$

where $e_o/e_i$ represents the voltage gain. Thus a gain of 1000 would be 60 dB, a gain of 200,000 would be 106 dB, a gain of unity would be 0 dB, etc. The Bode diagram, though constructed from only gain-frequency information, provides phase angle information as well. For the minimum phase circuits mentioned, the phase angle is related to the rate of change of the gain with frequency, and phase angles can be determined directly from the slopes of the Bode diagram. On the log-log plot a zero slope corresponds to a zero phase shift, a slope of -1 to a $90^o$ phase lag, -2 to a $180^o$ phase lag, etc. In terms of dB, a slope of -20 dB per decade (dB/dec) represents a $90^o$ lag, -40 dB/dec a $180^o$ lag, etc. The normal operation of an operational amplifier is inverting, and phase angles are expressed relative to this inversion or $180^o$ phase lag. Thus only the additional phase shift is reflected in the Bode diagram.

An interesting facet of the Bode diagram approach is that the transfer functions corresponding to the individual terms K and $\beta$ can be plotted individually, and then graphically added to produce the log-log plot or transfer function of loop gain. The Bode diagram for this circuit of Fig. 51 is shown in Fig. 53.

The criterion for stability, stated in terms of the Bode plot, is that a system will operate stably, provided that the slope of the loop gain transfer function at the unity gain axis is less than -2 (40 dB/dec). To maintain a sufficient margin for stability, slopes should not approach -2 too closely. The Bode diagram approach to the evaluation of circuit stability is very convenient to use and also provides very useful information about how a circuit can be modified for stable operation.

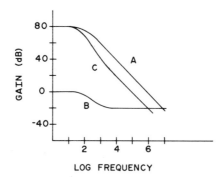

Fig. 53. The Bode diagram for the circuit of Fig. 51. A—amplifier gain,
B—load transfer function, C—loop gain. The slope of the loop gain at the
0-dB axis is 20 dB/dec, indicating a stable circuit [67].


3.   Evaluating the System Open Loop Gain

All methods of stability evaluation rely on a knowledge of the open loop
gain of the system.  Knowledge about the loop gain or about K and $\beta$ individually can be obtained in a variety of ways.  Normally K and $\beta$ can be determined separately by direct measurement or by calculations based on more
fundamental information.

a.  K, the Controller Gain.    In the simplest potentiostatic circuit the
controller is a single amplifier and the load is simply the electrolysis cell.
The controller gain, K, is then the open loop gain of the operational amplifier.  For some amplifiers the gain-frequency characteristics are available
directly from the manufacturer.  When this is not the case, sufficient information is supplied so that the over-all gain characteristics can be easily
determined.

To be compatible with a wide variety of feedback circuits most operational amplifiers are made to have a gain which decreases with a 20 dB/dec
slope.  The dc gain value and the unity gain (0 dB) crossover frequency for
the amplifier are generally given by the manufacturer.  Constructing the
gain-frequency plot then amounts to passing a line with a 20 dB/dec slope
through the unity gain crossover and terminating the line at the dc gain
value.  At lower frequencies the gain is usually constant and is given by a
zero slope line.  The point of intersection with the 20 dB/dec line is termed
the break frequency.  Actually this procedure gives an asymptotic representation of the gain curve and the actual plot passes smoothly from a zero slope
to the 20 dB/dec slope with increasing frequency.  At the break frequency the
gain is really 6 dB down or one-half of the asymptotic value.  This procedure
provides the gain-frequency characteristic for an amplifier which is typical
of the manufacturer's type.  A range is usually specified for each character-

istic and some uncertainty accompanies the gain-frequency plot derived in this manner.

Many operational amplifiers are designed to have gain roll-off values greater than 20 dB/dec. With such amplifiers higher gain values are available over most of the frequency region. This provides a higher loop gain and greater accuracy in some circuits. In many applications however this rolloff, if uncorrected, leads to instability. The roll-off rate of the amplifier generally is specified if it is greater than 20 dB/dec.

With chopper stabilized amplifiers, the specified dc gain is usually much larger than that of nonstabilized amplifier types. The gain-frequency character for the stabilized amplifiers differs from the typical case just given, primarily in the lower frequency types. This is due to the chopper amplifier gain (often 60 dB) which adds to the typical gain at very low frequencies. This difference is of little concern since the most important properties of the gain-frequency characteristics are at the higher frequencies, where the unity gain crossover frequency and the assumed 20 dB/dec slope provide the essential information.

b.  $\beta$, the Cell Transfer Function.  A reasonable estimate of the electrolysis cell transfer function can be obtained from simple calculations. These calculations have been shown by Booman and Holbrook [3] and by Schroeder and Shain [67]. They are based on an equivalent circuit representation for an electrolysis cell like that shown in Fig. 50. In the equivalent circuit, $C_{dl}$ represents the double-layer capacitance, $R_s$ the resistance between the counter and working electrodes, $R_u$ the portion of $R_s$ between the reference and working electrodes, and $R_r$ the internal resistance of the reference electrode. The ideal form for the cell transfer function has been given by Schroeder and Shain and is shown in Fig. 54.

c.  Measurement of Transfer Functions.  For potentiostat circuits involving several amplifiers the over-all controller gain can be determined from the measured or calculated gains of the individual components or units. The measurement of transfer functions has been used extensively by Booman and Holbrook who reported several useful procedures for these measurements. With the Bode diagram approach the feedback circuit can be viewed as a series of individual elements, each performing its open loop operation on the signal as it passes around the feedback loop. The operation of each unit is defined by the ratio of the signal at its output, $e_o$, to the signal at its input, $e_i$. Thus $e_o/e_i$, a frequency-dependent function, called the transfer function, defines the open loop operation of the individual unit. The loop gain of the system is then the product of the transfer functions of the unit operations comprising the feedback loop. Booman and Holbrook measured the transfer function of the various units in the control system, combined these to obtain the system open loop gain, and used the Bode criterion to evaluate the stability of the control circuit.

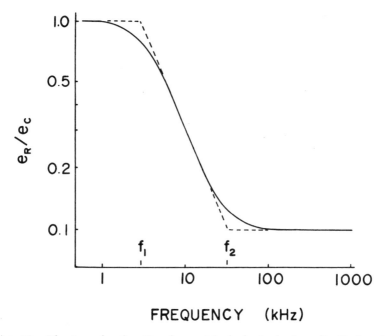

Fig. 54. The transfer function for an ideal electrolysis cell. Both actual (solid line) and asymptotic (dashed line) plots are shown for $R_T = 20\Omega$, $R_u = 2\Omega$, and $C_{dl} = 1\,\mu F$ (see Fig. 50); $e_r$ is the reference electrode potential and $e_c$ is the counter electrode potential [67].

The measurement of transfer functions provides a very powerful method of determining the open loop response since these measurements can be made while the control loop is essentially a closed loop operation. Booman and Holbrook were very explicit in defining the individual transfer functions of the operations comprising the system open loop gain, and they took several precautions to insure that their measurement techniques were sensitive to the open loop character of only one operation.

The direct measurement of transfer functions possesses several advantages over the calculation procedures. In calculating the transfer function it is quite difficult to account for loading effects of one operating unit on another. For example, in the simplest potentiostat the open loop gain of a single amplifier represents the transfer function of the controller. Yet this parameter is specified at no load or some specific load which may differ from the load in the real control circuit. Data concerning the output and input impedances of the amplifier could be used to assess any changes of the gain in real operation. But such an evaluation would be complex. Transfer function measurements which include loading effects can be made

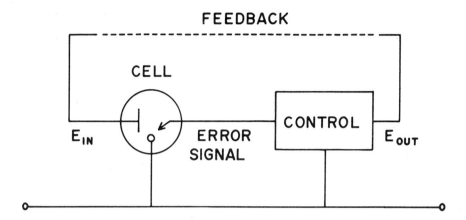

Fig. 55. The general schematic for a three electrode cell-controller system. (Reprinted [3, p. 1795] by courtesy of the American Chemical Society.)

under actual operating conditions or under simulated conditions where loading effects are quite close to the actual case.

Booman and Holbrook [3] separated the operations of their controlled potential coulometer circuit (Fig. 55) into two functioning units, the controller and the electrolysis cell. They described procedures for measuring the transfer function of each unit under the actual or simulated conditions. They found it convenient to measure transfer functions by use of a wave analyzer technique, introducing the high frequency signal directly into the section of the circuit being measured.

For measuring the controller transfer function, two circuits were used. The first, shown in Fig. 56, used the cell as the load while the second, Fig. 57, used a simulated cell. For both circuits the controller gain is measured directly. In the first circuit the amplifier input impedance does not load the reference electrode of the cell while in the second circuit this interaction is present.

For measurements of the electrolysis-cell transfer function, there is a further need for closed loop operation. The cell transfer function depends on the impedance within the cell. One of these, the double-layer capacitance, is strongly dependent on the working electrode potential. The cell has to be controlled at a fixed dc potential while the transfer function measurement is made.

The circuit used by Booman and Holbrook to determine the cell transfer function is shown in Fig. 58. In this circuit, resistors $R_1$ and $R_2$ were used to prevent any interaction between the controller output and the ac source.

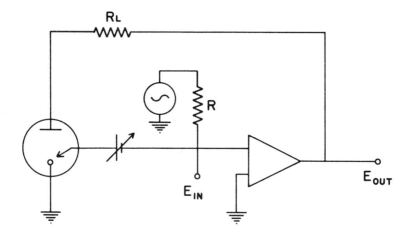

Fig. 56. Measurement arrangement for the control amplifier transfer function $E_{out}/E_{in}$ with a real cell as the load. (Reprinted [3, p. 1796] by courtesy of the American Chemical Society.)

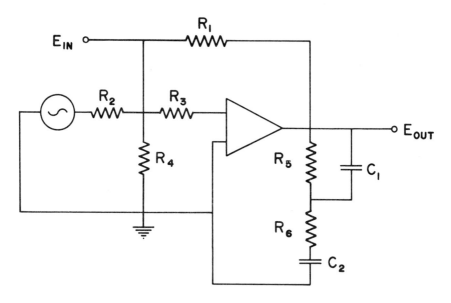

Fig. 57. Measurement arranged for the control amplifier transfer function $E_{out}/E_{in}$ with simulated cell impedance. (Reprinted [3, p. 1796] by courtesy of the American Chemical Society.)

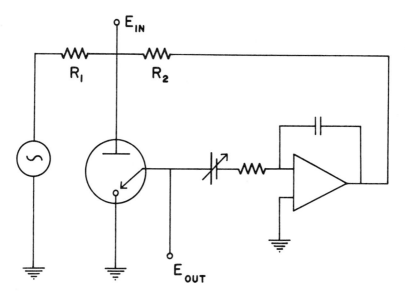

Fig. 58. The measurement arrangement for the cell transfer function at controlled potential. (Reprinted [3, p. 1796] by courtesy of the American Chemical Society.)

A large feedback capacitor was placed in the controller feedback loop, to reduce the controller bandwidth and to prevent the controller from responding to the extra signal introduced in the transfer function measurement. The controller still provided the dc control of the cell potential. Measuring the ratio of the ac signals at the counter and reference electrodes gave the transfer function directly.

The results obtained by Booman and Holbrook are shown in Fig. 59, along with the system loop gain.

### 4. Response and Accuracy of Potentiostats

The need for determining the loop gain of a control system goes beyond the need to evaluate its stability. The loop gain is important also in the determination of the system response and accuracy. The system open loop gain is a direct result of the system design, the types of circuits included, the components used, and the cell. These factors are under the direct control of the designer according to his needs for an intended application. As the desired system open loop gain is then a goal of the design, this system function also can aid in the design process by locating faults and indicating modifications. Thus, transfer function measurement and the evaluation of the system gain can be actively used in the circuit design process. Stability

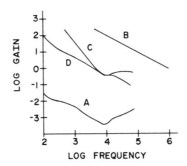

Fig. 59. The Bode diagram for the open loop response of the coulometer potentiostat circuit of Booman and Holbrook [3]. A — measured cell transfer function; B — amplifier open loop gain; C — system open loop gain before modification; D — system open loop gain after modifications to provide stability. (Reprinted [3, p. 1808] by courtesy of the American Chemical Society.)

is of course a necessary characteristic of any control system and has been a prime impetus to the study of operational amplifier control systems. However, the evaluation of the system open loop gain provides additional valuable information, the system response rate and the over-all accuracy of control.

The time dependent or transient characteristics of a closed loop system can be determined from its open loop transfer function by Laplace transformation methods. Time and the frequency operator $j\omega$ are a transform pair. The transformations are discussed in several places [71-73]. The relations are easily visualized in passive resistance-capacitance circuits where the time and frequency dependence are both well known. In amplifier circuits showing a steady 20 dB/dec rolloff, the rise time is, like an RC network, directly related to the system unity gain crossover frequency. Thus

$$\frac{e_o}{e_i} = 1 - \exp(-2\pi f K \beta t) \tag{31}$$

where $K\beta$ is the loop gain at frequency, f. However, $K\beta f$ is constant for a 20 dB/dec rolloff and equal to $f_0$, the unity gain crossover frequency. Equation (31) reduces to

$$\frac{e_o}{e_i} = 1 - \exp(-2\pi f_0 t) \tag{32}$$

and at $t = 1/f_0$, $e_o/e_i = 0.998$.

This relation shows the response to be an exponential function of time;

however, this holds only for the 20 dB/dec case. For system with larger rolloffs a more rapid response is observed but an overshoot is generally present. The control ratio approaches unity in a damped sinusoidal fashion. For systems with less than 20 dB slope, an overdamped response is seen and the approach to accurate control is smooth but slower than in the 20 dB case.

The prime consideration in determining response rate is the unity-gain frequency of the system loop gain. The accuracy of the control operation is related to the response rate and involves considering the response results from a different view. From the basic feedback equation it is seen that accuracy is related to the loop gain. Thus

$$\frac{e_o}{e_i} = \frac{K\beta}{1 + K\beta} = 1 - \frac{1}{1 + K\beta} \tag{33}$$

and whenever $K\beta$ is much larger than unity accurate control is obtained. The accuracy at any frequency can be determined directly from the open loop gain. The manner in which these measurements and evaluations are used has been demonstrated by Booman and Holbrook.

### 5. Modification of the Loop Gain

The determination of the system open loop gain will result in one of several possible conclusions. Basically, two situations arise, either the circuit will perform satisfactorily or it will not. In the latter case, the circuit must be modified. Hopefully, the nature of the loop gain itself will assist in determining what modifications must be made. When an unsatisfactory open loop gain is obtained it is necessary to reshape this characteristic. Booman and Holbrook [65] have provided the tools for this effort in their development of stabilization networks.

The nature and function of the electrolysis cell are such that only minor modifications can be made on its transfer function. The controller gain can, however, be shaped to match the cell transfer function and to produce an acceptable open loop gain. Reshaping the controller gain usually results in some loss in the magnitude of the loop gain. But if this action produces a stable, smooth response, the sacrifice is more than worth any loss in accuracy which results.

a. Obtaining the Optimum Controller Gain. The stabilization technique of Booman and Holbrook [65] involves such gain shaping. The desired form of the controller gain is determined by two factors, the cell transfer function and the optimum form for the system loop gain. The optimum loop gain would have a 20 dB/dec slope at and near the unity gain crossover frequency. Depending on the cell transfer function at near unity loop gain, the desired controller gain would be described by one of five possible gain frequency diagrams shown in Fig. 60. These forms result from the nature of the cell impedance in the frequency region near unity loop gain. Different shapes

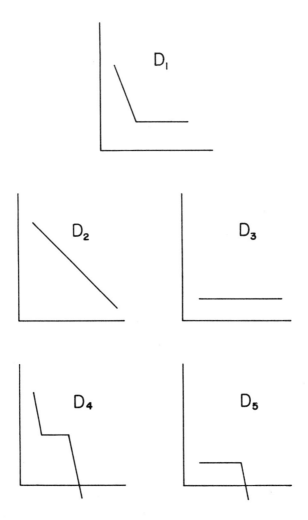

Fig. 60. The possible values of D, the controller open loop gain [Eq. (39)],
needed to produce optimum response with a particular cell. Values of $R_7$,
$C_2$, and $C_3$ (Fig. 48) are chosen to produce the desired form. (Reprinted
[65, p. 798] by courtesy of the American Chemical Society.)

arise when the impedance of the cell is reactive, resistive, or changing
from one to the other in this frequency region. The shaping technique used
by Booman and Holbrook causes the controller gain to complement the cell
transfer function. At frequencies where the cell transfer function is decreasing, the controller gain should be constant. Where the cell transfer function
is constant, the controller gain can decrease at 20 dB/dec. Alternately this

procedure could be viewed as making the phase shifts complementary. Increasing phase shift in the cell is offset by a decreasing phase shift in the controller. A 90° cell phase shift (20 dB/dec slope) requires a zero degree controller phase shift (zero dB/dec slope).

Four of the stabilization networks, when incorporated in the controller feedback loop, produce a gain curve showing two regions of 20 dB/dec rolloff separated by a constant gain (zero dB/dec slope) region. The amplifier loop gain is changed from its original form to a new shape. The values of the components of the stabilizing network are determined by the break frequencies of the desired controller gain and the size of the input resistors used in the controller. If the low and high break frequencies of the desired gain are $BF_1$ and $BF_2$ respectively, and R is the input resistance, then $R_7$ and $C_2$ and $C_3$ are given by

$$R_7 = K_1 R \tag{34}$$

$$C_2 = \frac{1}{2\pi R_7 (BF_2)} \tag{35}$$

$$C_3 = \frac{1}{2\pi R_7 (BF_1)} \tag{36}$$

where $K_1$ is the gain in the zero slope region of the desired gain.

The last type of controller gain desired is a constant 20 dB/dec slope (implying the cell impedance is resistive). The stabilization network required used only capacitor $C_2$ whose value is given by

$$C_2 = \frac{1}{2\pi F_3 K_2 R} \tag{37}$$

where $F_3$ is the frequency (where the system loop gain is 0.1), $K_2$ is the gain at $F_2$ ($= 0.1 F_3$), and R is the controller input resistor value.

Booman and Holbrook calculated the cell and system transfer functions for the potentiostat system of Fig. 48. The cell transfer function actually reflects the impedances in the cell itself and the other portions of the system from which the feedback signals are derived. For the circuit shown, the cell transfer function is the difference between two ratios. The first ratio, A, is the reference electrode follower output divided by the counter electrode signal. The second ratio, B, is the inverter output divided by the counter electrode signal. The net cell transfer function is proportional to A - B and is given by

$$A - B = \frac{1 + j\omega C_1 C_2}{1 + j\omega C_1 (R_1 + R_2 + R_L)} \tag{38}$$

where the symbols represent the components shown in Fig. 48.

The actual open loop gain is the product of this cell transfer function and the control amplifier open loop gain.

The desired open loop gain of the control amplifier is given by

$$D = \frac{-j2\pi F_3 K_3}{(A - B)} \tag{39}$$

where $K_3$ is the desired amplifier gain at frequency $F_3$.

The values of the cell transfer function, the system loop gain, and the desired control amplifier loop gain were calculated using a digital computer. Calculations were made at ten points per decade over a broad frequency range. The region of principal interest is bounded by $F_3$, the frequency at which the system loop gain is 0.1, and $F_1$ which is $0.01F_3$. This region contains the unity gain crossover frequency, the point of main concern.

The objective of Booman and Holbrook involved producing more than just stable circuits, but circuits which also produced optimum response for the components used. This goal required further consideration of the system parameters which affected the system performance. The size of the cell's uncompensated resistance, the total cell impedance, the feedback resistance of the current follower, and the cell's time constant were considered important to the system performance by the authors. They further assumed that these parameters could be controlled within certain limits.

The double-layer capacitance and the high frequency response of the controller were considered the principal factors in determining the maximum response available. The other system parameters were selected to achieve this maximum response with stable operation.

The optimum cell uncompensated resistance, $R_2$, is one which provides a cell time constant, $R_2C_1$ (defined at $T_1$), equal to the control system time constant, $T_2$. Thus

$$R_T = T_2/2\pi C_1. \tag{40}$$

The total cell resistance, $R_T$, is

$$R_T = R_1 + R_2 + R_2 \tag{41}$$

where $R_1$ is the compensated portion of the cell resistance and $R_L$ is the load resistance in series with the cell. Then $R_L$ is assumed equal to $R_T/2$ and $R_1$ is given by

$$R_1 = (R_T/2) - R_2. \tag{42}$$

The optimum value of $R_2$ is calculated by minimizing the rise time function, $\epsilon(T)$, of the double-layer capacitance voltage. Selected values of $R_2$, $R_T$, and $C_1$ are used to calculate the cell transfer function which is then used to determine the system open loop gain. The rise time expression

$$\epsilon(T) = \frac{1 + T_1 \exp(-T/T_1) - T_2 \exp(-T/T_2)}{T_2 - T_1} \tag{43}$$

is calculated for various $R_2$ values. For each total cell resistance, $R_T$, and double-layer capacitance, $C_1$, an optimum $R_2$ is obtained.

b. Obtaining Optimum Auxiliary Circuit Functioning. The proper functioning of the current follower is essential to the accuracy of the control system operation. This can be assured by requiring better performance from the current follower than from the control system as a whole. This involves keeping the current follower feedback resistor below a maximum value calculated by Booman and Holbrook. They assumed in calculating this value that the current follower was part of an inverter circuit (Fig. 61). The total cell impedance is the input resistance of the inverter. This inverter circuit will function well in the control system only when the gain of the current follower amplifier is large enough to hold its summing point near virtual ground. This requirement is expressed as

$$\frac{Z_c \times F_3}{F_2(Z_c + Z_f)} > 10 \tag{44}$$

where $Z_c$ is the total cell impedance, $F_3$ is the current follower amplifier unity gain crossover frequency, $Z_f$ is the maximum feedback resistance, and $F_2$ is the system unity gain frequency. The factor of ten provides a safety factor. This yields a maximum feedback impedance value of

$$Z_f = Z_c \left( \frac{F_3}{10F_c} - 1 \right). \tag{45}$$

Booman and Holbrook provided tables of the uncompensated resistance, the stabilizing network components, and maximum current follower feedback impedance for various values of total cell resistance and double-layer capacitance. They also gave the system unity gain crossover frequency, the response time, and the phase margin for each set of cell and circuit values. These were calculated using specific operational amplifiers as the controller

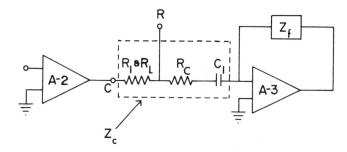

Fig. 61. Simplified schematic of the polarograph of Fig. 48 in the current follower mode, showing the roles of the cell impedance and the feedback impedance in the performance of the current follower. (Reprinted [65, p. 798] by courtesy of the American Chemical Society.)

and current follower, but the technique is applicable to any type of operational amplifiers.

### D. A Stabilizing Technique for Circuits Using

### Positive Feedback iR Compensation

As mentioned earlier, the control system accuracy and the electrochemical accuracy differ by the iR drop error in the cell which is due to the uncompensated resistance. When a signal is fed back to the controller to compensate for this error, more accurate control of the cell potential is possible [74, 75, 66]. Many of the workers who used this positive feedback form of iR drop compensation observed unstable operation of the control circuits.

The complexity of this problem led Brown et al. [76] to a new form of

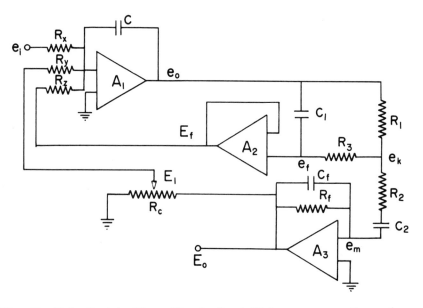

Fig. 62. Potentiostat with positive feedback iR drop compensation and with stabilizing capacitor, $C_1$, between counter and reference electrode of a simulated cell. Potentiostat components: $A_1$ — controller amplifier; $A_2$ — voltage follower; $A_3$ — current follower. Cell: $R_1$ — uncompensated resistance; $C_2$ — double layer capacitance; $R_3$ — the reference electrode resistance. Signals: $E_f$ — the normal cell potential; $E_1$ — the positive feedback signal; $E_0$ — the current output signal. (Reprinted [76, p. 1412] by courtesy of the American Chemical Society.)

circuit stabilization and to new techniques for evaluation system transfer functions.

When a positive feedback signal is used to compensate for the cell iR drop error, the transfer function of the system is drastically changed. This can be viewed as a change in the cell transfer function. In effect the positive feedback signal reduces the uncompensated resistance. This produces a cell transfer function with a 20 dB/dec decrease over a broad frequency range.

Full compensation produces a cell transfer function like that of a simple RC network, 20 dB/dec at all frequencies higher than the break frequency. This function combined with the common controller transfer function produces a 40 dB/dec slope in the high frequency region.

The technique of Brown et al. [76, 77] involved a reshaping of the cell transfer function. This was accomplished by introducing a capacitor between the counter and reference electrode terminals (Fig. 62). With this

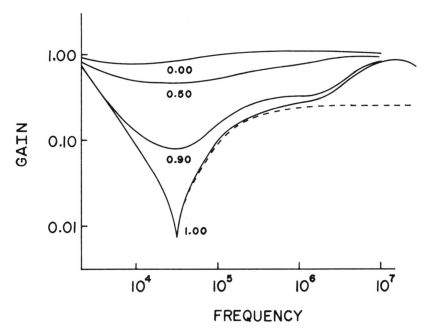

Fig. 63. Bode plot of the cell transfer functions (A + B) for the circuit of Fig. 62 with various amounts of iR compensation. A is the transfer function $E_f/e_o$ and B is the transfer function $E_1/e_o$. Circuit components: $R_1 = 25 \ \Omega$, $R_2 = 100 \ \Omega$, $C_2 = 1.00 \ \mu F$, $R_3 C_1 = 1.00 \times 10^{-6}$ sec, $C = C_f = 0$. $\beta$, the fraction of the resistance $R_2$ compensated by $E_1$ is given on each curve. (Reprinted [76, p. 1414] by courtesy of the American Chemical Society.)

additional component, the cell transfer function assumes the shape shown
in Fig. 63. The cell transfer function is affected by several parameters
but mostly by the natural break frequency of the cell, by the $R_3C_1$ time
constant, and by the amount of positive feedback compensation. Assuming
full compensation for the cell iR drop and using the basic cell parameters,
an optimum value of the $R_3C_1$ time constant can be calculated. For significantly smaller values of $R_3C_1$, the circuit will become unstable. The Bode
diagrams given in Fig. 64 demonstrate this fact, since, for $R_3C_1$ values
well below the optimum, the system open loop gain shows a 40 dB-or-
greater slope at unity gain. Brown et al. [76] obtained the system transfer
function by utilizing a technique similar to that employed by Booman and
Holbrook [65]. In addition, the transfer function for the system could be
calculated.

The eight voltages within the control system could be described by eight
separate equations. The various transfer functions of the system could be
expressed as a ratio of eighth-order determinants. When the values of

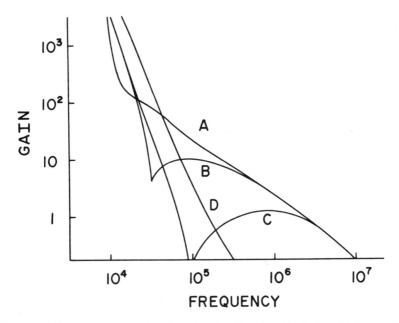

Fig. 64. The system transfer function for the circuit of Fig. 62 for various
$R_3C_1$ values. The $R_3C_1$ value calculated [76] for unconditional stability was
$4 \times 10^{-4}$ sec. $R_3C_1$ values: curve A, $10^{-5}$ sec; curve B, $10^{-6}$ sec; curve
C, $10^{-7}$ sec; curve D, 0. Circuit components: $R_1 = 25\Omega$, $R_2 = 100 \ \Omega$, $C_2$
$= 1.00 \ \mu F$, $C_2 = 1.00 \ \mu F$. (Reprinted [76, p. 1416] by courtesy of the
American Chemical Society.)

various system components and the characteristics of the amplifiers were chosen, the cell and system transfer function could then be generated. With the use of digital computation a large number of system variables could be studied to determine their effect on the open loop gain. The results of their calculation are represented as Bode diagrams of the system loop gain.

The introduction of the additional capacitor allows the use of full iR drop error compensation and thus provides an electrochemical accuracy equal to the control system accuracy. This accuracy is obtained below the frequency where the added capacitor begins to shunt the normal cell path. Above this frequency the accuracy diminishes. But it is in this high frequency region that the added capacitor aids in stabilizing the circuit. In general the electrochemical accuracy provided by this circuit is superior to that given by ordinary circuits, even those systems which show better control system accuracy at higher frequencies.

## REFERENCES

[1]. C. N. Reilley, J. Chem. Ed., 39:A853, A933 (1962).

[2]. D. D. Deford, Stabilized Follower Amplifier, Applications Bulletin, G. A. Philbrick Researches, Boston, Massachusetts.

[3]. G. L. Booman and W. B. Holbrook, Anal. Chem., 35:1793 (1963).

[4]. Bulletin Q25AH/1B1 Rev. 6, Philbrick Researches, Boston, Mass., Sept. 1, 1966.

[5]. G. L. Booman, Anal. Chem., 29:213 (1957).

[6]. M. T. Kelley, H. C. Jones, and D. J. Fisher, Anal. Chem., 31: 1475 (1959).

[7]. C. V. Evins and S. P. Perone, Anal. Chem., 39:309 (1967).

[8]. M. T. Kelley, H. C. Jones, and D. J. Fisher, Anal. Chem., 31:488 (1959).

[9]. R. C. Propst, Anal. Chem., 35:958 (1963).

[10]. F. B. Stephens and J. E. Harrar, U. S. Atomic Energy Comm. Rept. UCRL 7165 (1963).

[11]. R. P. Buck, Anal. Chem., 35:692 (1963).

[12]. R. W. Stromatt and R. E. Connally, Anal. Chem., 33:345 (1961).

[13]. H. C. Jones, W. D. Shults, and J. M. Dale, Anal. Chem., 37:680 (1965).

[14]. J. E. Harrar and E. Behrin, Anal. Chem., 39:1230 (1967).

[15]. M. T. Kelley, D. J. Fisher, and H. C. Jones, Anal. Chem., 32: 1262 (1960).

[16]. D. D. DeFord, Division of Analytical Chemistry, 133rd Meeting American Chemical Society, San Francisco, April, 1958.

[17]. C. Auerbach, H. L. Finston, G. Kissel, and J. Glickstein, Anal. Chem., 33:1480 (1961).

[18]. C. K. Mann, Anal. Chem., 33:1484 (1961).

[19]. W. L. Underkofler and I. Shain, Anal. Chem., 35:1778 (1963).

[20]. R. Annino and K. J. Hagler, Anal. Chem., 35:1555 (1963).

[21]. W. M. Schwarz and I. Shain, Anal. Chem., 35:1770 (1963).

[22]. R. R. Schroeder, Ph.D. Thesis, University of Wisconsin, 1967.

[23]. S. P. Perone and T. M. Mueller, Anal. Chem., 37:2 (1965).

[24]. H. C. Jones, W. L. Belew, R. W. Stelzner, T. R. Mueller, and D. J. Fisher, Anal. Chem., 41:772 (1969).

[25]. R. L. Meyers and I. Shain, Chem. Instr., 2:203 (1969).

[26]. J. L. Huntington and D. G. Davis, Chem. Instr., 2:83 (1969).

[27]. R. A. Durst, J. W. Ross, and D. N. Hume, J. Electroanal. Chem., 7:245 (1964).

[28]. J. R. Alden, J. Q. Chambers, and R. N. Adams, J. Electroanal. Chem., 5:152 (1963).

[29]. W. F. Kinard, R. H. Philp, and R. C. Propst, Anal. Chem., 39: 1556 (1967).

[30]. D. E. Smith and W. H. Reinmuth, Anal. Chem., 33:482 (1961).

[31]. D. E. Smith, Anal. Chem., 35:1811 (1963).

[32]. W. L. Underkofler and I. Shain, Anal. Chem., 37:218 (1965).

[33]. J. W. Hayes and C. N. Reilley, Anal. Chem., 37:1322 (1965).

[34]. U. Eisner, C. Yarnitzky, Y. Newirowsky, and M. Ariel, Israel Chem., 4:215 (1966).

[35]. R. F. Evilia and A. J. Diefenderfer, Anal. Chem., 39:1885 (1967).

[36]. R. DeLevie and A. A. Husovsky, J. Electroanal. Chem., 20:181 (1969).

[37]. M. S. Krause, Jr. and L. Ramaley, Anal. Chem., 41:1365 (1969).

[38]. G. W. Ewing and T. H. Brayden, Jr., Anal. Chem., 35:1826 (1963).

[39]. R. W. Murray, Anal. Chem., 35:1784 (1963).

[40]. D. G. Peters and S. L. Burden, Anal. Chem., 38:530 (1966).

[41]. D. Stein and H. A. Laitinen, Anal. Chem., 38:1290 (1966).

[42]. W. D. Shults, F. E. Haga, T. M. Mueller, and H. C. Jones, Anal. Chem., 37:1415 (1965).

[43]. H. B. Herman and A. J. Bard, Anal. Chem., 37:590 (1965).

[44]. D. T. Napp, D. C. Johnson, and S. Bruckenstein, Anal. Chem., 39:481 (1967).

[45]. J. M. Matsen and H. B. Linford, Anal. Chem., 34:142 (1962).

[46]. J. H. Christie, G. Lauer, and R. A. Osteryoung, J. Electroanal. Chem., 7:60 (1964).

[47]. G. W. O'Dom and R. W. Murray, Anal. Chem., 39:51 (1967).

[48]. H. L. Pardue and W. E. Dahl, J. Electroanal. Chem., 8:268 (1964).

[49]. A. J. Bard and H. B. Herman, Anal. Chem., 37:317 (1965).

[50]. J. W. Hayes, D. E. Leyden, and C. N. Reilley, Anal. Chem., 37:1444 (1965).

[51]. W. J. Blaedel and R. H. Laessig, Anal. Chem., 37:332 (1965).

[52]. M. J. D. Brand and G. A. Rechnitz, Anal. Chem., 41:1185 (1969).

[53]. M. J. D. Brand and G. A. Rechnitz, Anal. Chem., 41:1788 (1969).

[54]. R. P. Buck and R. W. Eldridge, Anal. Chem., 35:1829 (1963).

[55]. G. Lauer, H. Schlein, and R. A. Osteryoung, Anal. Chem., 35:1789 (1963).

[56]. C. F. Morrison, Anal. Chem., 35:1820 (1963).

[57]. D. M. Oglesby, S. H. Omang, and C. N. Reilley, Anal. Chem., 37:1312 (1965).

[58]. A. D. Goolsby and D. T. Sawyer, Anal. Chem., 39:411 (1967).

[59]. T. R. Mueller and H. C. Jones, Chem. Instr., 2:65 (1969).

[60]. R. Bezman and P. S. McKinney, Anal. Chem., 41:1560 (1969).

[61]. J. Schon and K. E. Staubach, Regelungstechnik, 2:157 (1965).

[62]. H. Gerischer and K. E. Staubach, Z. Electrochem., 61:789 (1957).

[63]. F. G. Will, Z. Electrochem., 63:484 (1959).

[64]. I. Shain, J. E. Harrar, and G. L. Booman, Anal. Chem., 37:1768 (1965).

[65].   G. L. Booman and W. B. Holbrook, Anal. Chem., 37:795 (1965).

[66].   G. Lauer and R. A. Osteryoung, Anal. Chem., 38:1106 (1966).

[67].   R. R. Schroeder and I. Shain, Chem. Instr., 1:233 (1969).

[68].   H. Chestnut and R. W. Mayer, Servomechanisms and Regulating System Design, Vol. I, 2nd ed., Wiley, New York, 1955.

[69].   A. M. Hardie, The Elements of Feedback and Control, Oxford, New York, 1964.

[70].   J. L. Bower and P. M. Schultheiss, Introduction to the Design of Servomechanisms, Wiley, New York, 1958.

[71].   V. S. Levadi, Calculation of Transient Response from Frequency Response, Experiment Station, College of Engineering, Ohio State Univ., Columbus, 1959.

[72].   W. B. Coulthard, Transients in Electrical Circuits, 2nd ed., Pitman, London, 1951.

[73].   R. V. Churchill, Operational Mathematics, 2nd ed., McGraw-Hill, New York, 1958.

[74].   D. Pouli, J. R. Huff, and J. C. Pearson, Anal. Chem., 38:382 (1966).

[75].   E. R. Brown, T. G. McCord, D. E. Smith, and D. D. DeFord, Anal. Chem., 38:1119 (1966).

[76].   E. R. Brown, D.E. Smith, and G. L. Booman, Anal. Chem., 40:1411 (1968).

[77].   E. R. Brown, H. L. Hung, T. G. McCord, D. E. Smith, and G. L. Booman, Anal. Chem., 40:1424 (1968).

B. Digital (and Hybrid) Instrumentation

Chapter 11

# INTRODUCTION TO THE ON-LINE USE OF COMPUTERS IN ELECTROCHEMISTRY

Robert A. Osteryoung

Department of Chemistry
Colorado State University
Fort Collins, Colorado 80521

## I. INTRODUCTION

The use of computers in electrochemistry has grown rather markedly in the last several years to the point where a considerable number of electrochemists now feel unable to operate unless they have a computer on-line to their experiment. The scope of this introductory chapter is limited; we shall be concerned with the use of a computer "on-line" — i.e., used for data acquisition and experimental control and operating only on the acquired data. We will not be concerned with "number crunching" or treatment of data obtained in an off-line situation. Since the author is not a computer expert, his comments are limited primarily to his own experiences and reflect his own prejudices. The use that we have made of computers in the electrochemical area has been for the computer system to serve as a programmable function generator, supplying voltages to the conventional potentiostatic electrochemical system and obtaining the result of a voltage change — i.e., measuring a current or charge — which is acquired and stored in the computer's memory. In developing this procedure, it is hoped that the neophyte reader will gain a broad picture of how to accomplish the task. It is the author's conviction that, when it comes to the on-line use of computers, the only real way to obtain the detailed picture is to actually "do it."

A general overview of the ideas involved in the on-line use of computers in chemistry may be found in two recent articles by Frazer [1, 2], of Livermore Radiation Laboratory, who is one of the pioneers in the use of computers in an analytical chemistry laboratory.

## II. BACKGROUND AND JUSTIFICATION

A computer permits one to do more accurate measurements and perform them much more rapidly than do usual methods. Our own experience may illustrate the point. Several years ago we were engaged in using a technique known as chronocoulometry [3]. The essence of this method is that a potential step is applied to an electrode, the charge passed as a function of time is determined, and the potential may then be stepped back to its original value with the charge-time behavior again being followed. Initially we carried out this experiment by employing operational amplifiers as the electrochemical instrumentation — potentiostats, current followers, integrators, and relay-operated switches attached to complicated timing circuitry to apply steps of given magnitudes for given times — and recorded the charge-time behavior on an oscilloscope, from whence it was photographed. Then data were taken off the picture, and the appropriate diagnostic plots were made. Outside of the difficulty in taking accurate data from a Polaroid print and the tedium in taking off sufficient data points to make the required plot, the procedure worked reasonably well. At that time we had in our laboratory a multichannel analyzer which had been purchased to carry out some radio-tracer work, but also with the vague idea that it could prove useful in electrochemical studies. The analyzer was eventually employed as a digital oscilloscope by being fed a voltage signal which was changing with time [4]. The analyzer has 400 channels, i.e., 400 words in its memory in which you can put information. However you can put into its memory only data. Using a voltage-to-frequency converter, for instance, and a time base which opens each channel, or word, for a fixed period of time, one counted the output of the voltage-to-frequency converter for a fixed time, and then moved on to another channel for another time period. The information stored in the memory of the analyzer was the number of counts, which were proportional to the voltage input. One could output this number onto punched paper tape and convert the tape into punch cards, feed the punch cards into a computer, and avoid pictures and tedium. However, a considerable amount of auxiliary electronics is required. Minor changes in the experimental procedure necessitated major changes in the electronics required to apply the voltage pulses and supply synchronization. This took time.

In any case, largely through the efforts of Lauer (see [4]), we realized that a small digital computer could do the same sort of thing with far more flexibility, less dependence on inflexible hardware, and complete control of parameters from a Teletype keyboard. Thus, we were led into the on-line computer world.

Where did the computer differ from the analyzer? In general, both performed analog-to-digital conversions, both stored data, and both could output that data. But, because the analyzer could store only data — all its "instructions" were hardware-wired — flexibility was lost. The computer could store both data and instructions; it could be programmed by "software" which could be changed by writing another program and putting that program into the machine.

With this background, it might be worthwhile to consider some of the building blocks required to make up an on-line computer system useful for carrying out certain kinds of electrochemical work.

### III.  SYSTEM BUILDING BLOCKS

#### A.  General

We will look at some of the building blocks which go to make up a computer system. It will be necessary to stay vague in terms of certain details; in other instances, particularly concerned with "interface," the link between the computer and the experiment, some detail will be required.

The computer itself, as purchased in a bare-bones form, consists of a central processor, some sort of fast memory generally referred to as core memory, and some means for the operator to communicate with the computer. The communication in most small and stripped-down machines may be via Teletype keyboard, paper tape, or switch registers on a console. The central processor is the place where the arithmetic and logical operations take place. As a rough analogy, the central processor, or CPU in the jargon, is the nervous system while the core memory is exactly that, a memory not a brain.

The link between the CPU and the means of communication involves an interface — a link of various hardware so that the electrical impulses out of the computer may eventually, via the communication device, appear as a letter or number which man can comprehend or at least try to. The simplest system, for example, will consist of a central processor, 4096 words of core memory, a Teletype, and paper-tape reader-punch. Systems such as this may cost $6000-10,000. Such a system can be used to do a variety of mathematical operations, but not a thing in the experimental world.

An excellent article by Butler [5] attempts to compare various small computers available in mid-1970. Although the state-of-the-art in the minicomputer area has already progressed much further since it appeared, the criteria discussed would be of interest to those trying to obtain information in this rather bewildering field.

What else do we need? Suppose we just want the computer to accept data into its memory; we can then operate mathematically on that data. Most experimental data — e. g. , the response of an electrochemical cell to a perturbation, or the output of a photomultiplier or NMR unit — can be obtained in the form of an analog or continuous voltage which changes with time. The computer is a digital instrument; it understands if a voltage is 0 or 5 V but nothing in between. So, to permit communication between our instrument's experimental readout and the computer, we must translate from analog language to digital language; the interpreter, strangely enough, is called an analog-to-digital converter (A/D). If one has a continuously changing voltage signal, say a sine wave, the converter samples the voltage at periodic intervals and tells the computer, in digital language, what the value of the voltage is at the instant of sampling; the computer can store this information. Thus, an input analog sine wave voltage would be broken up into a series of discrete information points. Suppose we want the computer to take data at fixed time intervals; the computer, from information held in its memory, can tell the A/D converter when to convert — i. e. , when to sample the voltage. In other words, it could tell the converter to "go" at fixed intervals of time if it knew what the time was. So we have to go out and buy the computer system a watch. If we have some sort of an oscillator which operates at a fixed frequency, we can make the computer count. In fact, we can build a clock into which we put the pulses from an oscillator; it can then compare the number of pulses with another number previously put into it by the computer. When the two match, the clock can tell the computer, "it is time for you to do something."

So, we now have an A/D converter and a "programmable clock" — what it is called if we can put information into it from the computer, which it can use to compare to the number of pulses out of the fixed oscillator. What else do we need? Suppose we want the computer to control something in the external world; perhaps we want, in some manner, to make it put out a voltage level into a potentiostat. Well, the computer internal tongue is digital, the potentiostat input has to be an analog voltage, and we obviously have to have a digital-to-analog converter (D/A). The computer can input digital information into this circuit, and out comes a voltage level. The computer output, in digital, is essentially quantized, hence a computer can generate a sine wave as a series of voltage levels which can be changed very rapidly to give the appearance of a sine wave if put through a D/A converter and external filters. What more do we need? Perhaps we want to turn something on or off; then we may need relays or electrical switches, such as FET's, which can be operated by the computer.

Finally, suppose we wish to see the output in a graphical form. We need a plotter or recorder of some sort which can accept digital information either directly from the computer or via a D/A converter if it is a pretty conventional X-Y recorder, for example. Even here it is not that straight-forward; let us consider what happens if we wish to plot a point. Suppose our recorder is a conventional analog recorder and we then use D/A con-verters. The computer inputs digital information to the converter which transforms it to an analog dc level — we are plotting one point, so the X- and Y-axis input voltages must be constant until the point is plotted and we need two D/A's. Now, the pen of the recorder takes time to move and we have to move it along, hopefully with the pen up so that we do not smear the paper. The computer may take about 1 μ sec to input the digital information into the D/A converter, and the D/A output attains its analog value soon after. But the pen may take a half-second to find its proper place on the paper — remember that both the X and Y inputs are changing. Obviously, then, the computer cannot input more information into the A/D converters for input to the recorder until the recorder has gotten to where it wants to be and puts the pen down to make the point. The recorder has a special attachment which, when it is in the proper position, not only plots but also signals the computer and says, "I'm done, I'm done." The computer, mean-while, could be doing other things or it could be just waiting around for the "Done" signal from the plotter. In any case, as you can see, when the computer interacts with the external world, there may be instances when it has to know if the world has reacted or not, because the computer acts a lot faster than any mechanical device, such as a typewriter, punch, relay, etc., and faster than many other electrical devices, such as A/D converters.

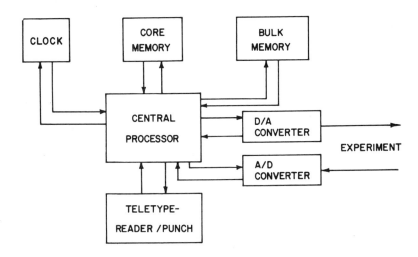

Fig. 1. Block diagram of computer system.

Are we done? Perhaps, but by the time we have all this together, we find that when we do our programming — when we instruct the machine how to do what we want and obtain the data — we may run out of core. Most core is very fast; we can put a word in and get it out (write and read) on the order of 1 $\mu$sec. But we do not always need a memory that fast and that expensive, so we may try to obtain some other kind of "bulk" memory which can contain many words, but which we may have to wait a bit to reach. Such bulk storage devices, magnetic disks or magnetic tapes, are very useful to the small computer in an on-line chore because if the data from the experiment is not coming in too fast, it can be transferred in large units called blocks onto the bulk storage device.

An outline of our system may be seen in Fig. 1. Further details of computer systems applied to various electrochemical problems may be found in papers by Lauer and collaborators [6, 7], Perone and co-workers [8-10], Kugo et al. [11], and our own recent work [12].

## B. Specifics

We have not paid heed to the detailed electronics, and it is totally beyond the intent of this chapter to discuss detailed electronics of interfacing. Nevertheless, it may be instructive to at least consider some requirements for communicating with the external world and some of the requirements on interfacing. As an example of this, suppose that we consider the clock. We wish a programmable clock, i. e., we wish to be able to input information to the clock, have it count and, after a given time, output a signal indicating that the prescribed amount of time has elapsed.

In Fig. 2, we give a general idea of the communication link between the computer and a pair of devices. The computer may wish to send data to the device, and it must exercise certain control functions on various devices; for example, in our discussion of the A/D converter, the computer must tell the device when to convert and it must be capable of accepting information — "Done," data, etc. — from the device. Since the external devices are in parallel in most useful systems, we must have some method of signaling to one particular device, and one alone, that it and it alone is the one we wish to activate. This is akin to a party-line telephone situation; if everybody gets on the line at once, no business can be transacted. The address lines connected to each device operate in such a manner that each device responds only when its address code is output by the computer. The information is digital in nature, that is, each line carries one of two voltage levels output by the computer. These voltage levels carry binary, or two-state, information. By proper connection, each device responds when its "number" appears on the line. For example, suppose we have a three-line digital address, as in Fig. 2a. We define the line as being a 0 state if 0 V is on it, and in a 1 state if -3 V is on it. Then consider a device with 3 address lines whose number is 101. If the voltage sequence -3 V, 0 V, -3 V

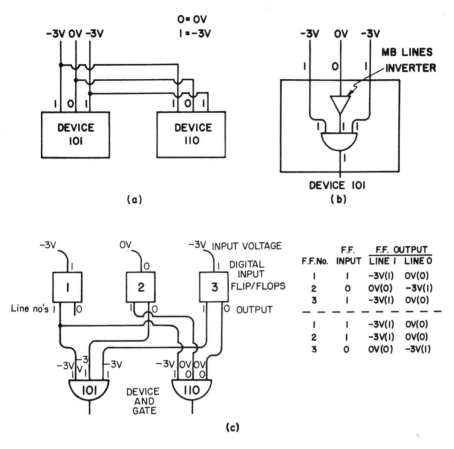

Fig. 2. (a) Block diagram: communication link to a device. (b) AND gate as device selector. (c) Flip/flop – AND gate as device selector; truth table for flip/flop.

is output, as in Fig. 2a, then the Device 101 is being signalled. Internal electrical elements, called gates, can be connected in such a manner that the device is activated if, and only if, the 101 is output. Device 110, next door, will not respond to this signal; its chimes haven't been rung. The nomenclature used here is, for the most part, that of Digital Equipment Corporation PDP family of computers, and in particular that of the PDP-8 series [13].

To explain this, we must understand the operation of an AND gate, the half-moon-shaped device in Fig. 2b. An AND gate is an electronic device whose voltage output will be "1" (or -3 V, or whatever voltage level 1 is) if, and only if, all the inputs to it are also 1's. If any one of n inputs is 0, then the output of the gate is 0. If the lines labeled MB lines in Fig. 2b are what we have called address lines, they are so coded going into the gate that only when 101 appears will the AND gate in the device have all 1's at its input. A simple way to do this, for example, might be to have an inverter — a device which converts 1 to a 0 voltage level and vice versa — in the gate input of the middle line of Fig. 2b, which leads to an AND gate. Thus, in Fig. 2b, the device AND gate would see all its inputs at -3 V if, and only if, the output from the computer were 101 because the inverter on the middle line would make the input to the AND gate of the device 111 only under these conditions. (The device with address 110 might have an inverter at the input to its AND gate at the third line.)

There are other ways of accomplishing the same thing; the output from the computer comes from a "register" which is usually a series of devices called "flip/flops." A flip/flop is an electronic device constructed to have certain characteristics (see Fig. 2c). This device has two outputs; the left output line is designated line 1, the right output line is line 0. The device operates in such a way that if the output of line 1 is a logical 1 (which is -3 volts), the output of line 0 is a logical 0 (or 0 volts). Similarly, if line 1 is a logical 0, or 0 V, then line 0 is a logical 1, or -3 V. An input which causes the device to change its state changes both outputs. The "1" state of the flip/flop is defined as having output line 1 at -3 V, which means output line 0 is at 0 V, or logical 1 and 0, respectively. Similarly, the "0" state finds output line 1 at 0 V, and output line 0 at -3 V, or logical 0 and 1, respectively. In effect, we "read" the device by looking at output line 1.

The 1 state of a flip/flop is achieved when a logical 1, or -3 V, is input; the 0 state is achieved when a logical 0, or 0 V, is input. (The flip/flop actually responds to pulses, ignoring pulses in one direction and responding to pulses in the other direction. It also has a "clear" input which may be pulsed to put the flip/flop in a defined state in an unambiguous manner.)

Suppose we "set" the computer to output 101 on its device address lines; these lines may be viewed as the output of a series of flip/flops, which constitute a register in the computer. If we connect the left line (output line 1) of flip/flop 1 and 3 to the input of an AND gate leading to device 101, and the right line (output 0) of flip/flop 2 to the AND gate, then the AND gate output (which must go to -3 V to activate device 101) will go to 1, or -3 V, only when all its inputs are at logical 1's, or -3 V. This will happen only when 101 (or -3 V, 0 V, -3 V) is input in some manner into the flip/flops. The table of Fig. 2c shows the result discussed here. Similarly, observe the connection to the AND gate of device 110. We connect the left lines of flip/flop 1 and 2, and the right output line of flip/flop 3, to the AND gate input. Its output will now go to -3 V only when 110 is input into the three flip/flops. Note that the input to the AND gate of device 110 is 101 when device 101 is activated. In effect, we have a parallel device-activating

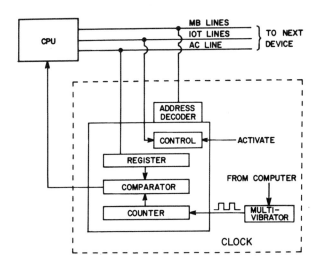

Fig. 3. Programmable clock-block diagram.

system which will signal one, and only one, device at a time. Other devices, with different addresses, will not respond to anything but their own address.

Figure 3 is a block diagram of the clock. The MB lines are address lines, hardwires from flip/flops so that the AND gates of the clock are activated only when the clock's address appears on the MB lines. In the DEC PDP-8 system, the "information" to be transferred appears on the lines designated AC; the AC is the accumulator, or main arithmetic register of the PDP-8. This information appears at the input of a holding register, but the register cannot accept the information until activated; this is because the input to a register, or a series of flip/flops, is via an AND gate as well. Observe Fig. 4. The control pulses (Fig. 4b), designated IOT to conform to PDP-8 nomenclature, are pulses designated 1, 2, and 4 which appear after the device address is generated and are separated in time. Any, all, or none of these pulses may be generated by the program. The "Clear" input of the flip/flop "clears" it — that is, it sets the flip/flop to the "0" state, in which the left output, designated 1, has 0 V appearing there. This means that the input designated 0 has -3 V on its output. Since we "read" the left line, the flip/flop is in the "0" state. The clear input will always cause the flip/flop to go to its 0 state. The set input responds to a 1 pulse at its input and sets the flip/flop into its 1 state — i.e., output 1 of the flip/flop goes to -3 V and output 0 to 0 V (the opposite of the cleared or 0 state). Suppose the data line into the flip/flop has a 0 on it — i.e., it is at 0 volts. If the MB lines are set to the device address, AND gate 1 will go to one output. Remember, the data line has a 0 on it. Now, AND gates 3 and 4 have one input — the output from AND gate 1, at 1. If a pulse, IOT 1, is applied to AND gate 4,

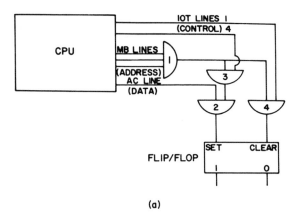

(a)

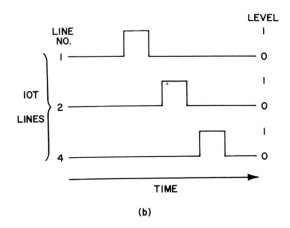

(b)

Fig. 4. (a) Communication – CPU to flip/flop. (b) IOT pulse sequence.

the output of the AND gate 4 will go briefly to 1 (the IOT pulse is from 0 to 1 and back to 0) and this will clear the flip/flop or set it to the 0 state. Now, IOT 4 is applied to the input of AND gate 3, bringing both its inputs to 1 and the output of AND gate 3 goes to 1. Now, AND gate 2 has a pulse input from 3 going to 1, but its other input, from data line, is 0, and the flip/flop stays at 0. Thus, a 0 is "transferred" from the data line to the flip/flop of the register. Suppose the data line contains a 1. Then the argument is as before, we clear the flip/flop, etc., and when the pulse from AND 3 is applied, bringing BOTH inputs of AND gate 2 to a 1 state, the output of AND 2 goes to 1 briefly, setting the flip/flop to a 1 state, resulting in a transfer of a 1 from the data line to the flip/flop.

We can thus transfer information from the computer accumulator to an external "register," both the accumulator and the external register being a series of flip/flops.

We can, under our software program instruction, load a number into the accumulator and, as described, transfer it into an external register. We can also go the other way — i. e., we can transfer information from an external register into the accumulator, and hence into the computer memory. We can see, therefore, that we have the capability to transfer information from the accumulator to the clock register shown in Fig. 3, or to any other register such as that in a D/A converter, etc. In a similar manner, using a device address output from the computer on the MB lines, we can activate the multivibrator on the clock. (The multivibrator period may be set manually in some instances.) Its pulses are counted by the counter register in the clock and when the contents of the counter are equal to the clock register containing the number input from the computer, the comparator circuit causes a pulse to be output by the clock. This pulse may be used to signal the computer to do something. This is a programmable clock, in that the clock register may be changed at will under software control to result in a series of clock pulses as a particular function of time.

The general area of commands which result in device addresses being output on the MB lines are INPUT/OUTPUT or I/O routines. It should now be obvious that the most useful small computers for on-line data acquisition and control are those which are designed to have a large I/O capability and to be reasonably simple to interface.

In the system described, the CLOCK output pulse is fed into an INTERRUPT input line of the computer. The INTERRUPT line interrupts whatever the computer may be doing — and in our operation the computer may either be sitting around waiting for the INTERRUPT or carrying out some calculation on previously obtained data — and causes the program to JUMP to some subroutine to service, or take care of, the INTERRUPT. It will also store the location of the program step it was executing when interrupted so that, when finished, it can resume its interrupted operation.

## IV. ILLUSTRATION

Perhaps at this point it might be worth going through the sequence of operations required to carry out a simple electrochemical experiment (Fig. 5). We wish to apply a voltage pulse to an electrochemical cell and measure the current-time transient resulting. We write a software program to do this. Reference may be made to Fig. 6, which is a flow diagram of what we have to do. Specifically, we wish to apply a voltage step to a 4-sec-old Hg drop and make a current measurement every 10 msec thereafter. We would initially set our multivibrator to have a 1-msec base period.

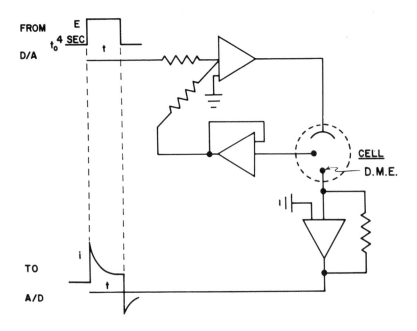

Fig. 5. Experimental diagram.

We start by loading the D/A converter with a number which will cause the voltage applied to the potentiostat to assume its initial value. This is an I/O command. Since we do not wish to take data forever, we designate some location in core as a counter and put a number into it which will be the number of data points we wish to take. After each point is taken, we decrement the counter and test to see if it has reached zero. We then activate a relay to knock the drop off. (If we desired, we might get fancy and use a birth detector which detected drop fall and started the clock and the experiment after a drop had fallen.) We load the clock register with the number 4000, since the time base is set to 1 msec and the experiment desires to apply a pulse after the drop is 4 sec old. We then load the clock register and start the multivibrator. This is essentially one command, and makes use of the IOT pulses to effect the loading of the clock register followed by starting the multivibrator. We might in some instances use a separate command to load and start if desired. We would then turn on the INTERRUPT. The PDP-8/I computer has only one interrupt; other small computers have a hierarchy of interrupts which permit a priority assignment. However, since several things may be on the INTERRUPT line at once, we would normally disable the INTERRUPT until we really wanted it to operate. At this time we might put the program into a "wait" loop; the computer is stalling until it obtains an INTERRUPT signal. When this is received, we

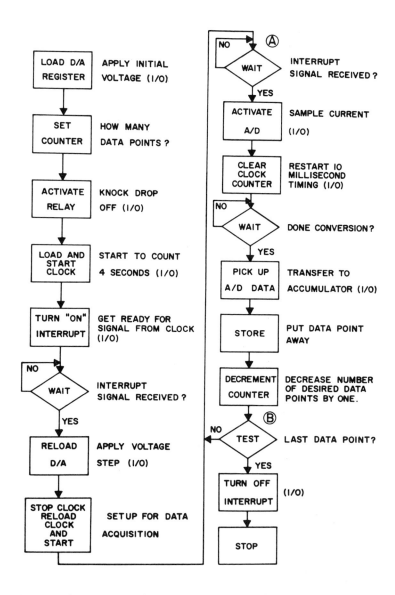

Fig. 6.  Flow diagram for experiment control and data acquisition.

can reload the D/A, thus applying a voltage step, stop the clock, reload
the clock register with 10, since we wish an interrupt every 10 msec, and
start the clock. Once again we wait for 10 msec for the INTERRUPT line
to be activated. When the INTERRUPT is received, we permit the clock to
keep going, but see that the counter starts over again. We then activate
the A/D, sampling the current at the output of the current-measuring
amplifier (Fig. 5), and again wait until a test of the A/D reveals it is done
converting.

The actual A/D conversion takes time; a 12-bit conversion — one that
gives 12 bits of binary information — may take as long as 40 $\mu$ sec, depending on the particular A/D converter used. This is typically much longer
than the computer's ability to cycle through an instruction, hence we must
wait and test the converter to see if the conversion is done. Once it is, we
transfer the data into the accumulator and put it into a previously designated
location in the core memory. We can then decrement the counter and see if
we have obtained the required number of data points; if so, we stop. If not,
we go back into the wait loop, waiting for the interrupt from the clock to say
it is time to take another current reading. Again, if we wished to be fancy
we might program the computer to take data on a logarithmic or square-root
time base, having it recalculate each time or, more likely, having the numbers stored and reloading the clock register after each INTERRUPT.

Things are not quite as simple as they seem; without detailing too
many problems, we must make sure, for example, that the time required
to go from point A on the flow chart to point B is less than the time period
between data points. If it is not, we may have serious problems. In the
sequence described, with the A/D converter described, about 50-60 $\mu$ sec
would be required to proceed from the point when the INTERRUPT is received to the point when one tests for the number of data points. This shows
one of the problems inherent in real-time programming; it is real time,
and one must be very aware of how long operations take. In the description
above, the determined time is not precise. For instance, some few microseconds are required in the steps from the point when the first INTERRUPT
is received, saying that 4 sec have passed, until the clock is reloaded to
start the 10 msec interval. The timing uncertainty is of the order of a few
microseconds. It is, however, very reproducible.

A large number of details have been omitted. We clearly must define
some core locations where we have sufficient space to place acquired data.
This could be done by establishing a block which would be reserved for this
purpose prior to starting the experiment.

Once we have the desired experimental data, obtained on-line, in core,
we can move to a stored program which operates on the acquired data. We
may perform the required mathematical analysis, plot out the information
in various forms on an oscilloscope or plotter, etc.

## V. SUMMARY

Why use a computer at all? The answer to this has really been the purpose of the material above. The computer, as applied in electrochemistry, can obtain experimental data both accurately and rapidly. Very precise timing circuitry permits obtaining data at precise intervals of time. One may obtain data, for example, at an initially high rate, and decrease the data sampling rate as the perturbation experiment proceeds. The type of experiment performed may be varied under computer control by using various switches or relays which operate under computer control. For example, experimental change from a controlled potential to a controlled current mode may easily be accomplished. The type of experiment performed may be readily changed. For instance, the experimental setup indicated in Fig. 5 can, by changing only software, carry out polarography (assuming a dropping-mercury electrode is employed), pulse polarography, square-wave polarography, and staircase voltammetry [7, 12]. Minor hardware changes — the addition of an integrator, for example — would permit carrying out chronocoulometric experiments [6, 7].

Signal averaging techniques may be easily applied; ensemble averaging should permit more accurate results obtained with considerably less tedium [8, 12].

One is faced today with a whole host of small computers from which to choose. Up to this point one of the major problems has been the need to program these small minicomputers in machine language. This is not really difficult, and indeed results in efficient use of available core, but it poses problems when personnel change. Making changes in someone else's machine-language program can be frustrating. The ability to use higher-level languages, Fortran, for example, and still program for data acquisition will ease this problem, probably at a cost of requiring more core to do the same job. However, in the past few years the cost of the minicomputer has dropped rather drastically. The use of medium- and large-scale integrated circuitry, only now really starting to appear, points to a continuing price reduction. The cost of additional increments of core and of tapes and disks is also decreasing. As hardware costs go down, and people costs go up, it seems reasonable to expect more attention to be paid to increasing the ease of changing programs at the cost of using more hardware.

It is the author's conviction that the only way to really appreciate the power of the small computer as applied to electrochemistry is to apply it. Many who have done so have commented on the problem of graduate students becoming so involved with the computer that they neglect the chemistry. As the use of this equipment spreads, this problem, hopefully, will disappear. It is probably true that whatever equipment in this area one chooses today, one will find something better tomorrow. Yet, it is probably true that the equipment will not really be obsolete. One will gaze longingly at someone else's miniminiminicomputer, no bigger than a breadbox, perhaps, and capable of doing an addition in 300 nsec, compared to the table-filling monster of today, taking perhaps 1.5 μsec for an addition. But, because the use to which the equipment is put does not change, the equipment will probably not become obsolete.

It is inevitable that the computer will become, in the '70's, what the multipurpose operational amplifier system was in the '60's — the tool, treasure, and plaything of the electrochemist.

## ACKNOWLEDGMENT

This work was supported in part by the Air Force Office of Scientific Research under Grant AFOSR-71-1995.

## REFERENCES

[1].    J. W. Frazer, Anal. Chem., 40:21A (1968).

[2].    J. W. Frazer, Chem. Instr., 2:271 (1970).

[3].    J. Christie, G. Lauer, R. Osteryoung, and F. C. Anson, Anal. Chem., 35:1979 (1963).

[4].    G. Lauer and R. A. Osteryoung, Anal. Chem., 38:1137 (1966).

[5].    J. L. Butler, Instr. Tech., 17(10):67 (1970).

[6].    George Lauer, Roger Abel, and F. C. Anson, Anal. Chem., 39:765 (1967).

[7].    G. Lauer and R. A. Osteryoung, Anal. Chem., 40:30A (1968).

[8].    S. Perone, J. E. Harrar, F. B. Stephens, and Roger Anderson, Anal. Chem., 40:899 (1968).

[9].    S. P. Perone, D. A. Jones, and W. F. Gutknecht, Anal. Chem., 41:1154 (1969).

[10].    W. F. Gutknecht and S. P. Perone, Anal. Chem., 42:906 (1970).

[11].    T. Kugo, Y. Umezawa, and S. Fujiwara, Chem. Instr., 2:189 (1969).

[12].    H. E. Keller and R. A. Osteryoung, Anal. Chem., 43:342 (1971).

[13].    PDP-8 Logic Handbook, PDP-8/L Handbook, Small Computer Handbook, Digital Equipment Corp., Maynard, Mass., 1968-70.

Chapter 12

APPLICATIONS OF ON-LINE DIGITAL COMPUTERS
IN AC POLAROGRAPHY AND RELATED TECHNIQUES

Donald E. Smith

Department of Chemistry
Northwestern University
Evanston, Illinois 60201

## I. INTRODUCTION

It is undoubtedly evident to most readers of this volume that the application of small on-line digital computers ("minicomputers") represents the major area of innovation in chemical instrumentation at the present time. This particular theme is advanced and supported by a deluge of papers, review articles, news releases, etc. Many feel that when the dust has cleared, the "computer revolution" will be viewed as a development in chemical measurements whose importance is matched only by the earlier advent of analog electronic instrumentation.

In the narrower context of electrochemical measurements, probably the only previous "breakthrough" in instrumentation which might match the eventual impact of on-line computers has been the introduction of operational amplifier methodology [1-7]. The latter advance provided a generalized concept in analog electronics for electrochemical measurements [4-7] and revolutionized the philosophy of such measurements from one based on numerous, single-purpose instruments, to a much more efficient one characterized by modular, multipurpose units [2, 5, 6]. When considering their effects on experimental electrochemistry, one soon realizes that operational amplifier techniques and the on-line computer represent something more than two separate significant advances whose benefits are essentially additive. These two developments, when used in combination, interact in a manner which is in effect symbiotic. For example, the technical success of operational amplifier instrumentation generated in many laboratories a "data explosion." In this situation the limiting step was processing adequately vast amounts of raw data. Clearly, the on-line computer's capability for information handling removes this otherwise formidable limitation on the advantages realizable from operational amplifier instrumentation. Conversely, operational amplifier electronics substantially enhances the benefits an electrochemist can realize from the availability of an on-line digital computer. This is effected partly because "op-amp" analog equipment simplifies the interfacing problem. It also makes possible special innovations which utilize the computer's capabilities in a much more effective manner than would be possible with classical techniques.

The present chapter deals with a special branch of electrochemical methodology in which one measures the electrical response of the interfacial impedance to small-amplitude periodic perturbations (usually sinusoidal). The most prominent member of this group of techniques is referred to as either ac polarography or faradaic impedance measurements, depending on the detection method employed [7-9]. Faradaic rectification measurements [7-13], second-harmonic ac polarography [7-9, 14], intermodulation polarography [7, 15], square-wave polarography [7-9, 16], and rf polarography [17] are among other members in this group. This chapter's main purpose, stated in the simplest possible form, is to illustrate how some of the general remarks advanced in the preceding paragraph apply to this special class of

techniques. In attempting to achieve this end, emphasis will be placed on:
(a) appropriate techniques for interfacing the computer to the ac polarograph;
(b) experimental routines which utilize the computer to automate the conventional approach to ac polarography and related techniques; (c) novel
experimental approaches made feasible by the computer whose implementation represents a fundamental alteration in experimental procedures; and (d)
some special computer-aided routines of primary interest in analytical
applications. Owing to space limitations, the presentation will be qualitative
in nature, deemphasizing specifics such as program listings, details of data
analysis routines, and detailed circuit diagrams. Other sources [7-9, 18]
may be consulted for such information. For the same reason the discussion
will be confined mainly to that aspect of ac polarography which is most relevant to the subject at hand, namely, the <u>measurement,</u> not its <u>interpretation.</u>
The latter problem has been treated extensively in recent reviews [7-9].

## II. BASIC ASPECTS OF THE AC POLAROGRAPHIC EXPERIMENT

At this point a short discussion of the nature of the ac polarographic
experiment is useful to assist those unfamiliar with the technique. The
nature of the material in the following sections is such that an adequate background should be provided by considering the technique from the viewpoint of
an electrical measurement.

### A. Nature of the Perturbation and Observables

Alternating-current polarography is basically a measure of the admittance
of an electrochemical cell when a small-amplitude sinusoidal potential
(usually 1-20 mV) is imposed between a working and reference electrode. The
solution conditions are normally those characteristic of conventional dc polarography where one employs an "inert" supporting electrolyte whose concentration substantially exceeds that of the electroactive species of interest. Such
solution conditions normally guarantee a situation where the steps which
primarily control the characteristics of the observables (rate-determining
steps) are one or more of those processes which occur at or very near the
working electrode-solution interface. Diffusive mass transport, heterogeneous electron transfer, homogeneous or heterogeneous chemical reactions
coupled to the electron transfer step, and charging of the electrical double
layer are among the relevant rate processes [7-9]. Similar rate processes at
the reference electrode are rendered unimportant either by not permitting
current flow through the reference electrode, as in a modern three-electrode
potentiostat [1-7], or by employing a reference electrode whose area and
electroactive species concentrations vastly exceed those at the working electrode. Likewise, polarographic conditions tend to minimize, but do not
always totally negate, the importance of steps involved in transport of elec-
tricity through the bulk of the solution whose attendant manifestations are the
bulk solution ohmic resistance and transport of electroactive species by ion
migration.

A conventional ac polarogram involves application of a potential containing the small-amplitude sinusoidal component of fixed frequency and amplitude, together with a dc component which is slowly scanned. One measures the resulting alternating component of the cell current whose frequency is the same as that of the applied potential (fundamental harmonic current), as a function of the dc component of the applied potential. The resulting plot of

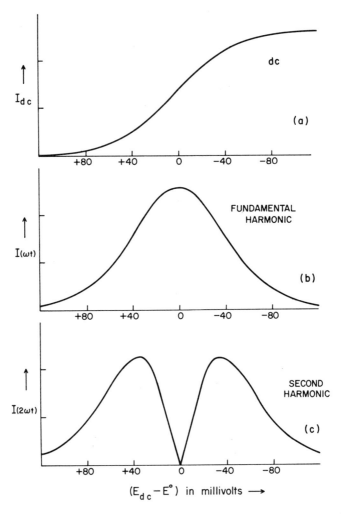

Fig. 1. (a) Dc, and (b) fundamental- and (c) second-harmonic ac polarograms with a diffusion-controlled system. (Ac polarograms are total current amplitude polarograms — i.e., "conventional" fundamental and second-harmonic polarograms.)

the fundamental harmonic alternating-current magnitude versus dc potential is the conventional ac polarogram. The response observed in such an experiment originates from two fundamentally different processes: (a) actual electron transfer associated with electrolysis of a particular electroactive species; and (b) a "virtual" charge transfer associated with charging the electrical double layer at the working electrode-solution interface. Current arising from electrolysis is referred to as the faradaic current, and in the vast majority of applications it is the component of interest. The faradaic current contributes a sharp, peak-shaped response to the ac polarogram which is coincident on the dc potential axis (abscissa) with the rising part of the conventional dc polarographic wave (see Fig. 1, a and b). The magnitude of the faradaic "wave" is normally directly proportional to the concentration of the electroactive species, the property which serves as the primary basis for quantitative analytical applications. Like the dc polarographic wave, the position of the ac polarographic peak on the abscissa is characteristic of the electroactive species, providing a basis for qualitative and quantitative analysis of multicomponent mixtures. Physical chemical applications of ac polarography are based on the fact that the magnitude, shape, and position of the faradaic peak is sharply dependent on the kinetic-thermodynamic properties of the various above-mentioned rate processes associated with electrolysis. The double layer charging current in most instances is primarily a function of the supporting electrolyte concentration and provides at most only secondary information regarding the electroactive species of interest. For most applications of ac polarography the charging current may be viewed as the major background or noise component whose existence limits the ultimate sensitivity and versatility of the method. Relative to the faradaic response, the charging current variation with dc potential is small, so that it contributes little structure to the ac polarogram, except in special cases (see tensammetric waves [8]). These points are illustrated in Fig. 2 which shows a typical ac polarogram, including the charging current contribution. For most conditions other characteristics of the double-layer charging current are those associated with charging a simple capacitor — e.g., the current increases in direct proportion to the frequency and leads the applied potential by 90° (approximately).

In addition to the double-layer charging current, a second "nonfaradaic" phenomenon also frequently contributes to making the observed ac polarographic results differ from the desired purely faradaic response. The ohmic resistance of the solution in combination with the cell current introduces an iR drop. This causes the alternating potential across the electrical double layer (the effective alternating potential as far as the faradaic impedance is concerned) to differ in amplitude and phase angle from the externally applied alternating potential. Since the magnitude of this disparity is directly proportional to cell current, the obvious outcome is a distortion characterized by a broadening and flattening of the ac polarographic wave and a shift in the phase angle, a parameter of importance in kinetic applications (see below).

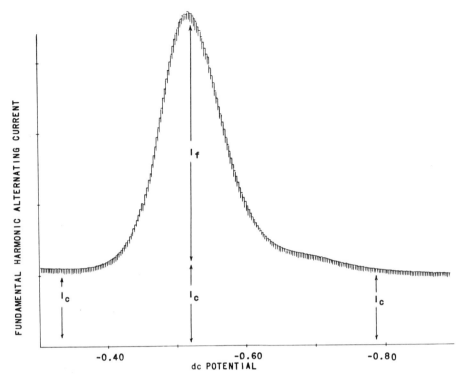

Fig. 2. Ac polarogram including double-layer charging current. System:
$1 \times 10^{-3}$ M (cyclooctatetraene) $Fe_2(CO)_5$ in dimethylformamide — 0.01 M
tetrabutylammonium perchlorate. Applied: Sinusoidal potential of 456 Hz
frequency and 20 mV peak-to-peak amplitude; dc scan rate of 25 mV/min.
Measured: Total alternating current at 456 Hz using sample-and-hold readout mode. $i_c$ = double-layer charging current. $i_f$ = faradaic current.

If one views the ohmic potential drop from a kinetic rather than electrical
viewpoint, one recognizes it as a manifestation of a rate process (bulk ionic
conduction) which usually exerts some contribution to the over-all rate,
notwithstanding the experimentalist's desires to eliminate this effect.

The foregoing two paragraphs were prefaced by the example of the so-
called "conventional ac polarogram" in which one measures the total fundamental harmonic ac amplitude as a function of applied dc potential. Those
familiar with ac measurements in any context will recognize that this is not
the only option open to the experimentalist. For example, within the framework of the fundamental harmonic response one can measure ac components
which are in phase and/or quadrature to the applied alternating potential
(phase-sensitive ac polarography). The phase angle may also be observed

directly. Thus, in addition to the "conventional," one may record any of several other types of ac polarograms such as: (a) in-phase current versus dc potential — the "in-phase ac polarogram"; (b) quadrature current versus dc potential — the "quadrature ac polarogram"; (c) phase angle versus dc potential — the "phase angle polarogram"; and (d) the in-phase versus the quadrature current components — the "complex plane polarogram." Reports of successful acquisition of each type of polarogram are found in the literature [7, 8, 19, 20] and each has been recommended for one purpose or another. Of the latter types, phase-selective detection of the in-phase fundamental harmonic current has received greatest attention. It has the advantage of greater sensitivity because the double-layer charging current, whose phase angle is nearly quadrature to the applied alternating potential, contributes much less to this polarogram.

In addition to the fundamental harmonic current, which is associated with the linear aspects of the cell admittance, the ac polarographic response includes minor current components which arise because of nonlinearities of the cell admittance. In an experiment such as we have been considering, where the applied potential is comprised of a single frequency component, "higher-order" current components are found at frequencies which are integral multiples (harmonics) of the fundamental frequency, as well as at zero frequency (dc). If one simultaneously applies two or more discrete frequencies to the cell, the nonlinear effects generate components at sum and difference frequencies [7, 15] as well as at the normal harmonics. Some electrochemists have devoted considerable effort to the measurement and interpretation of certain of these higher-order current components [7-15, 17, 21, 22]. These efforts have given birth to a variety of polarographic techniques which are closely related to conventional ac polarography. Some important examples are second harmonic ac polarography [7, 9, 14], intermodulation polarography [7, 15], and faradaic rectification measurements [7-13]. Some of these techniques are defined in Table 1. As is evident from the examples in Table 1, within the context of any particular higher-order current component one has the option of measuring the total, in-phase and/or quadrature current component or the phase angle. The primary stimulus for developing the capability to measure some of these higher-order current components is the fact that except at very high frequencies, by far the major contributor to the response is the faradaic process. The double-layer charging process behaves almost perfectly linear, contributing very little background ("noise") signal. Thus, despite the fact that the response is small, relative to the fundamental harmonic, observation of these higher-order currents actually represents a more sensitive and versatile probe into the faradaic process. This advantage is offset to some extent by measurement and interpretative complexities which combine to make most workers favor fundamental harmonic studies.

From the previous remarks, one should recognize that under ac polarographic conditions the cell current is comprised of a variety of observables,

TABLE 1

Ac Techniques Based on Observation of Faradaic Nonlinearity

| Technique | Applied ac potential | Observable |
|---|---|---|
| 1. Faradaic rectification measurements | $E = \Delta E \sin \omega t$ | Direct current induced by applied ac potential |
| 2. Second harmonic ac polarography | $E = \Delta E \sin \omega t$ | Total current component at frequency $= 2\omega$ (see Fig. 1c). |
| 3. Third harmonic ac polarography | $E = \Delta E \sin \omega t$ | Total current component at frequency $= 3\omega$ |
| 4. Intermodulation polarography | $E = \Delta E (\sin \omega_1 t + \sin \omega_2 t)$ | Total current component at frequency $= (\omega_1 - \omega_2)$ |
| 5. Rf polarography | $E = F(t) \sin \omega t$ where $F(t) =$ square wave | Total current component at frequency of square wave |
| 6. Phase-selective second harmonic ac polarography | $E = \Delta E \sin \omega t$ | In-phase or quadrature current component at frequency $= 2\omega$ |
| 7. Phase-selective intermodulation polarography | $E = \Delta E(\sin \omega_1 t + \sin \omega_2 t)$ | In-phase or quadrature current component at frequency $= (\omega_1 - \omega_2)$ |

only one of which is the conventional ac polarogram. Each observable contributes information of significance to the analyst or kineticist. Each is comprised of the desired faradaic component and nonfaradaic "noise" associated with the ohmic resistance and double-layer capacity. The relative contribution of the noise components depends markedly on the observable in question.

## B. Correction for Nonfaradaic Effects

Regardless of whether the application of ac polarography is analytical or kinetic in nature, careful, quantitative studies often demand some form of correction for effects of the ohmic resistance and double-layer capacity. An analyst may choose to avoid this task by generating from measurements on a series of standards, a calibration curve that reflects both the faradaic and nonfaradaic response. Such a calibration curve is usually nonlinear (effect of iR drop) and possesses a nonzero intercept, owing to the charging

current. Nevertheless, good analyses often may be effected using such
calibration curves, provided that the unknowns are run under precisely
identical circumstances (same cell geometry, dropping mercury electrode
capillary, ac frequency, and amplitude, etc.). This requirement may be
relaxed somewhat if analyses are run at low frequencies with highly con-
ducting solutions where one observes minimal nonfaradaic effects for
concentrations exceeding about 1 millimolar. While the analyst may sometimes use the calibration curve to avoid explicit correction for nonfaradaic
effects, the kineticist seldom has such an option. Kinetic-mechanistic
conclusions are usually based on ac polarographic data obtained over wide
ranges of applied ac potential and frequency. Much of this data will be substantially influenced by nonfaradaic noise.

Several basic approaches to compensating for nonfaradaic effects are
available. In principle, the most convenient involves the use of recently
developed analog instrumentation which under appropriate conditions can
provide automatic compensation [23-25]. Contributions of the ohmic potential drop are eliminated by introducing a positive feedback path in the
potential control network while charging current is reduced by subtractive
polarography [23]. The net result is an instrument readout which is inter-
pretable as a pure faradaic component. Although reasonably versatile, such
instrumentation is limited to frequencies below 10-20 kHz and to systems
where the electroactive component is not adsorbed significantly — i. e.,
where a priori separability of faradaic and double-layer charging current is
a valid assumption [9, 26, 27]. Partly because of these limitations and partly
because of the recent nature of the development, use of such instrumentation
is not widespread. Thus, presently most ac polarographic measurements
reported in the literature involve a readout reflecting nonfaradaic contributions which must be compensated by mathematical or graphical manipulation
to obtain the true faradaic response. Unlike the case with dc polarography,
such corrections amount to more than simple addition or subtraction because
the sinusoidal signals of ac polarography are characterized by two fundamental properties, amplitude and phase — i. e., the signals must be treated
as vector quantities. Thus, for example, in the case of instrumentation which
provides total faradaic current and phase angle, correction for ohmic resistance and double-layer charging current involves using mathematical relationships such as

$$i_c^* = i_{c,m}^* [1 + \cot^2 \varphi_M^*]^{1/2}, \tag{1}$$

$$R = \frac{V \cot \varphi_M^*}{i_c^*}, \tag{2}$$

$$i_c = \left(\frac{\Delta E}{V}\right) i_c^*, \tag{3}$$

$$\Delta E = [V^2 + (i_t R)^2 - 2V i_t R \cos \varphi_M]^{1/2}, \tag{4}$$

$$\sin \varphi_c = \frac{i_t R \sin \varphi_M}{[V^2 + (i_t R)^2 - 2V i_t R \cos \varphi_M]^{1/2}} \, , \tag{5}$$

$$Q = \varphi_M + \varphi_c \, , \tag{6}$$

$$\cos \varphi = \frac{i_t \cos Q}{[i_t^2 + i_c^2 - 2i_c i_t \sin Q]^{1/2}} \, , \tag{7}$$

$$I_f = [i_t^2 + i_c^2 - 2i_c i_t \sin Q]^{1/2}. \tag{8}$$

The notation employed in Eqs. (1)-(8) is defined in Table 2. These equations illustrate that to obtain the true faradaic current amplitude $I_f$ and phase angle $\varphi$, one must measure the quantities $I_t$, $i_{c, M}^*$, $\varphi_M$, and $\varphi_M^*$. Thus, compensation of nonfaradaic effects demands measurement of the current amplitude and phase angle in the presence and in the absence of the faradaic

TABLE 2

Notation Definitions for Eqs. (1)-(8)

| Symbol | Definitions |
|--------|-------------|
| V | Amplitude of externally applied alternating voltage |
| $\Delta E$ | Amplitude of effective alternating potential across interface with faradaic process |
| R | Ohmic resistance |
| $i_{c, m}^*$ | Double-layer charging current measured in absence of faradaic process |
| $i_c^*$ | Double-layer charging current in absence of faradaic process, corrected for ohmic potential drop |
| $i_c$ | Double-layer charging current in presence of faradaic process |
| $i_t$ | Total observed ac in presence of faradaic process |
| $I_f$ | True faradaic ac |
| $\varphi_M^*$ | Phase angle between externally applied alternating voltage and ac observed in absence of faradaic process |
| $\varphi_M$ | Phase angle between externally applied alternating voltage and total ac observed with faradaic process |
| $\varphi_c$ | Phase angle between externally applied alternating voltage and alternating potential across interface |
| $\varphi$ | True phase angle between faradaic ac and alternating potential across interface |

wave. A mathematically simpler situation exists if one employs phase-sensitive detection of in-phase and quadrature components of the alternating current where scalar subtraction replaces some of the vectorial relationships. The number of measurements required is the same.

The derivation and further details regarding application of Eqs. (1)-(8) are given elsewhere [7-9]. One point regarding the derivation which should be mentioned is that these equations are based on the assumption of a priori separability of the faradaic and double-layer charging processes [9]. When this assumption is invalid a more sophisticated and general approach developed by Sluyters-Rehbach and Sluyters [9], involving presentation of raw data in a complex plane format, is required to correct adequately for non-faradaic contributions. The latter approach entails mathematical relationships which are notably more cumbersome than Eqs. (1)-(8). In addition, data obtained over a wide frequency range are essential, whereas Eqs. (1)-(8) can be applied to a single-frequency experiment. Further details on this more general approach will not be presented here since a detailed lesson on compensation of nonfaradaic contributions is not the point of the present discussion. Rather, it is hoped that Eqs. (1)-(8) alone will help the reader appreciate the nontrivial nature of the data treatment routine whenever nonfaradaic compensation is necessary. It is particularly important to reflect on this problem in the context of a kinetic-mechanistic study where faradaic current amplitude and phase angle data are desired over a wide range of dc potential and frequency. Usually several hundred data points are required to characterize accurately the shape of the three-dimensional profile of current amplitude (or phase angle) vs. dc potential vs. frequency. Faced with the prospect of several-hundred-fold repetition of a calculation based on Eqs. (1)-(8), one soon appreciates the advantages of an on-line digital computer facility for electrochemical kinetic studies utilizing ac polarography. Although convenient, painless, real-time compensation of nonfaradaic effects is hardly the only benefit to ac polarography provided by an on-line computer facility (see below), it can be a major justification for acquiring such a facility where such corrections are done routinely. The need for repetitive, boring corrections for nonfaradaic effects has been a major reason for the use of the adjective "tedious" to characterize quantitative, kinetic studies using small amplitude ac measurements, such as ac polarography. In spite of this reputation, these techniques have found a prominent place in electrochemical methodology owing to the unusually rich and precise information content of the data. By freeing the experimentalist from the "tedium," the on-line computer should allow an even more impressive realization of the potentialities of ac techniques.

## III. THE COMPUTERIZED AC POLAROGRAPH

Figure 3 illustrates schematically an ac polarograph provided with an on-line computer. Actually, the schematic is sufficiently general that it

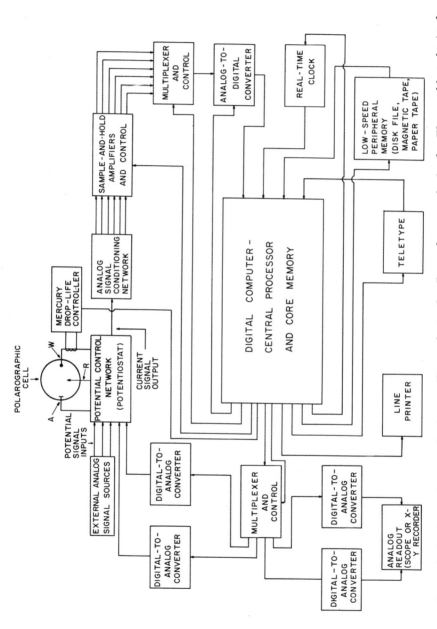

Fig. 3. Computerized ac polarograph: A — auxiliary electrode, R — reference electrode, W — working electrode.

could be taken as representative of any type of controlled-potential instrument coupled to a computer. The configuration permits computerized data acquisition, data analysis, and experiment control. The interconnections between the units are indicated with lines. Arrows show the direction of signal flow and, thus, implicitly designate inputs and outputs. The schematic is designed to illustrate a number of the useful features which are available with most commercially available laboratory computer systems. Some of these features are obviously "frills" which are not absolutely essential to set up an effective on-line computer operation. At the same time, one should recognize that not all available options are indicated, such as a card reader, a digital incremental plotter, etc.

The units depicted in Fig. 3, which one finds in conventional (noncomputerized) ac polarographs, are the external analog signal sources, potentiostat, polarographic cell, analog signal-conditioning network, analog readout, and mercury drop-life controller. External analog signal sources normally consist of a precision dc source which sets the initial voltage, a ramp generator which provides the "dc scan," and a sinusoidal oscillator. These signals are applied through a summing network to the potentiostat, which is a feedback control network whose purpose is to maintain the potential difference between the reference and working electrodes at a value equal to the sum of the potential signal inputs [1-6]. The current required to achieve potential control, which is the observable of interest, flows between the auxiliary and working electrodes. The potentiostat provides a current signal output which normally is a voltage proportional to the cell current [1-6]. The characteristics of the latter signal are usually modified appropriately by the analog signal conditioning network to yield an output which is compatible with the readout device. A typical potentiostat circuit is depicted in Fig. 4. Signal conditioning in ac polarography requires at least a high-pass filter to eliminate the dc component followed by rectification, and low-pass filtering to obtain a dc signal proportional to the ac signal amplitude. In careful quantitative work a more typical signal-conditioning network would involve some form of frequency-selective amplification to eliminate all undesired signal components (dc, unwanted harmonics, random noise pickup, etc.), combined with rectification in either the phase-selective [7] or full-wave mode. Circuit schematics of signal conditioning networks used in our laboratories for conventional and phase-selective ac polarography are illustrated in Figs. 5 and 6. In Fig. 3, multiple outputs are indicated for the analog signal-conditioning network. This implies the possibility of monitoring simultaneously several aspects of the ac polarographic response — e.g., simultaneous measurement of in-phase and quadrature fundamental harmonic current. Although this multiple readout option is seldom utilized in work reported in the literature, it represents a particularly appealing operation mode in the context of computerized ac polarography which will be discussed below. The mercury drop-life controller is a device designed to mechanically dislodge a mercury drop at a particular point after drop growth

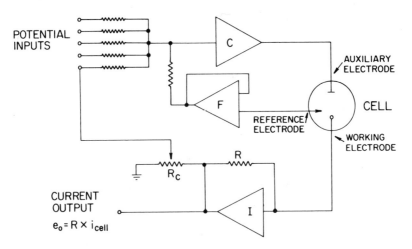

Fig. 4. Potentiostat: C — control amplifier; F — voltage-follower amplifier; I — current-measuring amplifier (current follower); $R_C$ — positive-feedback iR compensation control.

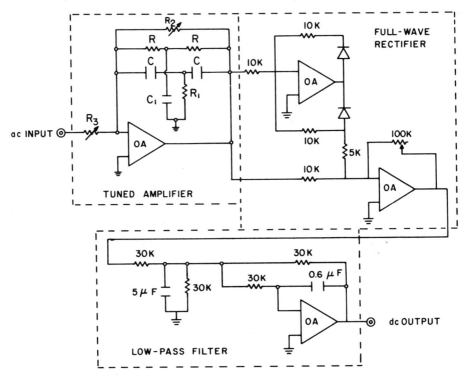

Fig. 5. Signal conditioning network for total ac measurement. OA = operational amplifier.

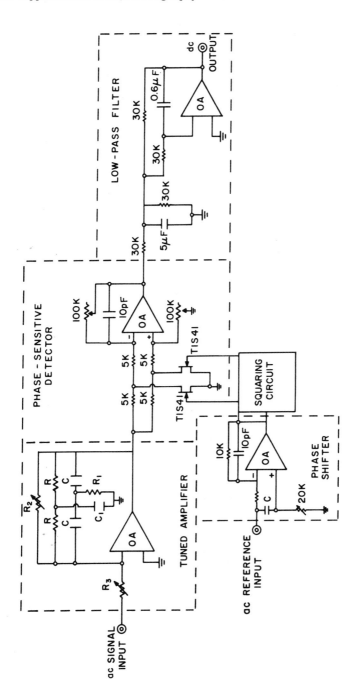

Fig. 6. Signal conditioning network for phase-selective current measurement. OA — operational amplifier.

has commenced. This usually is accomplished by a mechanical shock imparted by a solenoid, or the like, which is activated by a voltage pulse from a timing circuit [23]. Such a device has been used only sporadically with noncomputerized ac polarographs. However, the computerization of an ac polarographic experiment involving a dropping mercury electrode (DME) introduces the necessity of synchronizing the growth and fall of the DME with the computer's data acquisition cycle. Mechanical control of drop life represents an effective means of solving this problem so that the drop-life controller probably will become a more familiar commodity as on-line computer applications increase.

The devices in Fig. 3 which have not been discussed as yet are those normally associated with the on-line laboratory computer, regardless of the nature of the experiment. All items in question may be purchased, complete with interfacing, from the various vendors. The "command center" of the computerized instrument is represented in Fig. 3 as the digital computer with its central processor and core memory. It is this unit that stores and implements the program for the experimental sequence. Highly complex, sophisticated arithmetic operations may be performed within this unit for data analysis. Sophisticated timing sequences (aided by a real time clock) and associated driving signals for control of peripheral hardware and the chemical instrument also are generated within the confines of the computer. Arithmetic registers in the central processor are coupled to registers in peripheral equipment to enable input and output of digital information. Thus, the computer has the capability of accepting, transmitting, and manipulating data, as well as controlling the sequence of these events. The basic small laboratory computer usually is characterized by a core memory of 4096 (4 k) 12-18-bit words. Most core memories are expandable to at least 32 k words in 4 k increments. It is the relatively small core-memory capacity and word length which makes possible the low cost of these small computers, compared to the more familiar large scientific computers. Speed is not sacrificed substantially in most cases, as evidenced by typical cycle times of 0.3 to 2 $\mu$sec. Despite the seemingly small capacity and word length, a 4-k, 12-bit core memory is quite adequate for conventional ac polarographic measurements, provided one is willing to program in a low-order assembly language (as opposed to FORTRAN IV, for example). An 8-k core memory unit in our laboratory has been found adequate for most experimental routines discussed below, including those which generate relatively large numbers of data points per unit time. In addition, as indicated in Fig. 3, most problems arising from limited high-speed core memory capacity can be overcome through the use of a low-speed, high-capacity peripheral memory unit, such as punched paper tape which is a standard item in all small computers. More highly recommended than punched paper tape is magnetic tape or disk file memories. One of the small inexpensive disk file units provides storage space for 32-k 12-bit words per disk with a transfer rate of around 60 $\mu$sec per word. An inexpensive cassette style magnetic tape unit can provide, for

example, a 360-k, 16-bit word capacity per tape drive with a transfer rate
of about 1.5 msec per word. Thus, it is evident that when ultrahigh-speed
data acquisition and transfer are not essential, the use of low-speed peripheral memory units can provide the small lab computer with an effectively
unlimited memory capacity.

The sample-and-hold amplifiers and control, the multiplexer and control, and the analog-to-digital converter are the units in Fig. 3 which enable
transfer of data from the ac polarograph to the computer's central processor
and core memory. Because the computerized digital data-acquisition operation normally involves nonsimultaneous operation on the various signal
channels, the sample-and-hold amplifiers and control are required whenever
one desires the values of two or more signals at the same point in time.
Such a requirement obviously exists in ac polarography with a DME whenever
one wishes to follow two or more observables during the same experimental
run. Hardware provided by computer manufacturers enables sampling of
many analog signals simultaneously (typically ±50-200 nsec), after which
the sampled signals are "held" for a sufficient period to allow scanning of
the signal channels by the subsequent units in the data-acquisition system.
This operation is performed by the sample-and-hold amplifiers under computer control (the digital computer controls the timing and sequence of the
operation). Because the time for data acquisition is quite rapid (typically
10-2000 $\mu$sec per point), one often can do without the sample-and-hold
system in work with stationary electrodes where the ac polarographic
response varies slowly with time, relative to the situation with the DME.
During the time required to sequentially monitor all signal channels of interest, the stationary electrode response may vary negligibly so that
sequential data acquisition is, in effect, simultaneous. The multiplexer and
control is basically an array of electronic or electromechanical switches
that allows the computer to select which analog signal channel (multiplexer
inputs) is to be transmitted (multiplexer output) to the next stage in the data-
acquisition system. The switching rate with modern multiplexers is quite
rapid (e.g., 100 kHz) and they can control many channels (50-100 is common).
In fact, the economics of multiplexer design and construction apparently are
such that there is no advantage in building or buying a multiplexer that controls just a few channels. One is normally offered many-channel capabilities
and this should be considered a standard, built-in feature of the computerized
instrument. The analog channel selected by the multiplexer is delivered to
an analog-to-digital converter which transforms the analog signal into a
digital signal in the binary arithmetic format, the language "understood" by
the digital computer. The typical length (significant figures) of the binary
number generated ranges from 10 to 16 bits (binary digits). This implies
maximum resolution in the analog-to-digital conversion ranging from 1 part
per 1000 (10 bits) to 1 part in 65,000 (16 bits). Full-scale accuracies and
linearities exceeding 1 part per 10,000 are obtainable. The time required
for the analog-to-digital conversion ranges from 0.5 to 50 $\mu$sec in most

cases, although times in the msec range may arise as when a high-precision integrating digital voltmeter is the **A-D** converter. After A-D conversion is completed, data must be transferred into the computer memory, an operation which consumes another 1-100 $\mu$sec, depending on the system. The entire operation of channel selection, analog-to-digital conversion, data transfer to memory, and incrementation of a memory address register in preparation for the next data word may require 5-1000 $\mu$sec (except for the slow converter mentioned above). Finally, input impedances of analog-to-digital converters normally exceed 1 M$\Omega$ so that "loading" of the input signal source is seldom a problem. These facts lead one to the conclusion that the operations attending analog-to-digital conversion can be performed with accuracy and speed which exceed the needs of the classical ac polarographic experiment. The ac polarographer need not worry about degradation in signal fidelity once his signal is transmitted to the computer hardware, assuming certain simple rules are followed in interfacing (see below). Indeed, once the signal has been digitized the accuracy of the measurement has been essentially fixed. All subsequent operations on the digitized data can be effected without loss in accuracy.

In addition to accepting data from the output of the ac polarograph, the computer can transmit information to the polarograph, thus enabling experiment control. This aspect of the computer's function is represented in Fig. 3 by signal paths from the computer output to the potentiostat signal inputs and the mercury drop-life controller. Signals transmitted to the potentiostat signal inputs originate as digital words at the computer output, and are converted to the analog voltage format required by the polarograph in the digital-to-analog converter. Thus, voltages applied to the cell may be controlled directly by the computer program. The speed, accuracy, and resolution obtainable in digital-to-analog conversion is comparable to that quoted above for the inverse operation of analog-to-digital conversion. Output impedances of digital-to-analog converters are typically less than 1 $\Omega$, which negates most impedance matching difficulties. A control pulse from the computer to the mercury drop-life controller is derived from a single bit register in the computer. Logic levels of small computers are such that the voltage change accompanying a change in state of this register from the binary 0 to the binary 1 level (or vice versa) can be made sufficient to activate the mercury drop knocker. Any necessary buffer amplifiers may be considered an integral part of the mercury drop-life controller.

The teletype, line printer, scope display, and X-Y Recorder are the units depicted in Fig. 3 which allow information transfer between the operator and the computer. A Type ASR-33 Teletype is the standard item provided with small computers for programmed digital readout of information from the computer and entry of digital information to the computer by the operator. The teletype also provides a low-speed paper-tape punch and reader. The 10-character/sec maximum printing rate of the ASR-33 teletype represents a prohibitively slow step in many readout operations with small computers, particularly when real-time readout is desired. Although not a problem when computerizing the classical ac polarographic experiment, this speed limitation is annoying when one desires real-time readout in certain novel approaches to ac polarography described below. The readout speed limitation often can be circumvented through the use of a line printer. Reasonably fast (e.g., 430 characters per sec) models are now available at moderate cost. The scope display can be used as either an analog (graphical plot display) or digital (character display) readout device, whose speed limitations are essentially negligible. Combined with a light pen, the scope display allows the operator to input information to the computer. Unfortunately, the only available permanent record (photograph) provided by a scope display is somewhat unsatisfactory, particularly in the case where digital information is displayed. For analog recording a more desirable permanent readout can be obtained with an X-Y recorder. Recorders and oscilloscopes are basically analog readout devices, so that digital-to-analog converters must be inserted between them and the computer. Units are now available which include the digital-to-analog converters as internal components. Such devices are referred to as scope computer terminals and digital incremental plotters and are interfaced directly to the computer.

## IV. INTERFACING THE COMPUTER AND AC POLAROGRAPH

In considering the question of interfacing an ac polarograph and computer, it will be assumed that the polarograph already possesses the capability of driving a conventional recorder. This implies that ac-to-dc conversion, such as is effected by the circuitry in Fig. 5 and 6 has been achieved.

Interfacing a computer to any chemical instrument involves taking the necessary steps to ensure electrical compatibility of the components which interact at the digital-analog interface. For example, the magnitude, impedance characteristics, noise characteristics, and grounding characteristics of the input signal to the digital-to-analog converter must be consistent with converter requirements. Thus, if a **digital-to-analog** converter is designed to operate on signals in the range 0 to -10 V, it is not only important that the signal from the polarograph be of negative polarity, but that the largest signal excursions roughly approach the full-scale level (10 V) of the digital-to-analog converter. An input signal which never exceeds a few hundred mV will be converted with poor resolution in the same sense as one would observe in a voltage-to-position conversion of such a signal by a 10-V full-scale recorder. Similar considerations apply when considering the output of digital-to-analog converters.

Because computer manufacturers have designed their analog-to-digital and digital-to-analog converters to meet the average needs of laboratory instrumentation, interfacing is usually a relatively simple operation. However, one is seldom so fortunate that near-optimum interfacing can be achieved simply by a direct connection of appropriate instrument and computer terminals. Usually some additional steps are necessary. One may contract to have the computer vendor implement these extra steps (an expensive convenience) or one may adopt a do-it-yourself approach. The fast majority of interfacing operations fall in one of the following categories: (a) amplification; (b) inversion of signal polarity; (c) impedance matching; (d) filtering; (e) conversion of a floating signal to one referenced to ground; (f) synchronization of events in computer and polarograph. All these steps are readily achieved with the aid of the conventional operational amplifier circuits [28-30] illustrated in Figs. 7-12. All operational amplifier circuits illustrated are characterized by a very low output impedance so that when connected to the input of a digital-to-analog converter, impedance matching problems are negligible. A similar situation exists when operational amplifier circuits are employed to condition the output of a digital-to-analog

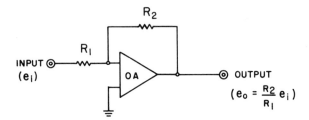

Fig. 7. Inverting amplifier. OA = operational amplifier.

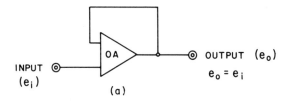

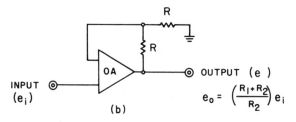

Fig. 8. Voltage-follower amplifiers: OA = operational amplifier. (a) Unity-gain follower. (b) Follower with gain.

converter, as all circuits shown either have or can be designed with an
adequately large input impedance.

The inverting amplifier (Fig. 7) may be used to invert signal polarity
and to modify the magnitude of a voltage signal. Figure 8 depicts the unity-
gain follower and the follower with gain. The former is an impedance
matching device which is used primarily to convert a high-impedance signal
source to one of low impedance. With appropriate operational amplifiers
(e.g., field-effect transistor input types), the follower will accommodate
signal sources with impedances characteristic of photomultiplier tubes or
glass electrodes, while its output can drive a load requiring 10-100 mA
(even larger currents can be furnished with the aid of a current booster
[28-30]). Few, if any, commercial or home-made ac polarographs produce
output signals with impedances sufficiently large to necessitate the use of a
unity-gain follower. However, this device would be useful if one wished to
interface a precise impedance bridge to a computer (see discussion of subtractor circuit below). In addition, the unity-gain follower would be an ideal
intermediary between a low-power logic circuit in a computer and a low-
impedance load such as the relay coils in the mercury drop-life controller.
The follower with gain (Fig. 8b) is a means whereby one may achieve
amplification (gain $\geq$ 1) without inversion of signal polarity as well as achieve
the advantages in impedance matching which characterize the unity-gain
follower.

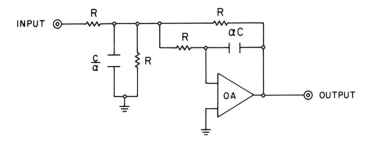

Fig. 9. Low-pass filter with second-order Butterworth response. $\alpha = 0.353$;
$f_O$ = cut-off frequency = $(2\pi RC)^{-1}$; OA = operational amplifier.

The circuit in Fig. 9 is a low-pass filter with a second-order Butterworth response. This response characteristic provides a sharper rolloff in
the gain-versus-frequency profile and a flatter response on the low-frequency
side of the rolloff than the commonly used single time-constant RC filter.
The Butterworth filter is quite effective when one wishes to eliminate high-
frequency noise from low-frequency or dc signals. In ac polarography the
Butterworth filter is profitably utilized to filter the ac "ripple" from the
outputs of full-wave and synchronous rectifiers in the current signal-

conditioning network (see Figs. 5 and 6). With proper choice of the cutoff frequency, the resulting dc signal is a precise representation of the instantaneous ac amplitude—i.e. the ac amplitude-versus-time response is not distorted by the action of the low-pass filter, even with a DME. Of course, the filter also effectively eliminates random high-frequency noise which can be troublesome when high-speed analog-to-digital converters are employed, unless one can perform ensemble averaging.

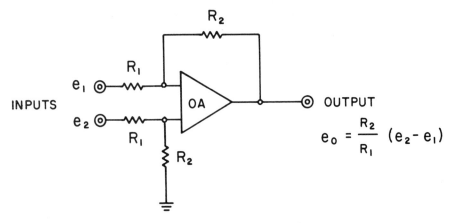

Fig. 10. Subtractor. OA = operational amplifier.

In some cases analog-to-digital converter inputs are single-ended rather than differential so that the input voltage must be referenced to ground. If the signal available from a polarograph is floating, such as when a bridge circuit is involved in the output [31], this output must be converted to one which is referenced to ground. The subtractor circuit in Fig. 10 is the standard means to this end. In addition to performing the function in question; the subtractor may be designed with substantial gain. If the signal source impedance is not too high (e.g., $\leq 1$ M$\Omega$, approx), the subtraction can also aid in impedance matching (by using sufficiently large resistors in the subtracter inputs). Alternatively, follower amplifiers may be interposed between the signal source and the subtractor inputs.

Most ac polarographic measurements are performed with a dropping-mercury electrode. As was already mentioned in the discussion of the computerized ac polarograph (Section III), the ac-versus-time oscillations attending this electrode necessitate appropriate procedures to synchronize the drop oscillations with the computer data-acquisition cycle. The steps to achieve this end should be viewed as the most crucial and complicated of the operations associated with interfacing (often interfacing involves only these steps—see next paragraph). The approach employed routinely in this writer's laboratory has been to control the mercury drop life mechanically and to utilize sample-and-hold circuits which "sample" the various signals of interest a few msec before the drop is dislodged and "hold" the sampled signal levels while the computer performs the analog-to-digital conversion on the sample-and-hold circuit outputs [23, 32]. Because our computer does not include a real-time clock, an analog timing circuit based on operational amplifiers has been employed to control these operations. Although quite adequate, the circuit diagram will not be given here as we prefer to recommend the use of a computerized real-time clock which is more precise and versatile. However, we do recommend the sample-and-hold circuit shown in Fig. 11. Although not as fast as the typical offering of the computer manufacturer, it is more than adequate in these regards for conventional ac polarography and it features much lower cost, variable gain, and often a lower dc offset drift.

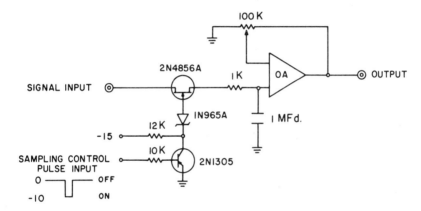

Fig. 11. Sample-and-hold circuit. OA ; operational amplifier.

Normally, the utilization of an analog-to-digital converter which accepts signals of only one polarity is no problem in ac polarography where output signal polarities are usually fixed. At most, one may have to employ an inverting amplifier (Fig. 7) to make the polarograph output compatible with the analog-to-digital converter. However, it is possible in certain types of investigations to encounter a situation where the ac polarographic signal changes polarity in the course of a dc-potential scan. This can occur only in phase-selective detection of ac polarographic currents. In the case of fundamental harmonic currents such behavior is observed with certain systems that exhibit negative impedance characteristics [33-35], although such encounters are rare. However, in phase-selective second-harmonic (or higher-harmonic) measurements, changes in signal polarity as one scans the dc potential are the rule rather than the exception [7, 19]. Aside from purchasing a more versatile analog-to-digital converter, one may solve the

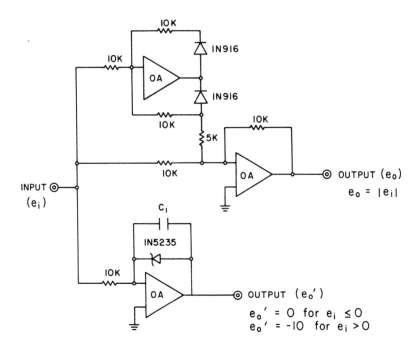

Fig. 12.  Absolute-value circuit with zero-crossing detector.

interfacing problem in such cases with the unit shown in Fig. 12, which is comprised of an absolute-value circuit (full-wave rectifier) and a zero-crossing detector. This circuit operates on the output of the signal-conditioning network shown in Fig. 6.  The absolute-value circuit will provide an output whose polarity is unidirectional and consistent with the require-ments of the analog-to-digital converter, while the output voltage magnitude equals that of the input signal. Information regarding the sign of the original signal is provided by the output state of the zero-crossing detector which is monitored by one of the multiplexer channels. An alternative to this technique would be to simply add a dc bias to the phase-detector output so that the resulting signal never changes polarity. The dc bias could be subtracted out in the data-analysis program. However, such a technique reduces by a factor of two the dynamic range and resolution of one's instrument, properties which are preserved by the circuitry in Fig. 12.

For most ac polarographs, some combination of the foregoing circuits should be sufficient to interface the polarograph to the computer. It should be evident from the foregoing discussion that the fortunate individual is one who possesses a typical operational amplifier ac polarograph with op-amp signal-conditioning circuits such as are shown in Figs. 5 and 6. Such instrumentation already provides the appropriate impedance matching, signal levels, and polarities. At most, one is faced with the construction of the components discussed earlier for synchronizing data acquisition with the DME along with the construction of the circuit in Fig. 12, although for most studies the latter will be unnecessary. If in addition to an op-amp polarograph the instrument capabilities already include mechanical drop-life control, the interfacing task may amount to installing a few wires between the computer and the polarograph. Incidentally, the discussion in this section should reveal the basis for statements made at the outset regarding the favorable interaction between operational-amplifier instrumentation and the computer.

## V. SOME BASIC EXPERIMENTAL ROUTINES WITH

## THE COMPUTERIZED AC POLAROGRAPH

We have considered the basic characteristics of the ac polarographic experiment, the small laboratory computer, and the combined computer-polarograph system, together with how the "marriage" is achieved. With this background it is now feasible to examine some examples of applications of the computerized ac polarograph. In this section we will consider, as examples, two computerized experiments, one whose objective is chemical analysis and one whose goal is electrochemical kinetic measurements. Both cases basically involve applications of the classical ac polarographic experiment, combined with a number of computer-enabled modifications which aid the operator and/or enhance data fidelity. These modifications do not represent fundamental departures from the basic experimental routine, but amount to nothing more than operations or judgments performed by the computer which were previously a task for the investigator. Experimental routines which fundamentally alter the nature and philosophy of the ac polarographic experiment are considered in the next section.

### A. A Routine for Chemical Analysis by ac Polarography

Figure 13 presents a flow diagram of a typical routine for computerized quantitative analysis by ac polarography. The program assumes that the measurement will be carried out with a dropping-mercury working electrode, that the potentiostat employed utilizes positive feedback for compensation of ohmic potential drop, and that the ac (fundamental or higher harmonics) is measured by phase-sensitive detection. Under such conditions nonfaradaic effects should be small or negligible for sufficient electroactive component concentrations ($>10^{-6}$ $\underline{M}$). Thus, the background signal correction included in the program is a simple scalar subtraction of this background as assessed by the computer at potentials where faradaic currents are negligible. This background may arise from small uncompensated charging-current contributions, amplifier offsets, etc.    Other features incorporated in the

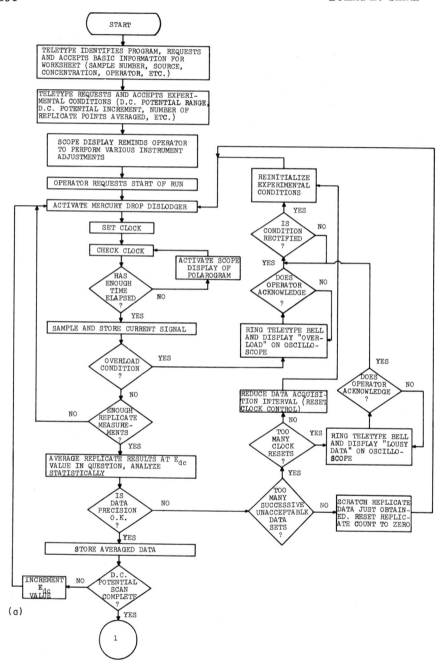

Fig. 13. Flow diagram for analysis by computerized ac polarography. (a) Experiment-initialization and data-acquisition stages. (b) Data-analysis stage.

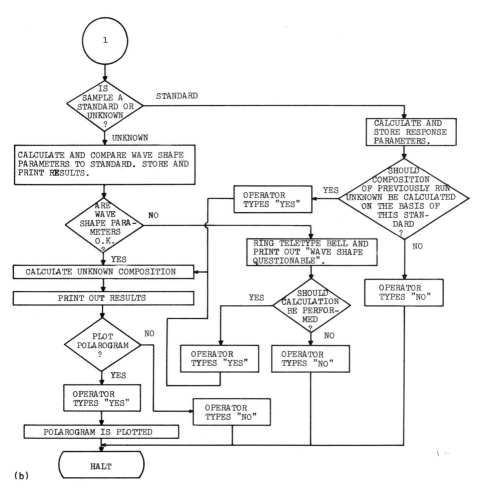

Fig. 13 (cont.)

experimental procedure include: (a) computer control of the dc potential
scan via the digital-to-analog converter output; (b) computer timing of data
acquisition and the mercury drop life; (c) real-time analog readout of the
polarograms on an oscilloscope display; (d) ensemble averaging of replicate
measurements at a particular dc potential; (e) several computerized tests to
sense and possibly correct polarograph malfunctions, along with an alarm
activation to alert the operator of problems; (f) calculation of the quantity of
unknowns in the desired units (e. g., % composition, parts per million, etc.)
by comparison with data from a standard; (g) the option to obtain a permanent analog plot of the polarogram.

The first three blocks in the flow diagram of Fig. 13 refer to preliminary operations which involve a "conversation" between the computer and operator with the aid of the teletype and scope display. Upon starting the program, the teletype first identifies the program to ensure that the correct one has been loaded. This is followed by a series of inquiries (teletype) and answers (operator) which amount to generating a worksheet for the sample in question. Simple data such as the date, operator's name, sample number, sample type (standard or unknown), sample origin manufacturer, type of product), solvent, supporting electrolyte, reference electrode, etc., may be requested and supplied. This is followed by inquiries and answers related to the experimental conditions to be employed in the experimental run—e. g., standard concentration, dc potential range to be scanned, dc potential increments to be employed in the "scan" (see below), time in drop life at which measurement is made, how many replicate measurements are to be averaged at each dc potential, etc. After the desired information has been typed the computer reminds the operator to perform certain critical adjustments (e. g., set the initial voltage). These messages are displayed on the oscilloscope. This could be done by teletype, but since the earlier teletyped information is relevant to the analyst's worksheet while the "reminders" are not, the scope display is preferred at this point.

Once all systems are adjusted the operator initiates the experimental run by pressing the appropriate key or keys on the teletype (e. g., type "go"). The exact operation required at this point is included with the reminders on the scope display. The computer initiates the experiment by mechanically dislodging the mercury drop and setting the real-time clock counting register to the beginning of its counting cycle (set clock). These two operations immediately bring the mercury drop and computer data-acquisition cycles into synchronization. The computer then "watches" the clock (check clock) until the time has been reached for data acquisition. During the actual experimental run, the vast majority of time (easily 99.9%) is devoted to waiting or "clock watching" between data-acquisition and analysis cycles. Although more complex, the data acquisition, analysis, and testing operations normally utilize only a few milliseconds. This waiting period is profitably utilized by periodically carrying out the program for real-time oscilloscope display of the polarogram. The flow diagram implies alternate clock checks and scope display renewals which are repeated with sufficient frequency that error in timing data acquisition is negligible and a flicker-free display is obtained, even with a nonstorage oscilloscope. (A brief flicker may be observed during the data-acquisition and analysis cycles.) When the clock signifies that the time for data acquisition has been reached, the current signal from the phase-sensitive detector is sampled, converted to a digital signal, and stored in memory. The data point is then tested for overload. This implies that signal levels are excessive, either from the standpoint of polarograph or analog-to-digital converter limitations. An overload condition may result from improper adjustment of instrument sensitivity or from instrument malfunction, such as potentiostat oscillation. Detection of overload results in repeated ringing of the teletype bell (the "alarm") to alert the operator and a display of an overload message on the oscilloscope. (A more sophisticated system could incorporate an attempt at automatic adjustment of polarograph gain and/or potentiostat stabilizing networks by the computer via computer-controlled switching networks.) Once the overload condition is corrected the operator can restart the data-acquisition routine.

The lack of an overload condition allows the program to proceed to a sequence of data tests which determine whether the desired number of replicate measurements have been obtained at a particular dc potential and whether the data ensemble is statistically acceptable. Such operations are most helpful in ensuring a noise-free polarogram when using a dropping-mercury electrode and in determining the existence of a damaged capillary. Even a well-conditioned DME exhibits an occasional erratic drop—i.e., a drop which dislodges prematurely due to vibration, etc. —particularly with certain nonaqueous solvents. In a more serious situation one may have a mistreated capillary which abruptly begins to behave very erratically. Finally, with mechanically controlled drop lives it is possible that improper choice of the data acquisition timing will lead to a condition where the drop life one is trying to enforce exceeds the natural drop life. Owing to the dc-potential dependence of the natural drop life, this may occur only over part of the dc potentials of interest, perhaps ensuing after the polarogram is partially run. In any event, the drop will dislodge before one desires and loss of synchronization of the DME and data-acquisition cycles is observed. All these sources of erratic behavior will be detected by the computer and appropriate action taken. The first of the data tests determines whether sufficient replicate measurements have been obtained at the dc potential in question. If not, the measurement operation is repeated until the desired data ensemble is acquired. The replicate data points are then subjected to a simple statistical analysis which determines the mean value and standard deviation and judges whether the standard deviation exceeds acceptable limits or whether one or more of the data points is statistically excludable ("bad point"). If data precision is judged unacceptable, the data collection operation is repeated at the same dc potential until a satisfactory set of replicate data is obtained or until the computer concludes that too many successive unacceptable data sets have been encountered. In the latter event the computer automatically readjusts (shortens) the data acquisition timing in the hope that the erratic data are caused by improper timing as described above. All experimental conditions are reinitialized, the new timing condition is printed out, and the polarogram is restarted. If this attempt is unsuccessful, one or more additional timing readjustments may be effected until good data are realized or the number of timing readjustments reaches a preprogrammed limiting value at which point the alarm is sounded and a "lousy data" message is flashed on the oscilloscope display. Once the problem is identified and corrected (probably by installation of a new capillary), the run can be restarted.

When an acceptable data set is obtained (this should be the prevalent occurrence), the average value is stored and, unless the scan is complete, the dc potential is incremented by stepping the output of the digital-to-analog converter and the data-acquisition operation is repeated at the new dc potential. (The fact that one "scans" the voltage range of interest by occasionally stepping the potential at the start of a new drop actually fits the constant-dc-potential assumption of ac polarographic theory [7] better than the conventional ramp voltage.)

When the potential range of interest has been scanned, the program enters the final data-analysis and calculation phase. The nature of the subsequent steps depends on whether the sample in question was an unknown or standard. In the usual case where an unknown is involved, the wave shape of the polarogram obtained is compared to corresponding characteristics of a previously run standard. Except in unusual cases (e. g., systems with significant adsorption or with coupled second-order chemical reactions) wave shape is a concentration-independent observable which can be used as a final test to evaluate the fidelity of the data just obtained. This test can ascertain more subtle problems with the polarographic run than the overload and erratic data tests. An improperly prepared solution or an unexpected excipient in the sample may introduce an interfering polarographic wave or may change the characteristics of the electrode reaction on which the analysis is based. Either situation might be detected by such a comparison of unknown and standard wave shape, a procedure which is seldom employed in analytical laboratories (except where disparities are qualitatively obvious) because of the time investment required. The computer can effect this comparison in an acceptably short period of time, regardless of the mathematical sophistication of the wave shape characterization. The detection of a wave shape which deviates significantly from the standard causes the teletype to print out an appropriate message which includes wave shape parameters (e. g., various moments of the peak, wave widths at various fractions of peak height, etc.). The operator, on the basis of his judgment, may instruct the computer to either abort or proceed with the calculation. In the latter event, or if the wave shape test was satisfactory, the unknown composition is calculated on the basis of results from the previously run standard by comparing peak current magnitudes. The operator may then request a plot of the polarographic data before the program is terminated. In the cases where the original sample was a standard, the calculation phase of the program first calculates and stores appropriate response parameters related to wave magnitude and shape. The operator is then questioned regarding whether the data just obtained should be used as a basis to perform calculations on previously run data for an unknown. A "yes" initiates the usual unknown calculation routine. A "no" terminates the operation, unless the operator requests a plot of the standard's polarogram.

The foregoing description of an experimental routine for computerized chemical analysis by ac polarography should be viewed as just one example of many operation sequences which can be implemented readily with a typical on-line laboratory computer system. One can expand or simplify the experimental routine in question, according to one's needs and imagination.

### B. A Routine for Electrochemical Kinetic Investigations

### by ac Polarography

A flow diagram of a program for computerization of a common type of electrochemical kinetic study by ac polarography is given in Fig. 14. The program is designed for use whenever the primary objective is the determination of the electrochemical rate parameters, $k_s$ and $\alpha$ [7-9], and when there is reason to expect that the electrode reaction in question corresponds to the simple quasi-reversible case [7]. Although it tests whether the observed response is consistent with the quasi-reversible mechanism, the program has no options for quantitative calculations when a more complicated mechanism is detected. It is assumed that simultaneous input signals provide the in-phase and quadrature components of the ac obtained at a particular point in the mercury drop life, and that such data are obtained from solutions prepared with and without the electroactive component of interest. These data on the total and background currents are then analyzed to mathematically compensate for nonfaradaic effects on the basis of the assumption of a priori separability of the faradaic and double-layer charging currents [26, 27]. Calculated faradaic response parameters are displayed in real time and are used for mechanistic verification and calculation of $k_s$ and $\alpha$ when the polarographic run is completed. It should be noted that in our laboratory we routinely utilize automatic analog compensation for nonfaradaic effects [23-25], which frees the computer from the task of mathematical compensation. However, aside from some extra time devoted to programming and a somewhat longer experiment time, the more classical approach invoked in this example may be preferred by many, and can hardly be viewed as inconvenient. Other operations implemented by this program include: (a) computerized elimination of the depletion effect; (b) detailed implementation of operator judgment in the final phase of data analysis; (c) computerized switching of the ac potential frequency; (d) a number of operations which are identical or similar to those used in the previously discussed program for analytical work—e. g., several automatic data quality tests.

As in the previous example, the first operations performed upon starting the program involve a "conversation" in which the computer requests and accepts via teletype certain essential information. Various experimental conditions are input which include, in addition to those mentioned in the flow diagram, all other parameters which are necessary to perform the theoretical calculations associated with data analysis (e. g., capillary characteristics,

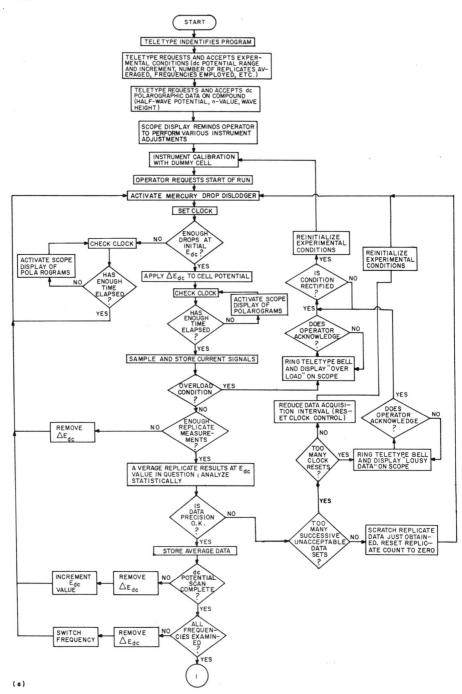

Fig. 14. Flow diagram for electrochemical kinetic measurements by computerized ac polarography. (a) Experiment-initialization and data-acquisition stages. (b) Data-analysis stage.

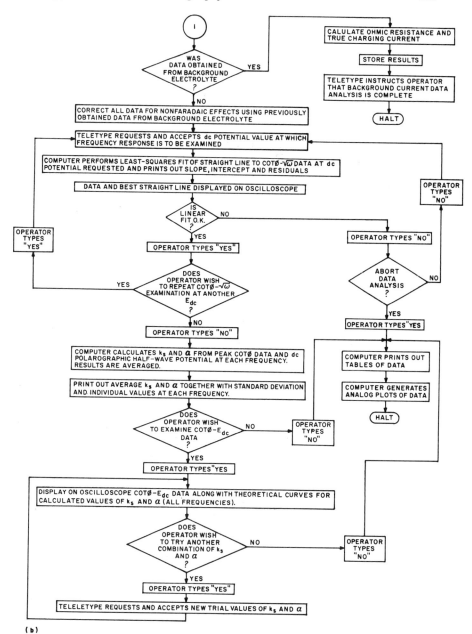

(b)

Fig. 14 (cont.)

temperature, etc.). The operator also is requested to input certain data obtained from a dc polarographic experiment which are necessary for the data-analysis operation. Appropriate reminders for the operator are then flashed on the oscilloscope display, followed by a calibration step. The latter involves inserting one or more precision dummy cells in place of the polarographic cell. The computer reads the dummy-cell current signals, requests the dummy-cell impedance characteristics (teletype), and uses these data to calculate the scaling factor(s) relating the true cell ac to the magnitude of the resulting digital signal. This calibration enables the computer to determine and readout absolute current magnitudes. This is essential in most kinetic studies, in contrast to the previous case (Fig. 13) where the use of relative signal levels is sufficient.

When the calibration step is completed, data acquisition is initiated by operator command (teletype). After dislodging the drop and setting the clock, the program tests to determine whether sufficient drops have fallen at the "initial dc potential value." This operation is part of a routine designed to eliminate any errors associated with the depletion effect [36]. The term "depletion effect" refers to the fact that electrolysis at a particular mercury drop slightly influences the initial surface concentrations "seen" by the subsequent drop—i.e., solution stirring when a drop falls is not 100% efficient in restoring the bulk concentration values at the capillary orifice. Because of mathematical difficulties, polarographic theory usually ignores this phenomenon, thus introducing a disparity between actual and assumed experimental conditions. Although relatively small in ac polarography, the depletion effect is readily detected [37] and can be troublesome in certain situations such as when highly viscous solvents are employed [38]. So-called "first-drop" measurement [36, 39] is one of the approaches utilized to eliminate the effect. The technique involves simply allowing several drops to fall at a potential where no electrolysis occurs, after which the potential is stepped to the value of interest and the polarographic measurement is performed on a single drop (the "first drop"). The entire operation is repeated to perform subsequent measurements. The growth and fall of several mercury drops at "inert" potentials adequately stirs the solution so that the desired, well-defined initial conditions apply for the drop utilized in the measurement. The relatively rare accounts of implementation of "first-drop polarography" describe manual operations. Automation of first-drop polarography requires an unusual applied potential function involving a dc-biased voltage impulse train of slowly increasing amplitude with impulse widths equal to and synchronous with the mercury drop life and a repetition period corresponding to an integral number of drop lives ($\geq 2$). Although analog generation of such a function is probably not impossible, it is inconvenient, as are the special timing operations required for automatic recording. However, the availability of an on-line computer makes the automation of this operation relatively trivial, requiring little more than a few additional instructions in the program. The first step in the relevant operation sequence is the previously mentioned test to determine whether sufficient drops (usually two or three) have fallen at an "initial potential" where no electrolysis occurs. A negative conclusion causes the computer to hold the initial potential for the normal period assigned to the drop life before dislodging the drop. Meanwhile a real-time oscilloscope display of data already obtained is maintained. This operation is repeated until sufficient drops have dislodged. At this point the computer adds to the initial potential an impulse of amplitude $\Delta E_{dc}$ which drives the dc potential applied to the newly formed drop into the region of interest for the duration of the drop life. At the appropriate point in the drop life, data points are obtained for the in-phase and quadrature current signals and subjected to the overload test. After data acquisition the $\Delta E_{dc}$ function is removed and several drops again are permitted to form at the initial potantial before the next measurement. Thus, the operations necessary to obtain measurements only on "first-drops" are effected routinely by the standard on-line computer system. Although the purchase of an on-line computer for ac polarography is not justified on the basis of its ability to automate "first-drop" measurements, it should be recognized that along with its more obvious and important functions, the computer affords innumerable minor conveniences such as this one.

As in the previous example, measurements are repeated at a particular dc potential (particular $\Delta E_{dc}$) until the desired number of replicate points are obtained. The data ensemble is averaged and tested for unacceptable scatter, etc. Too much scatter leads to the same operation sequence as in Fig. 13, except that direct operator intervention leads to a recalibration step to accommodate any instrument readjustments. Adequate data precision leads to storage of the average in-phase and quadrature currents which are utilized in the data analysis and in the real-time oscilloscope display of the polarograms. The value of $\Delta E_{dc}$ then is incremented so that subsequent measurements are performed at a new dc potential value. The $\Delta E_{dc}$ is incremented after each acceptable data ensemble is acquired, thus scanning the dc potential range of interest. When the scan is complete the applied frequency is switched automatically to a new value and the entire data acquisition operation is repeated. Eventually data are obtained at all frequencies of interest and the program proceeds to the data-analysis phase. One should recognize that modern lock-in amplifiers (phase-sensitive detectors) in the signal-conditioning path provide a highly selective (tuned) frequency response, yet accommodate automatic frequency switching without necessitating instrument recalibration.

The data-analysis section of the program for kinetic investigations is characterized by considerably more reliance on operator judgment than in the previous example. This is done primarily to demonstrate the capability for computer-operator interaction in the data-analysis phase. It should not be taken as an implication that totally automated data analysis is impossible in the more complex situation encountered in kinetic studies. It is possible to develop a program in which data analysis is under sole control of the computer, whereby the operator simply waits until the computer prints out the mechanism and rate constants or acknowledges failure to fit the data to the rate law(s) for the mechanism(s) considered. One can generally expect that the less the operator is involved in data analysis, the more complex and time-consuming will be the programming. The optimum program for any laboratory will depend to a large extent on the training of the personnel. If the researchers are intimately acquainted with ac polarographic theory, the types of responses to expect with various mechanisms, etc., a totally automated data analysis routine would be a waste of talent, not to mention some useful data which the computer might not treat properly because it was not programmed for all eventualities. It is usually best to combine in some manner the computer's numerical manipulation capabilities with human insight and instinct whenever the data-analysis operation is nontrivial. In effect we are espousing the same philosophy used by NASA when attempting to justify manned space flight over the unmanned option.

The path followed by the data-acquisition routine in Fig. 14 depends on whether the data were obtained on the background electrolyte (which should be run first) or on a solution containing the electroactive component of interest. In the former case, a simple routine is followed which involves calculation of the ohmic resistance (preferably from the high-frequency data) and the double-layer charging currents, and storage of these data. If the solution contains the electroactive compound, the first data-analysis step is the compensation of all data points for the nonfaradaic effects and storage of the resulting faradaic response data. At this point one begins a comparison of the faradaic data to the theory for a simple quasi-reversible mechanism. The analysis is based solely on the phase-angle response which has numerous advantages in simplicity and convenience over a corresponding operation involving current amplitudes [7]. The first step is based on the fact that with a quasi-reversible system, a linear response is expected in the $(\cot \varphi - \omega^{1/2})$ profile which follows the relationships [7-9]:

$$\cot \varphi = 1 + \frac{\sqrt{2\omega}}{\lambda} \tag{9}$$

where

$$\lambda = \frac{k_s}{D^{1/2}} (e^{-\alpha j} + e^{\beta j}) \tag{10}$$

$$\beta = 1 - \alpha \tag{11}$$

$$j = \frac{nF}{RT}(E_{dc} - E_{1/2}^r) \tag{12}$$

$$E_{1/2}^r = E^0 - \frac{RT}{nF} \ln\left(\frac{f_R}{f_0}\right)\left(\frac{D_0}{D_R}\right)^{1/2} \tag{13}$$

$$D = D_0^{\beta} D_R^{\alpha}. \qquad (14)$$

Notation definitions employed in Eqs. (9)-(14) are given in Table 3.

TABLE 3

Notation Definitions Employed in Eqs. (9)-(16)

| Symbol | Definition |
|--------|------------|
| $\varphi$ | Phase angle of fundamental harmonic current relative to applied alternating potential |
| $\omega$ | Angular frequency of applied alternating potential |
| $k_s$ | Heterogeneous charge-transfer rate constant at $E^{o}$ |
| $\alpha$ | Charge-transfer coefficient |
| n | Number of electrons transferred in heterogeneous charge-transfer step |
| F | Faraday's constant |
| R | Ideal gas constant |
| T | Absolute temperature |
| $E_{dc}$ | Dc component of applied potential |
| $E^{o}$ | Standard redox potential in European convention |
| $f_i$ | Activity coefficient of species i |
| $D_i$ | Diffusion coefficient of species i |

These relationships indicate that a linear ($\cot \varphi - \omega^{1/2}$) plot of unity intercept should be observed at all dc potentials where the faradaic current is measurable. Only the slope varies with changing $E_{dc}$. Deviations from this behavior, such as a nonlinear response, suggest a more complicated electrode reaction mechanism. To test whether the desired quasi-reversible mechanism is operative the operator responds to the teletype request for an $E_{dc}$ value at which the ($\cot \varphi - \omega^{1/2}$) profile is to be examined. The computer then effects a least squares fit of a straight line to the data, followed by a printout and display of the relevant results. From the display of actual data points and the best straight line fit, the operator must determine whether or not the fit is satisfactory. A negative operator response (e. g., because the data indicate a curved ($\cot \varphi - \omega^{1/2}$) plot so that a straight line fit is meaningless) gives the operator the choice of stopping the data-analysis routine or repeating the straight line fit at another dc potential. An aborted data analysis leads to a digital and analog "data dump" routine whereby faradaic current data are printed and plotted to enable subsequent study of the results. Whether or not the linear fit is satisfactory, the operator has the option of repeating the operation at another $E_{dc}$ value. If after examining the $(\cot \varphi - \omega^{1/2})$ response at various $E_{dc}$ values the operator concludes that the quasi-reversible mechanism is operative, the computer can be directed to calculate $k_s$ and $\alpha$. For the sake of illustration, it is assumed that $k_s$ and $\alpha$ are calculated from peak $\cot \Phi$ data at each frequency. The results are averaged and the average values printed out along with the associated standard deviations. This calculation is based on the fact that with the quasi-reversible mechanism, the peak $\cot \varphi$ magnitude and position ($E_{dc}$ value at peak) follow the relationships [7]

$$[\cot \varphi]_{peak} = 1 + \frac{(2\omega D)^{1/2}}{k_s [(\alpha/\beta)^{-\alpha} + (\alpha/\beta)^{\beta}]} \tag{15}$$

$$\left[E_{dc}\right]_{peak\,\cot\varphi} = E_{1/2}^r + \frac{RT}{nF}\, \ln\left(\frac{\alpha}{\beta}\right), \tag{16}$$

which allow convenient calculation of $k_s$ and $\alpha$, provided $E_{1/2}^r$ is known ($E_{1/2}^r$ is given by the dc polarographic half-wave potential unless $k_s$ is very small). If, upon obtaining the printout of $k_s$ and $\alpha$, the operator is quite confident that all is well, he can direct the computer to terminate data analysis, which leads to the data dump routines. However, if the operator wishes to double-check or to attempt further refining of the calculated rate parameters, the program will provide on request an oscilloscope display of all data along with the theoretical response for the calculated $k_s$ and $\alpha$. The presentation is given in the format of $(\cot \varphi - E_{dc})$ plots. Figure 15 gives a typical example of such data when a reasonably satisfactory fit is obtained. Figure 15 may be taken as suitably representing the appearance of the display in question. On inspecting this display, the operator may be satisfied and terminate the run, or he may request a new display using new values of $k_s$ and $\alpha$ for the theoretical curves. This may be repeated until the operator finds the rate parameters yielding the most satisfactory theory-experiment fit. There are numerous reasons why this latter option should be invoked in most cases. While consistent with mathematic simplicity, the use of peak $\cot \varphi$ data to calculate $k_s$ and $\alpha$ has disadvantages such as: (a) most of the acquired data are ignored; (b) some of the $\cot \varphi$ peak potentials may be difficult to locate, such as when $[\cot \varphi]_{peak}$ is very small (near unity) or very large and $\alpha$ values calculated under such conditions may erroneously bias the average value. These and other factors suggest that in all but ideal cases the initially calculated $k_s$ and $\alpha$ should be viewed as first approximations. The process of examining the $(\cot \varphi - E_{dc})$ display and observing the theory-experiment fit obtained for various trial $k_s$ and $\alpha$ combinations allows the operator to make use of all data obtained and to invoke his judgment

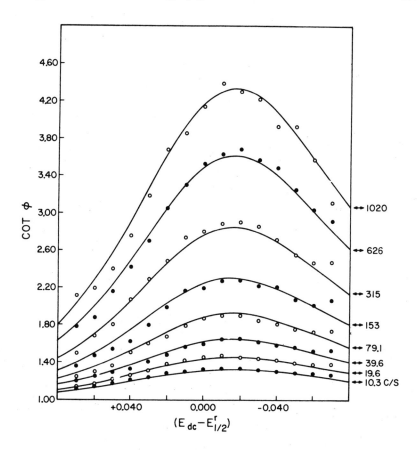

Fig. 15. Typical cot $\varphi$ - $E_{dc}$ profile with quasi-reversible system. System: 3.36 mM TiCl$_4$ in 0.022 M$_2$H$_2$C$_2$O$_4$, T = 25 $^\circ$C. Applied: Sinusoidal potential of 10 mV peak-to-peak amplitude, frequencies given on figure. Measured: Cot $\varphi$ vs. $E_{dc}$, solid lines represent theory for $k_s = 4.6 \times 10^{-2}$, cm sec$^{-1}$, $\alpha = 0.35$, and D = 0.66 $\times$ 10$^{-5}$ cm$^2$ sec$^{-1}$, points are experimental data. (Permission to reprint was granted by the American Chemical Society.)

regarding which data should be given greatest weight in the rate parameter calculation, etc. One can even cope with the existence of moderate perturbations on the desired quasi-reversible mechanism with such a procedure, e.g., if a coupled chemical reaction is affecting the response, the calculation of $k_s$ and $\alpha$ may be based on high-frequency cot $\varphi$ data where the chemical kinetic effects are small [7].

The program represented by Fig. 14 is one of many examples which can be advanced for purposes of kinetic studies by ac polarography. Despite its lack of generality, we feel that the program illustrates many of the capabilities of an on-line computer system which are relevant to kinetic studies.

## VI. COMPUTER-ENABLED INNOVATIONS IN AC POLAROGRAPHY

The previous section dwelled on more obvious examples of on-line computer applications in ac polarography. The computer was used as a means to improve efficiency without significantly altering the basic experiment—i.e., the nature of the input perturbations and the measured observables conformed to the "classical" procedure. However, this writer feels that the on-line computer is likely to play a much more profound role than implied by these earlier examples, in the future development of the experimental technique in question. In this larger role the computer will be responsible for changes in basic experimental procedure and philosophy in much the same manner as it is already influencing infrared [40] and NMR [41] spectroscopy. In the present section we shall discuss briefly some techniques that fit this category. In each case a change is made in some basic aspect of the ac polarographic experiment to more effectively utilize one or more of the computer's special abilities and to substantially enhance the information flux.

### A. Multiplexing in ac Polarography

One of the important features of the computerized ac polarograph represented in Fig. 3 is the capability of handling, essentially simultaneously, numerous analog signal channels. This capability is practically wasted when one employs the classical approach to ac polarography—e.g., no more than two input channels are utilized in the operations discussed in the previous section. In this writer's laboratory a major portion of our effort in investigating on-line computer applications has been devoted to finding approaches whereby the ac polarographic experiment can be modified to better utilize the computer's multichannel data-acquisition capability. Two techniques have arisen from this work. Both involve the application of multiplexing principles to ac polarography.

Multiplexing in electronic operations generally refers to the transmission of more than one type of information (signal) over a single "line". This is followed by separation and analysis of the "mixed" pieces of information at the receiver. Separation is usually based on the time and/or frequency characteristics of the information. In a certain sense, the first aspect of multiplexing (several information types passed over one "line") always exists when the classical ac polarographic experiment is performed, but the second step (analysis of more than one information type) is not attempted. That is, the cell current signal under ac polarographic conditions contains information related to the dc response of the faradaic impedance (dc polarographic current), to the linear aspects of the faradaic impedance's ac response (fundamental harmonic current), and to the nonlinear characteristics of the faradaic impedance (higher harmonic currents). However, usually only one current component is monitored. Indeed, in the classical context the measurement of dc, fundamental harmonic, and second harmonic currents are each viewed as separate experiments and, in some laboratories, each measurement involves a separate instrument. Because the kinetic-mechanistic information provided by these various current harmonics is frequently complementary, significant stimulus exists for measuring each in fundamental electrochemical studies. Rather than approach these measurements as separate techniques, we are applying the computer's multichannel data acquisition capabilities to effect a form of multiplexing in which dc fundamental harmonic ac (in-phase and quadrature), and second harmonic ac (in-phase and quadrature) currents are measured simultaneously [42]. This operation simply requires several analog signal conditioning networks similar to those shown in Figs. 5 and 6 which operate on the cell current signal to deliver the five signal components to the computer's multiplexer. An instrument flow diagram is shown in Fig. 16. Although our work with this approach is in its infancy, tests to prove its feasibility are complete [42] and applications have begun. Because this is one of two modes of multiplexing in ac polarography which we have considered, it is referred to as "ac polarography in the harmonic multiplex mode" to avoid ambiguity.

Harmonic multiplexing represents a minor departure from the conventional ac polarographic experiment in that it involves no alteration of the input signal, but simply utilizes more of the information transmitted by the cell current signal than usual. However, a single experiment provides little detailed information regarding the frequency response characteristics of the faradaic impedance. As in conventional fundamental harmonic studies, repetitive experiments are required, each at a different frequency, until the response is obtained at sufficient "points" to define the frequency response profile. This time-consuming operation can be eliminated in principle by another form of multiplexing. The latter involves simultaneous application of multiple discrete frequency components to the cell—i.e., one uses an input signal of the form

$$E(t) = E_{dc} + \Delta E \sum_{n=1}^{N} \sin \omega_n t \tag{17}$$

where N = total number of discrete frequency components.

This is accompanied by measurement of the cell current signals generated by each applied frequency. We refer to this technique as "ac polarography in the frequency multiplex mode." In principle, one could envision the simultaneous measurement of the fundamental and second harmonic ac polarographic signals, as well as intermodulation components [7] associated

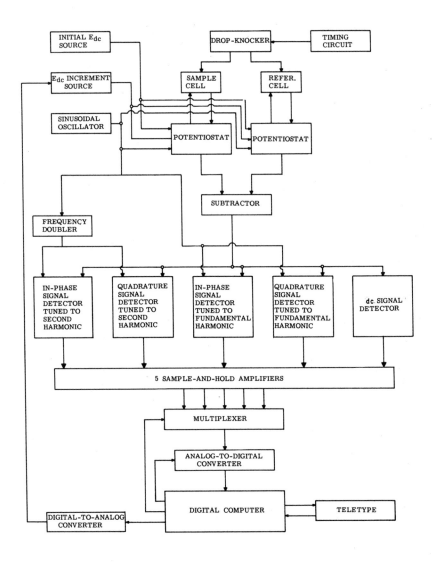

Fig. 16. Flow diagram of instrument for harmonic multiplexing in ac polarography.

with the various frequencies. However, the technical problems attending
separation and measurement of numerous second harmonic and intermodulation components from each other and from the larger fundamental harmonics
are serious with significant numbers of input frequencies. Consequently, for
purposes of the present discussion, "frequency multiplexing" will refer to
the simultaneous application of multiple frequencies to the polarographic cell
and the measurement of the fundamental harmonic response at each frequency.

Within the framework of the existing theoretical base of ac polarography, it is readily shown [43] that the fundamental harmonic current
components at each frequency are precisely the same as if the measurements
had been performed with only that particular frequency applied to the cell,
provided that one honors certain easily accommodated restrictions on the
amplitude of the applied alternating potential. Thus, conventional theoretical
interpretation may be applied to the data in frequency multiplexing. The only
other apparent limitation to this measurement concept is that the frequency
separation between the various components must be sufficient for their resolution by available tuned circuitry. Because the tuned amplifier-phase
detector signal conditioning approach offers very narrow bandpass, this also
represents a minor limitation. In principle, it is entirely possible to simultaneously measure the response at a sufficient number of frequencies to
define the fundamental harmonic frequency response throughout the range of
interest. This would enable one to complete an entire ac polarographic study,
including real-time data analysis by the computer, in the time required to
scan the dc potential range of interest—i.e., in the time required to run a dc
polarogram.

An instrument has been constructed in our laboratory to measure the
total and in-phase faradaic current components at four frequencies simultaneously [18, 43]. Although four discrete frequencies are not sufficient to
give the frequency response over the entire domain of interest (usually 8-12
frequency points are required), the use of this number of frequencies was
considered adequate to test the fidelity of the method. A flow diagram of the
instrument, which also features analog compensation of nonfaradaic effects
[23-25], is given in Fig. 17. Our experience with frequency multiplexing
has been gratifying. Tests with numerous systems have validated the concept
that data obtained with small-amplitude multiple-frequency inputs are identical to results obtained by the classical "one-frequency-at-a-time" approach.
Some typical experimental data are shown in Fig. 18 [43]. As a consequence
of these results, ac polarography in the frequency multiplex mode is now our
standard approach to ac polarographic measurements.

The most obvious advantage of either frequency or harmonic multiplexing is the grossly reduced time for data acquisition. Both techniques replace
redundancy in experimental measurement with redundancy in electronic circuitry to perform simultaneously many experiments in the classical sense.
Numerous distinct gains attend the multiplexing procedures other than just

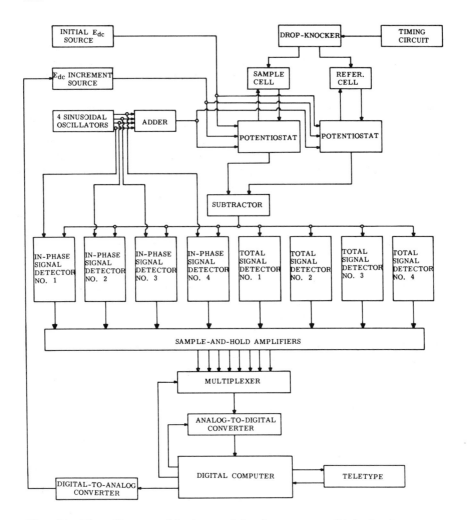

Fig. 17. Flow diagram of instrument for frequency multiplexing in ac polarography.

getting the experiment done in a hurry. Many advantages which lead to the expectation of enhanced measurement precision and, possibly, accuracy can be envisioned. For example, if an entire ac polarographic measurement operation can be finished in 15 minutes, rather than in a day or two, the requirements associated with solution stability are grossly reduced. Further, the significance of sources of response drift other than solution decomposition is also minimized. Among these are thermostat temperature drift, drift in electronic detectors and signal sources, slow leakage of $O_2$ into the cell,

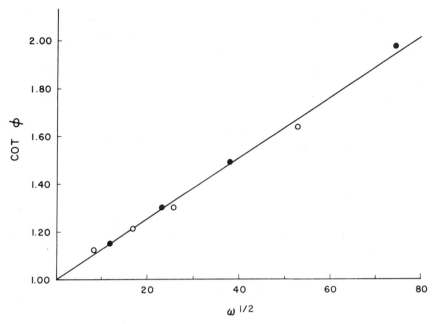

Fig. 18. Cot $\varphi$ - $\omega^{1/2}$ results obtained by ac polarography in the frequency multiplex mode. System: $2 \times 10^{-3}$ M $Cd^{+2}$ in 1.0 M $Na_2SO_4$ at 25 °C. Applied: Sinusoidal potentials of 10 mV peak-to-peak amplitude; four discrete frequency components applied simultaneously (figure shows results of two runs). Measured: Phase angles calculated from simultaneously measured total and in-phase ac components at each applied frequency; points represent experimental data. Solid line represents theoretical dependence for $k_s$ = 0.15 cm $sec^{-1}$, $\alpha$ = 0.31. (Previous results obtained in our laboratory by standard methods, $k_s$ = 0.15 cm $sec^{-1}$, $\alpha$ = 0.30.)

slow solvent evaporation owing to inadequate saturation of the $N_2$ train, and other annoyances which plague the polarographer from time to time. The fact that measurements at all frequencies are made simultaneously under identical conditions should reduce sources of data scatter which can hide features of mechanistic significance. The most evident disadvantage of multiplexing in ac polarography is the cost of the rather extensive analog circuitry required. At one time this problem precluded any attempt to develop multiplexing methods in our laboratory. However, low-cost operational amplifiers of reasonably high quality are now available which suppress the cost factor to an acceptable level. For example, the signal conditioning network shown in Fig. 6 may be constructed for less than $200. Thus, compared to the initial investment in an on-line computer, the additional investment required to implement frequency or harmonic multiplexing is relatively minor.

### B. The Use of Autocorrelation and Fourier Transform Techniques

Frequency and harmonic multiplexing in ac polarography adapt the experimental procedure to achieve efficiency by taking advantage of the enormous data acquisition capabilities of the on-line computer. Adaptations of the classical ac polarographic procedure also can achieve measurement efficiency by utilizing the computer's unique computational abilities. The example to be given emphasis here is just one of the existing possibilities.

An obvious extension of ac polarography in the frequency multiplex mode would be to employ an input signal characterized by a broad band spectrum such as "white" noise or some form of pseudorandom noise. The response of a polarographic cell to such inputs can be converted to the usual frequency response presentation by well-known mathematical operations. A crosscorrelation is performed [44, 45] of the input waveform (applied potential) and the cell output-current waveform. The result under appropriate conditions is the system's impulse response [44, 45, 47]. Fourier transformation of the latter yields the frequency response profile (amplitude and phase angles)—i. e., the conventional ac polarographic response is obtained. This is essentially identical to the operation performed in Fourier transform infrared spectroscopy [40] where an optical interferometer performs the crosscorrelation, while an on-line computer effects the Fourier transformation. In the context of ac polarography, such an operation requires a data acquisition system with the ability to sample simultaneously the input and output signals and repeat the operation at a rate at least four times as fast (3-dB error point) as the highest frequency of interest [45]. A reasonably fast on-line computer system obviously is required for the crosscorrelation and Fourier transform calculations. Fortunately, the state of the art is consistent with these requirements. Digital data acquisition at 50 kHz rates is not uncommon with relatively long word lengths (12-15 bit). Shorter word lengths accommodate MHz rates [45]. The calculations involved in Fourier transformation or crosscorrelation can be achieved in approximately 1 sec per 1000-point data set using software approaches based on special algorithms such as the Cooley-Tukey fast Fourier transform method [48]. More significant is the fact that hard-wired processors are now available [49] which perform such calculations approximately 20 times as fast (in 25-40 msec per 1000 points). Thus, with existing hardware one can perform an experiment with a white noise input to a polarographic cell based on the DME in which only a few tenths of a second would be required for data acquisition and analysis. Real-time readout of data in the conventional ac polarographic format could be achieved via a display oscilloscope. Adequate time would exist to store the frequency response data on high-capacity, low-speed memory (disk or magnetic tape) to allow repetition of the operation with each and every mercury drop. Such an experiment has not been reported, although a close analog has been described in which the system impulse response is obtained directly and converted to the frequency response spectrum by

Fourier transformation with an off-line computer [46].

At this point a number of questions arise regarding the relative accuracy (and other merits) of direct methods for obtaining the frequency response spectrum versus indirect methods using noise or impulses. The question of whether or not white noise inputs are preferable to impulses also is relevant. We will avoid these questions at this time because they are somewhat out of context with the main topic and because at this particular time supporting experimental data are unavailable. Furthermore, the main point of interest in this section's discussion is valid regardless of how these questions are answered—i.e., in its application to ac polarography, the on-line computer will stimulate significant basic changes in experimental approach. Perhaps this process of evolution will ultimately yield offspring which are unrecognizable from their ancestor. Time will tell.

## VII. SOME SPECIAL COMPUTERIZED ROUTINES OF INTEREST

## IN ANALYTICAL APPLICATIONS OF AC POLAROGRAPHY

Analytical applications of ac polarography are less demanding of electronic instrumentation primarily because a measurement at one frequency on the low side of the usable frequency spectrum (10-20 Hz) usually suffices. This situation enables one to employ with ease certain computer routines whose implementation at the higher frequencies employed in kinetic applications might prove difficult. A few simple examples of such analytically oriented routines are given in this section.

### A. Square-Wave Polarography

The foregoing discussion abounds in examples where the on-line computer serves as the source and control element for the dc component of the potential applied to the polarographic cell. The possibility that the computer might play the same role with regard to the applied alternating potential was not raised. This concept was given little emphasis because computer-generated output signals are necessarily characterized by voltage levels which change in discrete, discontinuous "steps." Thus, the computer can only approximate smoothly varying analog signals such as the sinusoidal input normally used in ac polarography. This approximation may be satisfactory at low frequencies, but leaves much to be desired in the case of high-frequency sine waves. However, there is one modification of ac polarography for which the abrupt-state changes characterizing a computer-generated voltage are ideally suited. The technique in question is square-wave polarography [16].

Square-wave polarography involves the use of a small-amplitude square-wave voltage in place of the usual sinusoidal potential. Alternating-current detection is effected with the aid of special gating circuits that sample the current signal near the end of each square-wave half-cycle. The difference between currents sampled during the positive and negative square-wave half-cycles is a measure of the ac response. This technique has proven quite effective in suppressing contributions of the undesired double-layer charging current. The charging current is characterized by sharp impulses which are coincident with state changes in the square-wave input and which decay rapidly during a square-wave half-cycle. On the other hand, the faradaic current decays relatively slowly under the same conditions. Thus, by sampling currents late in square-wave half-cycles, relatively small charging-current contributions are realized and an extremely sensitive measurement of the faradaic response is obtained.

From this rather brief description (see refs. [7, 8, 16] for details) it should be evident that the electronic operations associated with square-wave polarography are such that it seems to have been designed with an on-line computer in mind. A square-wave input voltage whose frequency is precisely controlled by a real-time clock is readily obtained at the computer's digital-to-analog converter output. Current sampling at the end of the square-wave half-cycles can be achieved with standard sample-and-hold circuits whose operation also is controlled by the real-time clock. A flow diagram for these basic operations is given in Fig. 19. One concludes that any analytical laboratory endowed with an on-line computer and a potentiostat (even this might be unnecessary—subsection B, below) needs only the appropriate software to possess a square-wave polarograph.

## B. Direct Digital Control

The use of a digital computer to effect automatic control is well known in many areas [50, 51]. A computer can be operated as a feedback controller where it generates a control signal at a digital-to-analog converter output in response to an error signal which is the difference between a desired input level and an actual input level as observed at an analog-to-digital converter. Meanwhile, the computer also plays its usual role as a data acquisition device. Such applications of direct digital control are most common in situations where high-speed response is not essential—e.g., in a chemical process stream, a machine tool operation, etc.

In the context of ac polarography, one could envision the use of direct digital control whereby instead of using the conventional analog potential control network (see Fig. 4), one simply connects the three electrodes directly to appropriate computer terminals and allows the computer to perform all functions associated with the electrochemical experiment. In a typical configuration, a digital-to-analog converter output (controller output) is connected to the auxiliary electrode, the reference electrode is connected to an analog-to-digital converter input channel (the feedback signal), and the working electrode is connected to ground through a resistor whose voltage drop (cell current signal) is monitored by another analog-to-digital converter

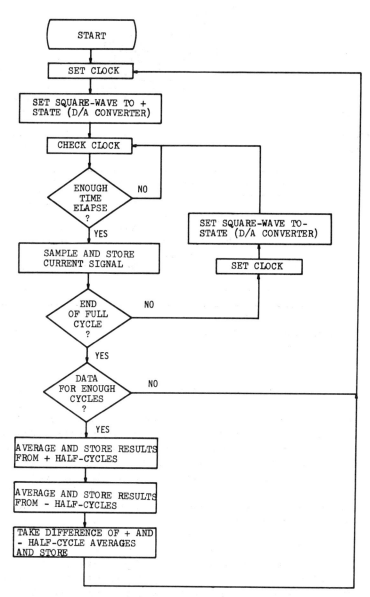

Fig. 19. Flow diagram of program for ac potential control and data-acquisition in square-wave polarography.

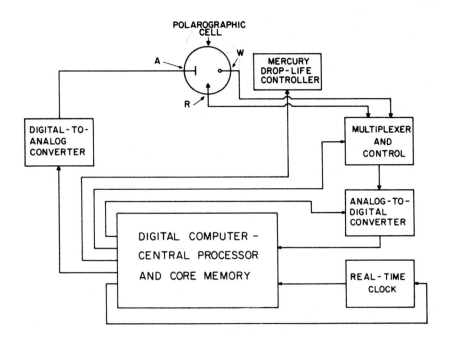

Fig. 20. Schematic of instrument using direct digital control. See Fig. 3 legend for definitions.


channel. Figure 20 provides a schematic of such a system. The digital equivalent of the feedback signal from the reference electrode is compared to a desired signal which is stored in memory. The magnitude and time response of the error determine the control signal delivered to the auxiliary electrode. One could program any desired mode of control—e. g., proportional, derivative, integral, various combinations of the latter, or any other mode that is consistent with accuracy and stability [52]. Of course, positive feedback compensation of ohmic potential drop should be included in the control program. With direct digital control the operation is a sequence of discrete steps involving signal sampling, error calculation, and response, which must repeat at a rate that is much larger than the required rate of change of the controlled variable. This requirement is a stringent one for electrochemical relaxation methods, but it probably can be met with existing computer systems for the case of low-frequency signals. With regard to signal and impedance requirements, available hardware is consistent with the needs for such operations, with the possible exception of analog-to-digital converter sensitivity where higher-than-normal sensitivities might be required.

The response of many electrochemists to the suggestion of direct digital

control in electrochemical experimentation might be, "Why bother?" After
all, the acquisition of an operational amplifier potentiostat is rather trivial
and inexpensive (an adequate potentiostat can now be constructed for less
than $50). Why utilize the computer's time in the control function, thereby
limiting its ability to perform other functions of which it is more capable?
The latter philosophy is certainly the one followed in this writer's laboratory.
Nevertheless, we acknowledge that there are situations where electrochemical
experimentation by direct digital control might be preferable. Consider, for
example, an analytical services laboratory that possesses an on-line computer facility but lacks an individual who is knowledgeable in electrochemical
instrumentation. In such an operation the "activation energy" associated with
evaluating, purchasing, constructing, and/or operating an analog potentiostat
may be sufficient to retard any interest in polarographic analysis. Thus, the
ability to perform a polarographic experiment by simply connecting three
wires from an already available computer to an electrochemical cell, loading
a program, and running the experiment may be essential for the use of polarographic techniques in such a laboratory operation.

### C. Digital Simulation of Analog Signal Conditioning Operations

As a final note we wish to point out that whenever a computer's data-
acquisition cycle is much faster than variations in signals of interest one
may allow the computer to perform most conventional signal-conditioning
operations. Every analog signal-conditioning function (e.g., low-pass or
high-pass filtering, tuned amplification, lock-in amplification, etc.) can be
approximated by a series of digital operations—i.e., by a computer program.
The approximation is excellent whenever digital data points are sufficiently
closely spaced on the time base that they accurately describe the shape of the
analog waveform. Once such data have been acquired, the computer can
operate on them in an appropriate manner to simulate a desired analog filter
action, or some other action [53]. For example, an ideal lock-in amplifier
consists of an ideal electronic multiplier and a low-pass filter [7, 19]. A
digital computer can effect lock-in amplification by the following operation
sequence: (a) digital data acquisition of the signal of interest and a reference
waveform for an appropriate number of cycles; (b) point-by-point digital
multiplication of the two waveforms; (c) digital filtering of the product data;
(d) multiplication of the result by a constant (amplification).

Although we do not recommend that one purchase a digital computer
merely to serve as a lock-in amplifier, Butterworth filter, or the like, such
applications should not be overlooked if a computer is available. Whenever
an ac polarographic experiment does not keep a computer particularly busy,
as in an analytical application, one can afford to allow the computer to digitally
acquire the data points required to define the current signal waveform and
perform the signal conditioning. One is then in a position to forego purchase
(or use) of tuned amplifiers, lock-in amplifiers, low-pass filters, and the
other analog circuits normally associated with the modern ac polarograph [7].

## ACKNOWLEDGMENT

    All of the experimental results and concepts described in this chapter
which originated in the author's laboratory were made possible through the
financial support of the National Science Foundation. Without such support,
the empirical basis for this chapter would not exist.

## REFERENCES

[1].    G. L. Booman, Anal. Chem., 29:213 (1957).

[2].    D. D. DeFord, Division of Analytical Chemistry, 133rd Meeting,
ACS, San Francisco, Calif., April, 1958.

[3].    M. T. Kelley, D. J. Fisher, and H. C. Jones, Anal. Chem., 31:1475
(1959); 32:1262 (1960).

[4].    C. N. Reilley, J. Chem. Ed., 39:A853, A933 (1962).

[5].    W. M. Schwarz and I. Shain, Anal. Chem., 35:1770, 1778 (1963).

[6].    G. Lauer, H. Schlein, and R. A. Osteryoung, Anal. Chem., 35:1789
(1963).

[7].    D. E. Smith, in Electroanalytical Chemistry, Vol. 1 (A. J. Bard,
ed.), Marcel Dekker, New York, 1966.

[8].    B. Breyer and H. H. Bauer, in Chemical Analysis, Vol. 13 (P. J.
Elving and I. M. Kolthoff, eds.), Wiley-Interscience, New York, 1963.

[9].    M. Sluyters-Rehbach and J. H. Sluyters, in Electroanalytical Chemistry, Vol. 4 (A. J. Bard, ed.), Marcel Dekker, New York, 1970.

[10].    P. Delahay, M. Senda, and C. H. Weis, J. Phys. Chem., 64:960
(1960).

[11].    P. Delahay, M. Senda, and C. H. Weis, J. Amer. Chem. Soc.,
83:312 (1961).

[12].    G. C. Barker, in Transactions of the Symposium on Electrode
Processes (E. Yeager, ed.), Wiley, New York, 1961.

[13].    R. de Leeuwe, M. Sluyters-Rehbach, and J. H. Sluyters, Electrochim.
Acta, 14:1183 (1969).

[14].    T. G. McCord and D. E. Smith, Anal. Chem., 40:289 (1968).

[15].    J. Paynter, Ph.D. Thesis, Columbia Univ., New York, 1964.

[16].    G. C. Barker and I. L. Jenkins, Analyst, 77:685 (1952).

[17].    G. C. Barker, Anal. Chim. Acta, 18:118 (1958).

[18].    B. J. Huebert, Ph. D. Thesis, Northwestern University, Evanston,
Illinois, 1971.

[19]. D. E. Smith, Anal. Chem., 35:1811 (1963).

[20]. J. W. Hayes and C. M. Reilley, Anal. Chem., 37:1322 (1965).

[21]. T. G. McCord and D. E. Smith, Anal. Chem., 41:1423 (1969).

[22]. T. G. McCord and D. E. Smith, Anal. Chem., 42:2, 126 (1970).

[23]. E. R. Brown, T. G. McCord, D. E. Smith, and D. D. DeFord, Anal. Chem., 38:1119 (1966).

[24]. E. R. Brown, D. E. Smith, and G. L. Booman, Anal. Chem., 49:1411 (1968).

[25]. E. R. Brown, H. L. Hung, T. G. McCord, D. E. Smith, and G. L. Booman, Anal. Chem., 40:1424 (1968).

[26]. P. Delahay, J. Phys. Chem., 70:2373 (1966).

[27]. K. Holub, G. Tessari, and P. Delahay, J. Phys. Chem., 71:2612 (1967).

[28]. Burr-Brown Research Corp., Handbook of Operational Amplifier Applications, Burr-Brown Research Corp., Tucson, Arizona, 1963.

[29]. Philbrick Researches, Inc., Applications Manual for Computing Amplifiers, Nimrod Press, Boston, Mass., 1966.

[30]. A Comprehensive Catalog and Guide to Operational Amplifiers, Analog Devices, Cambridge, Mass., Jan. 1968.

[31]. M. Ishibashi, T. Fujinaga, and A. Saito, Rev. Polarog. (Kyoto), 7:41 (1959).

[32]. E. R. Brown, Doctoral Dissertation, Northwestern University, Evanston, Illinois, 1967.

[33]. R. Delevie and A. A. Husovsky, J. Electroanal. Chem., 22:29 (1969).

[34]. B. Timmer, M. Sluyters-Rehbach, and J. H. Sluyters, J. Electroanal. Chem., 19:73 (1968).

[35]. D. E. Smith and H. R. Sobel, Anal. Chem., 42:1018 (1970).

[36]. J. Heyrovsky and J. Kuta, Principles of Polarography, Academic Press, New York, 1966, pp. 96, 99-103.

[37]. D. E. Smith, unpublished work, Northwestern University, Evanston, Illinois, 1966.

[38]. J. F. Coetzee, J. M. Simon, and R. J. Bertozzi, Anal. Chem., 41:766 (1969).

[39]. J. Kuta and I. Smoler, in Progress in Polarography, Vol. 1, (R. Zuman and I. M. Kolthoff, eds.), Wiley-Interscience, New York, 1962, p. 43.

[40].  M. J. D. Low, Anal. Chem., 41:97A (May 1969).

[41].  T. C. Farrar, Anal. Chem., 42:109A (April 1970).

[42].  D. E. Glover and D. E. Smith, Anal. Chem., submitted for publication.

[43].  B. J. Huebert and D. E. Smith, Anal. Chem., submitted for publication.

[44].  J. S. Bendat and A. G. Piersol, Measurement and Analysis of Random Data, Wiley, New York, 1966.

[45].  R. L. Rex and G. T. Roberts, Hewlett-Packard J., 21:1 (Nov. 1969).

[46].  A. A. Pilla, J. Electrochem. Soc., 117:467 (1970).

[47].  D. K. Cheng, Analysis of Linear Systems, Addison-Wesley, Reading, Mass., 1959.

[48].  V. W. Cooley and J. W. Tukey, Math. Comp., 19:297 (1965).

[49].  Raytheon ATP Computer, Bull. No. SP330A, Raytheon Co., Santa Ana, Calif., 1970.

[50].  J. T. Tou, Digital and Sampled Data Control Systems, McGraw-Hill, New York, 1959.

[51].  A. Poll and D. J. Ray, Control Engineering, 17:75 (April 1970).

[52].  D. E. Smith, C.E. Borchers and R. J. Loyd, in Physical Methods of Chemistry, Part I, 4th ed. (A. Weissberger and B. W. Rossiter, eds.), Wiley, New York, 1970.

[53].  S. A. White and T. Mitsutomi, Control Engineering, 17:58 (June 1970).

Chapter 13

ENHANCEMENT OF ELECTROANALYTICAL MEASUREMENT
TECHNIQUES BY REAL-TIME COMPUTER INTERACTION

S. P. Perone

Department of Chemistry
Purdue University
Lafayette, Indiana 47907

## I. INTRODUCTION

This chapter is intended to describe some ways in which the small, dedicated, laboratory computer can provide enhanced measurement capability for electroanalytical techniques. The work described is based primarily on three recent publications by the author and co-workers [1-3]. The particular electroanalytical techniques to which computerization has been applied are stationary-electrode polarography (SEP) [4] and the closely related

423

derivative voltammetry [5, 6]. However, the principles and methodology described should be generally applicable to other electroanalytical techniques and, perhaps, to analytical experimentation in general.

Certainly, one very important way in which the on-line digital computer can improve the capabilities of electroanalytical techniques is simply to provide automated experimentation, data assimilation, and straightforward data processing. This has been demonstrated amply already [7-9]. However, these approaches ultimately are limited by the inherent limitations of the particular measurement techniques. To take full advantage of the dedicated computer, one should utilize its capabilities for rapid "intelligent" feedback and incorporate the computer into the experimental control loop, as shown in Fig. 1. Thus, the computer could monitor the experiment, process the data as the experiment progresses, and "intelligently" modify the course of the experiment to provide optimum measurement conditions. Of course, the "intelligence" is related to the programming skill and the experimental intuition of the programmer. Moreover, the transfer function for "intelligent" response will depend on the degree of sophistication and number of calculations and decisions which must be made in "real time"—i.e., between successively acquired data points; also important are the computer's speed, and hardware arithmetic and logical capabilities. (A more quantitative discussion of response factors is given below.)

The important feature of this latter approach—real-time computer optimization of analytical measurements—is that a measurement technique can be generated which would be unattainable without the aid of an on-line computer. Moreover, in the process of developing this application, the experimenter must necessarily investigate systematically those experimental

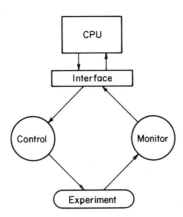

Fig. 1. Block diagram of laboratory instrumentation with computerized feedback loop.

parameters which most critically determine the computer optimization of the measurement. This can be done only with the computer in the control loop. However, the results of these investigations then provide the foundation for the design of new instrumental methods which might be implemented independent of an on-line computer. These aspects of laboratory computer applications—real-time computer interaction with instrumentation, and computerized experimental design of interactive instrumentation—will be discussed below.

## A. Response Factors for Real-Time Computer Interaction

Any computer operations executed during the lifetime of an experimental event can be considered "real-time operations." However, in the most demanding situation, a specified sequence of operations must be completed within the time interval between real-time clock pulses. Thus, for the rest of our discussion here, we will consider real-time computer functions as those which must be executed between clock pulses, assuming that each clock pulse represents a point in time where the computer must devote its attention to the experiment, either for data acquisition or control or both. Also, this discussion will be oriented toward a single channel data acquisition system, although extension to multi-channel data acquisition should be obvious.

There are many reasons for executing real-time computer operations. For example, the simplest and most fundamental task of the computer in real-time must be to handle acquired data. This involves storing the data as well as bookkeeping operations to keep track of where the data goes and how much has been acquired. In addition, the experiment may require that some computational tasks be executed in real-time. For example, real-time smoothing of acquired data can be carried out. One approach is to apply an n-point moving average to the data as they are acquired. More sophisticated smoothing algorithms can also be carried out in real-time, such as digital filtering or least squares smoothing. For more detailed discussions of data smoothing methods, the reader is referred to references [10, 11].

We should now consider the characteristics of real-time computer control of experimental systems. Figure 1 represents schematically the configuration of an experimental laboratory computer system. Three of the elements in the diagram are typical of conventional laboratory experimentation. These include: the fundamental aspects of the experiment itself; the monitoring instrumentation; and the control instrumentation. Ordinarily, it is the scientist who supervises the implementation of these other elements. He can also interact with the experiment to a certain extent by observing the output of one experiment and modifying conditions for the next so that more nearly optimum experimental conditions will exist. Moreover, the scientist could interact with any given experiment in real-time if the time scale of the experiment was long enough and the monitor output provided the appropriate correlations to allow intelligent decisions regarding the modification of experimental conditions to improve the experiment. The fundamental limitation here, of course, is the response time of the human as a feedback element.

Figure 1 shows a feedback path through the digital computer. Because the computer has a memory, because it can execute computational and logical operations, and because it can execute data acquisition and control functions, it can act as a pseudo-intelligent feedback element in chemical instrumentation. Moreover, because of its speed, computerized feedback control can be accomplished on a microsecond or millisecond time scale. Thus, an advantage of several orders of magnitude in the time scale for real-time experiment interaction over human real-time functions can be gained with a computerized feedback loop.

Response characteristics, transfer functions, and stability requirements for instrumentation incorporating computerized feedback loops can be defined by the application of conventional Laplace transform and z-transform analysis [12, 13]. Because sampled-data systems provide discontinuities in the monitoring and/or control loops, the response and stability characteristics of these systems can be quite different than for pure analog systems. Although a detailed discussion of system analysis for sampled-data systems and periodic controllers is beyond the scope of this article, a discussion of response limitations for computerized feedback can be presented.

The response limitations for computerized real-time interaction cannot be analyzed similarly to analog electronic devices. The digital computer can be considered as an active electronic element in instrumentation, but the electronic characteristics which it displays as a feedback element in instrumentation are unique. For example, the operational amplifier can be "programmed" to differentiate, integrate, add, subtract, or generate a combination of mathematical functions, by providing the appropriate input and feedback elements such as resistors, capacitors, inductors, transistors, etc. Likewise, the digital computer can be programmed to accomplish these mathematical functions. The fundamental difference, however, is that the analog devices generate essentially instantaneous functions whereas the computer must execute a sequence of programmed steps in order to provide the mathematical results. Whereas with an analog operational amplifier, the "power" of that device is related to the voltage gain, the "power" of the digital computer as an instrumental element is directly related to the computational time available. That is, because the computer can function only by executing a programmed sequence of discrete steps, the computer can accomplish a given computational or control objective only if it has sufficient time available. The greater the amount of time available, the greater the power of the computer to execute computational or intelligent feedback operations.

The amount of computer power available in real-time will depend on the stimulus frequency f; this is the frequency at which the computer is "poked" by the experiment requesting some external service. In the simplest situation, f is the data acquisition frequency—the rate at which digitized data are made available to the computer from the experiment. At a minimum, the computer's service, as each datum is made available from the digital data acquisition system, involves inputting the datum, saving it in memory, and some bookkeeping. Typical <u>minimum service time</u>, $\tau_M$, may be 20 or 30 $\mu$sec. The <u>real-time computer power</u> can be equated to the <u>available computational time</u>, $\tau_A$, between stimuli:

$$\tau_A = \frac{1}{f} - \tau_M. \tag{1}$$

To provide any useful feedback functions—where some computational evaluation of the data, logical decisions, and possible experimental control operations may be carried out in real-time—the value for $\tau_A$ must be large enough to accommodate the program execution required.

Certain additional factors related to the system response must be considered. First of all, the discussions here are presented in the context of <u>dedicated</u> systems, where the computer is interfaced to a single instrument. This implies that response to stimuli from the particular experiment is the highest priority activity of the computer. Thus, it is assumed that the execution of the real-time service program begins "immediately" upon request from the interfaced instrument. In fact, there is some <u>minimum response time</u>, $\tau_0$, required. The magnitude of $\tau_0$ depends on the computer hardware features, particularly the cycle time, and also depends on the specific instruction sequence used to identify and/or respond to a stimulus. However, regardless of whether the computer's <u>Interrupt System</u> or <u>program-controlled response</u> is used, $\tau_0$ may be several <u>microseconds</u>. (Moreover, for time-shared systems, $\tau_0$ may be many milliseconds.) This must be included in estimating $\tau_M$.

Another relevant consideration is that the values for $\tau_M$ or $\tau_A$ used in any calculations should be the "worst case" values. That is, where alternative program pathways exist, assume that conditions will always require the longest path.

## B. Some Sample Calculations

Consider now how one might use the system response analysis described above for the design of a particular experimental application involving real-time computer interaction. First, the essential elements of the minimal experimental service program—including the required input/output and bookkeeping instructions to be executed for each stimulus—must be established. The time required for these operations plus the minimum response time, $\tau_0$, correspond to the <u>minimum service time</u>, $\tau_M$. The programmer can then calculate $\tau_A$ for the specific data acquisition or service frequency, f, required in his application. Then, he must establish $\tau_R$, the <u>real-time computational time</u> required to provide the required calculations, decisions, and experimental control operations to allow computer interaction with the experiment. The linear combination of the two programming segments results in a <u>total real-time service time</u>, $\tau_T$, where

$$\tau_T = \tau_M + \tau_R . \tag{2}$$

If the programming is such that $\tau_R \leq \tau_A$, the proposed application is feasible. Alternatively, for a given value of $\tau_T$, one can calculate the maximum data acquisition frequency allowed. This can be obtained by setting

$$1/f = 1/f_{max} = \tau_T . \tag{3}$$

A specific example should illustrate the above discussion. Assume the computer system has a basic machine cycle time of 2.0 $\mu$sec, and most instructions are some integral multiple of this value. For a particular application, program-controlled service is used requiring a worst-case response, $\tau_0$, of 6.0 $\mu$sec. The additional minimal service programming requires 8 cycles, 16.0 $\mu$sec. Thus, $\tau_M$ is 22.0 $\mu$sec. The desired data acquisition frequency, f, is 10 kHz. Therefore, the value of $\tau_A$ is computed from Eq. (1) to be 78.0 $\mu$sec. Thus, the real-time interactive programming has to have a worst-case execution time, $\tau_R$, less than 78.0 $\mu$sec. This allows real-time programming requiring no more than 39 machine cycles. Alternatively, if the programmer determines $\tau_R$ and $\tau_T$, the limiting data acquisition frequency can be determined from Eq. (3). Thus, for $\tau_R = 140$ $\mu$sec, and $\tau_T = 162$ $\mu$sec, the limiting frequency is about 6173 Hz.

It should be noted here that, as a feedback element, the digital computer is a relatively slow device. In fact, it can often be assumed that analog instrumentation controlled or measured by the computer does not limit the overall system response. Thus, the overall system response and transfer function is often limited by the computer response characteristics discussed here. However, because the computer can provide "intelligent" feedback, the experimenter can trade speed for sophistication of interaction.

## II. REAL-TIME COMPUTER CONTROL IN
## STATIONARY-ELECTRODE POLAROGRAPHY

Stationary-electrode polarography (SEP), despite its many desirable characteristics for automated analysis [7], is seriously limited for application to mixtures. Because of the continuous nature of the experiment, currents from easily reducible species continue to flow and contribute to, distort, or mask currents measured for more difficultly reducible species. The nonideal aspect of the technique is the fact that a continuous linearly varying potential is applied to the electrolysis cell, regardless of the composition of the sample. If the linear sweep were discontinuous, stopping briefly after each reduction step to allow the more complete dissipation of the easily reducible species in the diffusion layer around the electrode, the interference with reduction steps for more difficultly reducible species would be considerably diminished. However, such a discontinuous or "interrupted-sweep" experiment would require some foreknowledge as to the composition of the mixture—and this is not likely in real analytical situations.

TABLE 1

Values of Current Function Ratios $\dfrac{\chi(at)_p}{\chi(at)_E}$  $\dfrac{\chi'(at)_p}{\chi'(at)_E}$, and $\dfrac{\chi''(at)_p}{\chi''(at)_E}$

as Functions of $(E - E_{1/2})^a$

| $n(E - E_{1/2})$, mV | $\dfrac{\chi(at)_p}{\chi(at)_E}$ | $\dfrac{\chi'(at)_p}{\chi'(at)_E}$ | $\dfrac{\chi''(at)_p}{\chi''(at)_E}$ |
|---|---|---|---|
| -150 | 1.813 | 6.044 | 15.01 |
| -175 | 1.991 | 7.983 | 24.08 |
| -200 | 2.153 | 10.18 | 35.03 |
| -225 | 2.299 | 12.54 | 50.26 |
| -250 | 2.436 | 15.02 | 68.00 |
| -275 | 2.563 | 17.63 | 88.90 |
| -300 | 2.671 | 20.22 | 115.6 |
| -325 | 2.788 | 22.99 | 144.5 |
| -350 | 2.896 | 26.16 | 172.6 |
| -400 | 3.096 | 32.28 | 241.0 |
| -450 | 3.304 | 38.90 | 330.0 |

[a] See Refs. [4, 5] for definition of symbols. $\chi(at)$, $\chi'(at)$, and $\chi''(at)$ are theoretically tabulated current functions for the 0-, 1st- and 2nd- derivative SEP curves, respectively.

The work of Perone et al. [1] illustrated how one can take advantage of the control capabilities of the on-line digital computer to overcome the resolution problems of SEP. Their approach was to allow the computer to interact with the experiment in real time to generate an interrupted-sweep experiment which was effectively "sample oriented." Some details of that work will be presented here.

### A. Sample-Oriented Analysis—Interrupted-Sweep Approach

#### 1. Resolution Limits

The theories of conventional SEP [4] and SEP with derivative readout [5,6] allow the accurate prediction of resolution limits. This can be done by calculating the theoretical ratio of the current functions at the peak and at some potential beyond the peak. Using the mathematical approaches outlined previously [4-6] and considering only reversible systems, this has

been done for zero-, first-, and second-derivative measurement [14]. The
results are shown in Table 1. [The peaks chosen for the first- and second-
derivative current functions are the largest ones in each case—at
$n(E - E_{1/2})$ = 18.8 and 14.4 mV, respectively.] Note that the interference
with a succeeding reduction diminishes considerably with derivative measurements and with increased potential separation of reduction steps. Resolution
increases with the order of the derivative; however, even with the derivative
measurement, the resolution is not particularly good when reduction steps
are closer together than about 300/n mV.

Thus, considering arbitrarily a separation of 300/n mV and equal n and
D values, it would be possible to resolve, with 5% contribution of the first
reduction step to the second, concentration ratios of 1:7, 1:1, and 5.8:1 for
the zero-, first-, and second-derivative measurements, respectively. These
calculations are exclusive of any other background contributions, with the assumption that they are negligible or can be measured independently for correction.

It would, of course, be possible to correct mathematically for the
interference caused by overlapping reduction steps. However, a limit is
reached with this approach when the interference is so great as to preclude
even the recognition of a succeeding reduction step. In any event, this
approach has been considered [2] and will be discussed later.

2.  Details of Interrupted-Sweep Experiment

The interrupted-sweep experiment involves a computer-controlled
potentiostat [1] and real-time analysis of fast-sweep derivative polarographic
data. (A system block diagram is shown in Fig. 2.) The computer continuously monitors the experimental output, is instantaneously aware of the
occurrence of reduction steps, and can interrupt the linear potential sweep

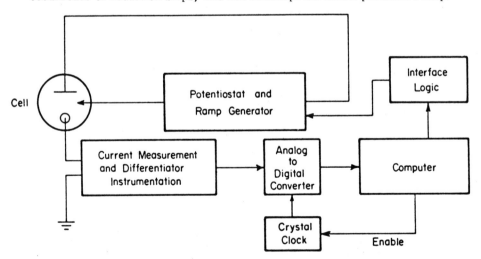

Fig. 2. SEP-1 system block diagram. (From Ref. [1].)

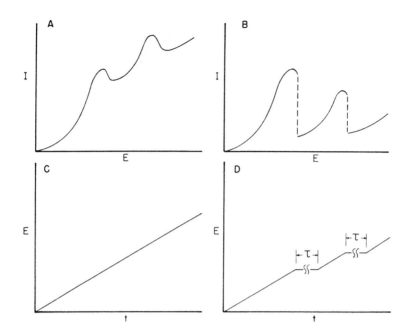

Fig. 3. Comparison of normal and interrupted-sweep stationary electrode polarography. (a) stationary electrode polarogram (without interrupt); (b) stationary electrode polarogram (with interrupt); (c) applied cell potential (without interrupt); (d) applied cell potential (with interrupt). (From Ref. [1].)

at an appropriate potential cathodic of each peak. The interrupt potential is held for a length of time computed to allow sufficient depletion of the electroactive species in the diffusion layer, and then the sweep is restarted. The interrupt delay time $\tau'$ is calculated in proportion to the magnitude of the reduction step, with the restriction that $\tau'$ not be so long as to cause convection processes to occur or to allow significant electrolysis of the next electroactive species. Thus, the controlling potential function—which is basically a series of ramp-and-hold steps—will be different for each experiment, depending on the sample mixture and composition. The experiment is custom tailored to the sample—i.e., sample oriented. A simplified comparison of the continuous- and interrupted-sweep experiments is shown in Fig. 3. A flow chart of the computer-controlled experiment is given in Fig. 4. The computer used in this work was a Hewlett-Packard Model 2115A with 16-bit word size, 8192-word core memory, 2.0-$\mu$sec cycle time, and a hardware-extended arithmetic unit. The data-acquisition system included a 10-bit, 33-$\mu$sec-conversion-time analog-to-digital converter (ADC). The timing was provided by an external 10-MHz crystal clock scaled down to 1 kHz, which was the maximum data-acquisition frequency used for all experiments. A detailed discussion of interfacing and control logic is given in Ref. [1].

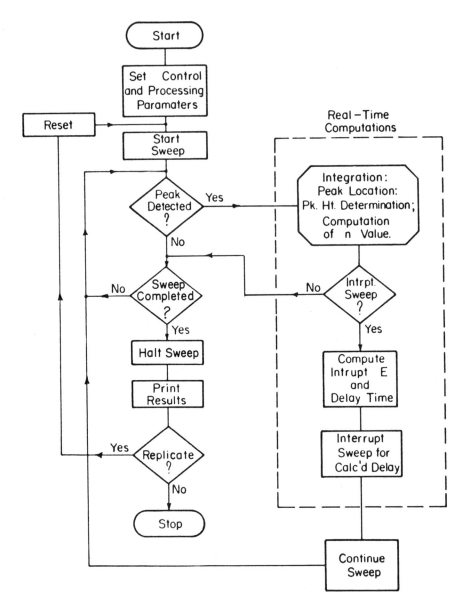

Fig. 4.  Flow chart for real-time computer-optimized SEP.  (From Ref. [1].)

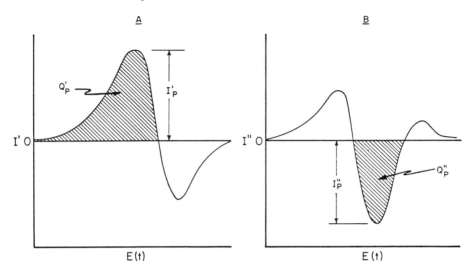

Fig. 5. Analytical measurements from first- and second-derivative voltammetric data. (a) first derivative; (b) second derivative. (From Ref. [1].)

The information taken in by the computer is extracted from either the 1st- or 2nd-derivative signal. Only the negative region of the derivative signal is seen by the ADC, and the measuring circuit is arranged so that only the largest peak in each case is the correct polarity. The result is shown in Fig. 5. The computer is programmed to look for sharp peaks —above an arbitrary threshold—and to measure and store peak heights, peak areas, and peak locations in real time.

When a complete peak is observed—i.e., one which goes above threshold, goes through a sharp maximum, and then comes back down below threshold—the computer executes the maximum real-time programming, calculating the n value, the potential at which the sweep should be interrupted $E_D$, and the delay time $\tau'$. (The maximum total real-time programming time $\tau_T$ is about 800 $\mu$ sec.) Then the computer allows the sweep to continue (if necessary) until the desired interrupt potential $E_D$ is reached; the sweep is interrupted at this point for time $\tau'$; the sweep is then reinitiated with the computer looking for the next reduction step, ready to reexecute a similar interrupted sweep. (The delay time $\tau'$ is computed in proportion to the peak height, with the restriction that $\tau'$ be between 100 and 1000 msec.)

## B. Advantages of Real-Time Calculations

### 1. Uses of Integral Data

It was possible to integrate the derivative peaks (Fig. 5) seen by the computer in real time and to use these integral data for analytical purposes, for calculating appropriate interrupt potentials, and to provide diagnostic

information. The area under a peak is directly proportional to the peak
height and, therefore, to the concentration of electroactive species. Thus,
the peak integral $Q_p$ is a concentration-dependent output. Moreover, should
the peak location routine in the program fail because the derivative peaks
are too noisy, broad, or small, the peak integrals will still be taken and
will provide a useful, reliable source of analytical information. In addition,
a peak-area threshold is incorporated into the program, whereby the area
of a given peak must exceed some arbitrary value before the computer will
recognize a signal excursion as a bona fide reduction peak. This is a very
useful processing parameter.

In the case of 1st-derivative readout, the value of the integral of the
observed peak is equivalent to the peak height of the conventional SEP. It
has been shown previously that for a reversible system the ratio of the
1st-derivative peak height $I_p'$ to the conventional peak height $i_p$ is related to
n [6] as given by Eq. (4),

$$\frac{I_p'}{i_p} = 13.2 \text{ nv,} \tag{4}$$

where v is the scan rate in V/sec. Thus, a determination of the ratio of the
derivative peak height to the peak integral can lead to an evaluation of n, and
the computer is programmed to do this in real time so that the information
is available for interrupted-sweep decisions. This information is also useful
for qualitative identification of species or for providing an error diagnostic
if wrong n values are obtained for known systems.

A similar relationship exists between the n value and the ratio of the
2nd-derivative peak height to the peak integral. The measured integral is
equivalent to the difference between the positive- and negative-going 1st-
derivative peaks. The relationship can be calculated from previous theoretical data [5] and is given in Eq. (5) (note the error in Eq. (2) of Ref. [1]:

$$\frac{I_p''}{I_{p_1}' - I_{p_2}'} = 21.2 \text{ nv.} \tag{5}$$

Thus, the n value can be obtained from either the 1st- or 2nd-derivative
measurements. For reversible processes, the computed n value is accurate.
For irreversible processes, the n value at least reflects the broadness of
the peak, and this is useful for the interrupt-potential calculations discussed
below.

## 2.   Selection of Interrupt Potential

In the interrupted-sweep experiment, the potential $E_D$, selected to be
held during the delay period $\tau'$, is critical. The objective is to select a
potential which is cathodic enough to adequately deplete the electroactive

species in the diffusion layer; that is, the potential should be chosen such that the concentration ratio $C_O/C_R$ approaches zero at the electrode surface.

The obvious problem is that selecting a value of $E_D$ cathodic enough to truly deplete the electroactive species at the electrode would eliminate the possibility of observing a succeeding closely spaced reduction. Thus a compromise must be reached, and a knowledge of the n value for the reduction step on which the delay is made is useful in selecting the appropriate interrupt potential.

In this work, interrupted-sweep experiments were run with the interrupt potential ($E_D$) selected by the computer after it had observed a complete derivative peak—i.e., when the derivative signal is going through zero. The $E_D$ is selected relative to the potential $E_Z$, at which the derivative signal goes through zero. The computer uses information provided initially by the operator in order to calculate $E_D - E_Z$. That is, the operator initially specifies $n(E_D - E_Z)$; the computer determines n and $E_Z$, and then selects $E_D$. Experiments are reported below where the operator-selected value of $n(E_D - E_Z)$ was varied for a series of runs to observe the effect of $E_D$ on quantitative resolution.

The influence of $E_D$ on the surface concentrations of species in the redox couple O and R is shown in Table 2. The calculations for Table 2 are based on the Nernst equation and reversible behavior. Also, it should be noted that $E_Z - E_{1/2}$ is $-28.5/n$ mV for the 1st-derivative measurement, and $-66.5/n$ mV for the 2nd-derivative measurement.

## C. Results

Two different two-component systems were studied in this work. The first system consisted of Tl(I) and Pb(II) in 1.0-M NaOH electrolyte. The half-wave potentials for these two species are separated by approximately 280 mV. The second system studied was that of Pb(II) and Cd(II) in a 2-M ammonium acetate-acetic acid electrolyte. The $E_{1/2}$ separation for this case

TABLE 2

Interrupt Potential Required for Specified Surface Ratio $C_O/C_R$

| $E_D - E_{1/2}$ | $C_O/C_R$ | $E_D - E_{1/2}$ | $C_O/C_R$ |
|:---:|:---:|:---:|:---:|
| 0 | 1/1 | $-100/n$ | 1/50 |
| $-28.5/n$ | 1/3 | $-118/n$ | 1/100 |
| $-59.1/n$ | 1/10 | $-177/n$ | 1/1000 |

TABLE 3

Peak Derivative Measurements of Smaller Components with and without
Computer Interaction for Binary Mixtures

| Mixture | Condition | $E_D - E_Z$, mV | % Std, $I_p'$ | % Std, $I_p''$ |
|---|---|---|---|---|
| 30:1 Tl:Pb | W/o interrupt | — | 50.8[a] | 92.6[b] |
| | Interrupted sweep | 0/n | 89.9 | 100.8 |
| | | -20/n | 92.8 | 100.8 |
| | | -100/n | 97.9 | — |
| 10:1 Pb:Cd | W/o interrupt | — | 45.3[c] | 91.9[d] |
| | Interrupted sweep | 0/n | 104.4 | 100.0 |
| | | -20/n | 104.8 | 86.9 |
| | | -30/n | 102.3 | — |
| 100:1 Tl:Pb | W/o interrupt | — | No peak[e] detected | 88.7[f] |
| | Interrupted sweep | -40/n | 70.8 | 101.0 |
| | | -100/n | 83.5 | 102.9 |
| 1000:1 Tl:Pb | W/o interrupt | — | No peak[g] detected | No peak[h] detected |
| | Interrupted sweep | -40/n | No peak detected | 78 |
| | | -100/n | No peak detected | 98 |

Predicted errors: (a) -48.4%, (b) -5.2%, (c) -45.6%, (d) -8.2%, (e) -160%, (f) -17%, (g) -1600%, (h) -170%. (Predicted errors based on Table 1 and known concentration ratios.)

was approximately 150 mV. All runs were made at a scan rate of 1.00 V/sec. Data points were taken at 1- or 2-mV intervals, and both the first and second derivatives of the reduction currents were observed for each system.

Various experiments were applied to the two systems. These included normal SEP and interrupted-sweep SEP with computer-selected interrupt potential. The values of $(E_D - E_Z)$ employed varied from $0/n$ mV to values which resulted in noticeable charging-spike interference. The delay time $\tau'$ for the first peak in each example was always near to or equal to 1000 msec, since the first peaks were always quite large. The results of these tests are

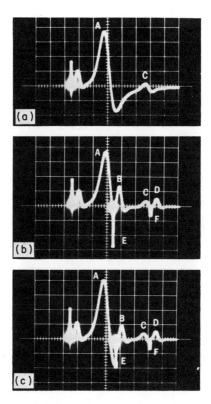

Fig. 6. First-derivative curves for 30:1 Tl(I)-Pb(II) system [$3.21 \times 10^{-4}$ M Tl(I), $1.05 \times 10^{-5}$ M Pb(II), 1.0 M NaOH]. Upper trace: without interrupt; Middle trace: with interrupt, $E_D - E_Z = -10/n$ mV; Lower trace: with interrupt, $E_D - E_Z = -40/n$ mV. The signals observed are: A, Tl(I) signal; C, Pb(II) signal; B and D, charging spikes; and E and F, current decay occurring during interrupt. (From Ref. [1].)

summarized in Table 3. Also included in these tables are the theoretical estimates of overlap error based on the data from Table 1.

The 1st-derivative data for the Tl (I) -Pb(II) system show continued improvement with increasing $(E_D - E_Z)$. However, beyond $(E_D - E_Z) = -100/n$ mV, some distortion apparently is caused by the charging spike of the re-started sweep overlapping slightly with the Pb(II) signal.

The error due to overlapping reduction signals shown in the 2nd-derivative data for the Tl(I)-Pb(II) system is small even without the interrupted sweep. (This is predicted, of course, from Table 1.) The improvement with the interrupted sweep is significant, however. The experimental effects of the interrupted sweep for both the first and second derivatives can be visualized in Figs. 6 and 7.

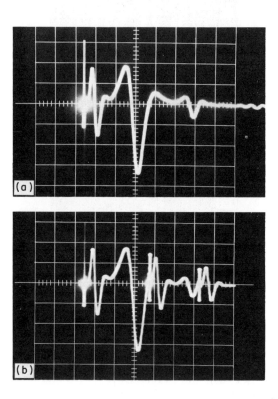

Fig. 7. Second-derivative curves for 30:1 T1(I)-Pb(II) system [3.21 × 10$^{-4}$ M Tl(I), 1.05 × 10$^{-5}$ M Pb(II), 1.0 M NaOH]. Upper trace: without interrupt; Lower trace: with interrupt, $E_D - E_Z = -10/n$ mV. (From Ref. [1].)

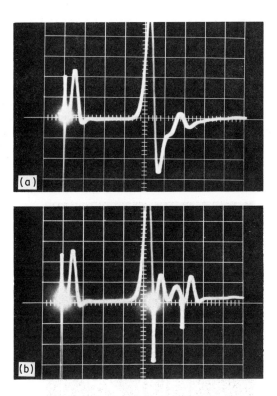

Fig. 8.   First-derivative curves for 10:1 Pb(II)-Cd(II) system $[1.05 \times 10^{-4}$ M Pb(II), $1.11 \times 10^{-5}$ M Cd(II), 2 M NH$_4$OAc, 2 M HOAc]. Upper trace: without interrupt; Lower trace: with interrupt, $E_D - E_Z = 0/n$ mV. (From Ref. [1].)

The first-derivative data for the Pb(II)-Cd(II) system, using the interrupted-sweep experiment, show some contribution to the Cd(II) peak from the charging spike, even with small values of $(E_D - E_Z)$. This interference, illustrated in Fig. 8, results in a small positive error for the second peak.

The width of the charging current spike for both the first and second derivatives is observed to be approximately 100 mV. Thus, if $E_D$ is to be set cathodic of $E_Z$, the end of the first signal peak and the start of the second signal peak must be at least 100 mV apart. This is a significant limitation on the interrupted-sweep experiment.

The second-derivative data for the Pb(II)-Cd(II) system show better over-all results than the first-derivative data. This was to be expected, on

the basis of earlier studies [5, 7] and the data of Table 1. It was observed
that for both systems the peak integral data $Q_p'$ and $Q_p''$ show greater error
than the peak height data. This is not unexpected, because the error from
the peak base is missed when peak overlap occurs. Thus, peak areas should
not be the primary source of analytical data.

The results for Tl(I)-Pb(II)-system 100:1 and 1000:1 mixtures in 1.0-$\underline{M}$
NaOH show a considerable improvement in quantitative resolution when the
interrupted-sweep experiment is employed. For the 100:1 mixture, the second peak is undetectable with a 1st-derivative readout (see Fig. 9). With
the interrupted sweep, however, not only is the second peak detectable, but
the measured value comes up to 84% of the correct value for 1st-derivative

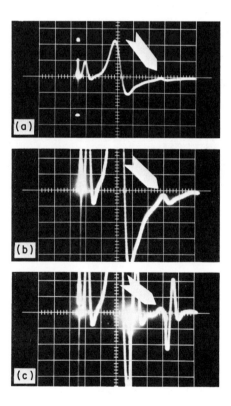

Fig. 9.   First-derivative curves for 100:1 Tl(I)-Pb(II) system [$2.68 \times 10^{-4}$
M Tl(I), $2.63 \times 10^{-6}$ M Pb(II), 1.0 M NaOH]. Arrow shows Pb(II) peak.
Upper trace: without interrupt; Middle trace: without interrupt, sensitivity
increased 5×; Lower trace: with interrupt, $E_D - E_Z = -40/n$ mV, sensitivity same as middle trace. (From Ref. [1].)

readout, and 100% for 2nd-derivative readout. For the 1000:1 mixture the second peak is undetectable, even with a 2nd-derivative measurement. By employment of the interrupted-sweep experiment with 2nd-derivative read-out, however, reliable and quantitative detection could be obtained.

### D. Observations

The work described here was intended to demonstrate that an on-line digital computer could be used to optimize an experimental measurement technique by real-time interaction with the experiment. The results clearly show a dramatic improvement in quantitative resolution of overlapping reduction signals, provided that a minimum $E_{1/2}$ separation of about 150 mV is present. Thus, the optimized measurement is subject to at least this one severe limitation, but nevertheless appears quite useful. Most importantly, the principle of real-time computer-optimized measurements in electro-analysis was demonstrated by application to a real system.

### III. COMPUTERIZED EXPERIMENTAL DESIGN
### OF INTERACTIVE INSTRUMENTATION

The experimental method described above illustrates the application of an on-line digital computer to generate, evaluate, and optimize a new electro-analytical approach. The general-purpose laboratory computer is well suited for this task because of the ease with which programmed control functions can be modified during the development of experimental control characteristics. However, having arrived at an optimum set of experimental control features, to continue dedicated use of an on-line computer for routine application of the technique might not be economically feasible. Thus, a more practical approach should be taken to adapt the technique developed with the general-purpose computer system for routine laboratory application. Jones and Perone [3] described the incorporation of the optimum parameters determined from an earlier work [1], summarized above, into a specialized instrument designed to generate the interrupted-sweep experiment without the need for an on-line computer. In this later work [3] they demonstrated the value of computer-controlled experimentation for the design of interactive experimental techniques that can then be hardware-implement, and also demonstrated that many programmed control operations can be converted easily to hardware logic and analog functions by using medium-scale integrated circuit (MSI) modules [15].

The optimum parameters selected earlier [1] were incorporated into a device in which few manual operations were needed and which, as a result, would implement the technique for most, but not all, of the cases studied. Detailed descriptions of the instrumentation and approach were presented [3]. Essentially the same functions as in the computerized technique were provided, but some changes were made. Only the 2nd-derivative peak height

$I_p'$ was used to extract quantitative information. The first-derivative zero crossing $E_Z'$ was used for qualitative identification, rather than the peak center of the 1st or 2nd derivative which was used in the computer-controlled technique. This was because of the ease with which the zero crossing could be detected with a hardware comparator. The peak width of the second derivative was used as an indication of n [5], rather than the ratio of the peak height to its integral as in the earlier work [1]. This was because the 2nd-derivative peak-width measurement was much easier to accomplish with hardware and gave very reproducible results. The interrupt-time delay was proportional to the second-derivative peak height, as it was for the computerized approach, and was again limited to between 100 and 1000 msec. Because the earlier work had shown that adequate resolution could be obtained up to at least 100:1 mixtures for $E_D - E_Z$, no provision was made in the hardware device for adjusting $E_D - E_Z$ in normal automated operation.

When the hardware-instrumentation approach [3] was compared with the computerized experimentation of the earlier work [1], several observations were made. First, the analytical results and limitations were essentially identical for the two approaches. The only exception was that the hardware instrumentation failed for a 1000:1 mixture. This was because the hardware device used a slightly less sensitive—though more reliable—peak detection method. Also, quantitative resolution of 1000:1 mixtures would require minimum peak separations of about 250/n mV and a value of $E_D - E_Z$ of -100/n mV [1]. Because the hardware device was designed for completely automated operation, $E_D$ was set to always equal $E_Z$, and this allowed application to any system where peaks were separated by 150 mV or greater, up to peak ratios somewhat greater than 100:1.

The limitations mentioned above reflect the objective of hardware development. The device was built to handle most analytical situations with a minimum of operator manipulations. Moreover, the cost was only about 5% of the computerized system. The device is much simpler to operate than the computerized instrumentation because only the initial cell potential, amplifier gain, and sweep mode (with or without sweep interrupt) need to be specified by the operator before an experiment. Nevertheless, it provides truly interactive instrumentation. Although the feedback is less flexible or "intelligent" than in the computerized system, it is optimized. In addition, the hardware device offers one other distinct advantage over the computerized approach. The sweep rate of the computerized system was limited by the time necessary to perform the real-time calculations. This amounted to about 800 μsec between data points, which limited the data rate to about 1 kHz. If data are taken at each mV during the sweep, this limited the sweep rate to about 1 V/sec. In the hardware device, this limitation does not exist as the hardware can perform several logical, arithmetic, storage, and control operations in parallel; thus fewer real-time steps are required. With proper modifications to the potentiostat, current follower, differentiators, and filters, the interrupted-sweep experiment could be run at sweep rates of

100 V/sec. With appropriate scaling of $\tau'$, this would allow much shorter experimental times and might be useful for the analysis of unstable systems.

The most important point here, though, is that the hardware instrumentation could not have been designed readily without the prior investigative study [1] using the on-line general-purpose computer. The optimum design parameters—as well as the very feasibility—of the interrupted-sweep technique were evaluated with the computerized experimentation.

## IV. COMPUTERIZED RESOLUTION OF CLOSELY SPACED

### PEAKS IN SEP

The experimental work described above demonstrated how computerized instrumental interaction could improve the quantitative resolution capabilities of SEP. However, that approach fails for peaks separated by less than about 150 mV. An alternative approach must be used to handle electroanalytical samples where SEP peaks are more severely overlapped. One approach taken has been presented by Gutknecht and Perone [2].

The approach involved extracting the analytical information from SEP data using mathematical deconvolution techniques. An empirical equation was developed which describes the general stationary electrode polarogram for a wide variety of electroactive species. The function is fit to a number of standard polarograms, and the constants of the function, as specifically determined for each species, are stored in computer memory. Upon analysis of an unknown mixture, these constants are used to regenerate the standard curves, a composite of which is then fit to the unknown signal. In the fitting process, account is taken of overlap distortion as well as experimental fluctuation of peak potentials.

The small computer was used to perform several different functions in the development of the on-line electroanalytical system. Primary among these were experimental control, timing, synchronization, and data-acquisition functions. In addition, with the aid of an oscilloscopic display system, off-line simulation studies were carried out to evaluate empirical equations developed for later on-line data processing. Finally, the computer was used to process SEP data acquired on-line for qualitative and quantitative information. The processing approach proved especially valuable for the analysis of mixtures of similar concentrations where the overlap was so severe as to preclude visible recognition of the individual signals.

A discussion of the numerical deconvolution approach is beyond the scope of this chapter, and the reader is referred to the original work for details [2]. However, a brief summary of the results of that work can be presented.

It was shown that for mixtures of similar concentrations of In(III) and Cd(II) in 1.0-$\underline{M}$ HCl, with a peak potential separation of 48 mV, it was

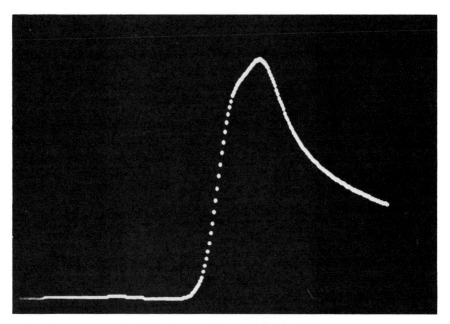

Fig. 10. SEP curves for real 1:1 In(III)-Cd(II) system, peak potential
separation 48 mV [3.84 × 10⁻⁵ M In(III), 3.95 × 10⁻⁵ M Cd(II), 1.0 M HCl].
Voltage range shown: -0.300 → -0.800 V vs. S.C.E.; maximum peak current: 3.7 μA. (From Ref. [2].)

possible to detect and quantitatively resolve the overlapped peaks with relative errors the order of 1 to 2%. A polarographic trace for a 1:1 mixture is
shown in Fig. 10. The approach was applicable to mixtures with concentration ratios as great as 10:1.

To establish the limiting peak separation which could be handled by the
deconvolution technique, synthetically generated polarograms of In(III)-Cd(II)
mixtures were used where peak separations were varied. The limiting peak
separation was found to be about 40 mV. Mixtures with concentration ratios
as great as 5:1 could be qualitatively identified and quantitatively resolved
with about 1 to 2% relative errors. By contrast, simple simultaneous equation calculations led to relative errors on the order of 10 to 35%. Moreover,
the visual detection of the two individual peaks was not possible, as shown in
Fig. 11.

It should be obvious that the numerical deconvolution approach and the
computerized interaction approach are complementary in many ways. The
latter is applicable to widely spaced peaks with large concentration ratios;
the former is applicable to very closely spaced peaks, but can handle more
limited peak ratios. Both approaches represent a significant enhancement of
measurement capabilities in SEP.

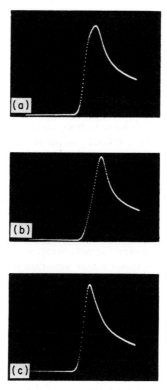

Fig. 11. SEP curves for synthetic 1:1, 1:5, and 5:1 mixtures of In(III)-Cd(II).
Peak potential separations: 38-42 mV; voltage range shown: -0.300 → 0.800
V vs. S.C.E. Upper trace: $4.80 \times 10^{-5}$ M In(III), $4.94 \times 10^{-5}$ M Cd(II),
1.0 M HCl; maximum peak current: 4.8 μA. Middle trace: $0.960 \times 10^{-5}$ M
In(III), $4.94 \times 10^{-5}$ M Cd(II), 1.0 M HCl; maximum peak current: 3.0 μA.
Lower trace: $4.80 \times 10^{-5}$ M In(III), $0.988 \times 10^{-5}$ M Cd(II), 1.0 M HCl; maxi-
mum peak current: 3.5 μA. (From Ref. [2].)


## V. CONCLUSIONS

Certain observations should be made here. First of all, real-time computer interaction with experimentation is not the answer to all measurement
problems. In some cases, in fact, it makes the problem worse. For
example, the interrupted-sweep approach is completely inappropriate for
SEP measurements of closely spaced reduction peaks. One is tempted to
generalize and state that the interactive approach fails when the interaction
distorts the fundamental processes of interest. It was shown here how one
might use the numerical analysis capabilities of the small computer for
solution of these measurement problems.

A second point is that one may not need to devise a real-time interaction scheme to achieve computerized optimization of experimental measurements. A perfectly adequate approach might involve an iterative method where the computer is programmed to analyze the data from a completed experimental run; make decisions regarding modification of controlled parameters for improved measurements; and then reinitiate the experiment under new conditions.

A third point to be made here is that, as demonstrated above, it may not be necessary to require a digital computer for implementation of real-time interactive instrumentation. However, the investigation of the approach and the establishment of the optimum mode of interaction are greatly facilitated by the on-line digital computer. Subsequent hardware implementation of the approach can be straightforward and economical.

A final observation to be made here is to attempt to define in general the experimental situations where real-time computerized interaction is advantageous and/or necessary for optimization of measurements. These situations seem to include those where separate dynamic experiment-associated chemical or physical processes occur which interfere with the measurement of interest at a particular time during the experiment. If the interfering processes can be independently evaluated by real-time computations, computerized interaction may be advantageous. If post-mortem analysis of unoptimized experimental measurements does not provide adequate information for subsequent experimental modifications, real-time computer interaction may be necessary for optimization.

It would be presumptuous on the part of this author to describe specifically how other measurement techniques might be optimized by real-time computer methods. Only the experienced worker skilled in the particular analytical method has the appropriate understanding and intuition for proper experimental design. However, several analytical methods have been recognized as being amenable to optimization by real-time computer measurements. These include gas chromatography [16], kinetic and other clinical methods of analysis [17, 18], as well as coulometric analysis [19]. Undoubtedly, many such applications will be developed in the near future.

## ACKNOWLEDGMENT

The support of the National Science Foundation Grants No. GP-8677 and GP-21111 is gratefully acknowledged.

## REFERENCES

[1].    S. P. Perone, D. O. Jones, and W. F. Gutknecht, Anal. Chem., 41:1154 (1969).

[2].    W. F. Gutknecht and S. P. Perone, Anal. Chem., 42:906 (1970).

[3].   D. O. Jones and S. P. Perone, Anal. Chem., 42:1151 (1970).

[4].   R. S. Nicholson and I. Shain, Anal. Chem., 36:706 (1964).

[5].   S. P. Perone and T. R. Mueller, Anal. Chem., 37:2 (1965).

[6].   C. V. Evins and S. P. Perone, Anal. Chem., 39:309 (1967).

[7].   S. P. Perone, J. E. Harrar, F. B. Stephens, and R. E. Anderson, Anal. Chem., 40:899 (1968).

[8].   G. Lauer, R. Abel, and F. C. Anson, Anal. Chem., 39:765 (1967).

[9].   G. Lauer and R. A. Osteryoung, Anal. Chem., 40:30A (1968).

[10].  A. Savitzky and M. J. E. Golay, Anal. Chem., 36:1627 (1964).

[11].  P. R. Bevington, Data Reduction and Error Analysis for the Physical Sciences, McGraw-Hill, New York, 1969.

[12].  J. E. Barnes, Jr., "Sampled Data Systems and Periodic Controllers," in Handbook of Automation, Computation and Control (E. M. Grabbe, S. Ramo, and D. E. Wooldridge, eds.), Vol. 1. Wiley, New York, 1958.

[13].  W. L. Luyben, Instrum. Technol., 18:58 (1971).

[14].  Paul E. Reinbold, M.S. Thesis, Department of Chemistry, Purdue University, Lafayette, Indiana, 1968.

[15].  John S. Springer, Anal. Chem., 42:23A (1970).

[16].  R. G. Thurman, K. A. Mueller, and M. F. Burke, J. Chromat. Sci., 9:77 (1971).

[17].  G. E. James and H. L. Pardue, Anal. Chem., 41:1618 (1969).

[18].  G. P. Hicks, A. A. Eggert, and E. C. Toren, Jr., Anal. Chem., 42:729 (1970).

[19].  F. B. Stephens, F. Jakob, L. P. Rigdon, and J. E. Harrar, Anal. Chem., 42:764 (1970).

# AUTHOR INDEX

Numbers in brackets are reference numbers and indicate that an author's work is referred to although his name is not cited in the text. Underlined numbers give the page on which the complete reference is listed.

## A

Abel, R. 140[24], 179, 424[8], 447
Adams, R.N. 185[11, 16, 18],
  192[11], 214, 215, 301[28], 348
Alberts, G.S., 77[15], 83, 101
Albery, W.J., 219[2-11], 221[17],
  240, 243[2, 3], 260
Alden, J.R., 301, 348
Anderson, R.E., 358[8], 367[8],
  368, 424[7], 428[7], 440[7], 447
Annino, R. 298, 348
Anson, F.C., 140[7, 21, 24, 32-35],
  117, 152[21], 178, 179, 180,
  210[33, 34], 215, 354[3], 358[6],
  367[6], 424[8], 447
Ariel, M., 303[34], 348
Ashley, J.W., 69, 100
Auerbach, C., 185[2], 214, 219[15],
  240, 294, 348

## B

Barclay, D.J., 140[33, 34], 180
Bard, A.J., 65[4, 5], 69[5], 71[5],
  100, 140[13, 29], 179, 185[8, 9],
  219[12, 13, 14], 230[12], 240,
  243[4, 5], 245[4, 12, 13], 256[12],
  257[12], 260, 313, 314, 349
Barker, G.C., 140[18, 31], 145[18],
  152[18], 179, 180, 370[12, 16, 17],
  375[12, 15], 415[16], 416[16], 430
Barradas, R.G., 5[8], 6[8], 44
Baticle, A.M., 140[41], 180

Bauer, H.H., 145[47], 180, 370[8],
  371[8], 373[8], 375[8], 379[8],
  399[8], 404[8], 416[8], 430
Behrin, E., 293, 347
Belew, W.L., 299[24], 301[24],
  348
Bell, R.P., 79[21], 88[21], 101
Bendat, J.S., 414[44], 422
Bertozzi, R.J., 402[38], 421
Berzins, T., 140[14], 145, 154[14],
  179, 180
Bevington, P.R., 78[18], 79[18],
  82, 90[18], 91[18], 101
Bezman, R., 314[60], 349
Birke, R.L., 105[10], 109[10], 117,
  140[17], 154[17], 179
Blaedel, W.J., 314, 349
Blount, H.N., 77[17], 90[32], 101
Bolzan, F.A., 140[31], 180
Booman, G.L., 135[29], 138,
  185[3, 6], 212, 281, 286[3],
  289[3], 291, 293, 315, 316, 322,
  325, 333, 335, 338, 339, 346[76],
  347, 349, 350, 370[1]. 371[1],
  377[24], 381[1], 411[24], 420,
  421
Bonnemay, M., 140[15], 179
Borchers, C.E., 418[52], 422
Bower, J.L., 328[70], 350
Brand, M.J.D., 314, 349
Brayden, T.H., Jr., 309, 348
Breyer, B., 145[47], 180, 370[8],
  371[8], 373[8], 375[8], 379[8],
  399[8], 404[8], 416[8], 430

449